Ulrike Grabski-Kieron, Ingo Mose,
Anja Reichert-Schick, Annett Steinführer (Eds.)

European rural peripheries revalued

RURAL AREAS

Issues of local
and regional development

LÄNDLICHE RÄUME

Beiträge zur
lokalen und regionalen Entwicklung

Edited by / herausgegeben von

Prof. Dr. Ulrike Grabski-Kieron
(Universität Münster)
Prof. Dr. Ingo Mose
(Universität Oldenburg)
PD Dr. Anja Reichert-Schick
(Universität Trier)
Dr. Annett Steinführer
(Thünen-Institut Braunschweig)

Band / Volume 1

LIT

European rural peripheries revalued

Governance, actors, impacts

edited by

Ulrike Grabski-Kieron, Ingo Mose,
Anja Reichert-Schick, Annett Steinführer

LIT

Cover photos:
Upper left corner: Isle of Barra, Outer Hebrides, Scotland – Photo: Annett Steinführer
Upper right corner: Grünhof, Rothenklempenow, Mecklenburg-Vorpommern, Germany – Photo: Anja Reichert-Schick
Bottom left corner: Gearrannan Blackhouse Village, Isle of Lewis, Outer Hebrides, Scotland – Photo: Ingo Mose
Bottom right corner: Tuscany, Italy – Photo: Ulrike Grabski-Kieron

We gratefully acknowledge the financial support of ERASMUS+ Higher Education – Jean Monnet

Gedruckt auf alterungsbeständigem Werkdruckpapier entsprechend
ANSI Z3948 DIN ISO 9706

Bibliografische Information der Deutschen Nationalbibliothek
Die Deutsche Nationalbibliothek verzeichnet diese Publikation in der Deutschen Nationalbibliografie; detaillierte bibliografische Daten sind im Internet über http://dnb.d-nb.de abrufbar.

ISBN 978-3-643-13050-1

Verlagskontakt:
Fresnostr. 2 D-48159 Münster
Tel. +49 (0) 2 51-62 03 20
E-Mail: lit@lit-verlag.de http://www.lit-verlag.de

Auslieferung:
Deutschland: LIT Verlag Fresnostr. 2, D-48159 Münster
Tel. +49 (0) 2 51-620 32 22, E-Mail: vertrieb@lit-verlag.de

E-Books sind erhältlich unter www.litwebshop.de

Preface

The publication of this volume was inspired by a workshop of international scholars held in Oldenburg, Germany, by the German Rural Geographers Working Group, together with the Jean Monnet Chair "Europeanisation and sustainable spatial development" of the University of Oldenburg in November 2014. As such the book comprises several of the presentations given at the workshop in an updated form, together with a number of invited contributions by authors from across Europe. This collection is meant to provide an overview of the state of rural peripheries on the European scale, which is lacking at present. At the same time, the book is pioneering a new series of publications, "Rural areas: Issues of local and regional development", which is edited by the German Rural Geographers Working Group with a major focus on rural development from both a local and regional perspective.

Financial sponsoring for the publication of this volume was provided by both the Jean Monnet Chair "Europeanisation and sustainable spatial development" at the University of Oldenburg and the Local, Regional and State Development/Spatial Planning Working Group, Institute of Geography, University of Münster. The editors are very grateful for this valuable support. They also wish to thank Matthias Kling for his assistance in the editing process and Claudia Schroer (both of the Institute of Geography, University of Münster) for technical support.

Ulrike Grabski-Kieron, Ingo Mose, Anja Reichert-Schick, Annett Steinführer

Münster, Oldenburg, Trier, Braunschweig

June 2016

Contents

Notes on the contributors

Kristine Beurskens (née Müller) studied Geography, Political Science and Sociology at the Humboldt University Berlin (Germany), the University of the District of Columbia (USA) and the University of Amsterdam (The Netherlands). She holds a PhD (Dr.) in Geography (2012). She is based at the Leibniz Institute for Regional Geography in Leipzig (Germany), where her work is focused on questions around borders, informality and social inequality as well as on the development of qualitative research methods in social and spatial research.

Mary Cawley studied Geography and History at the National University of Ireland (NUI) Galway (Ireland). She holds a PhD (1979) in Geography from the University of London (England) and has been a Fulbright Scholar at the Pennsylvania State University (USA) (1984–85; 1993). Her research focuses on rural change and development with particular reference to population, services and tourism and she has led many projects on these themes. She is Emeritus Senior Lecturer in Geography at NUI Galway and a member of the Steering Committee of the International Geographical Union Commission on the Sustainability of Rural Systems.

Samuel Depraz studied Geography at the Ecole Normale Supérieure of Lyon and the University of Montpellier (France). His PhD, completed in 2005, focused on the transformation of post-socialist rural areas in Eastern Germany and Hungary. Since then, he holds a position of maître de conferences at the University of Lyon. He has been a visiting researcher at Münster University (Germany). His current studies deal with the incidence of nature protection measures on local development and the ongoing gentrification process of favoured rural areas in France.

Christoph Creutziger studied Geography, Political Science and Sociology at the University of Münster (Germany). He is based at the Leibniz Institute for Regional Geography in Leipzig (Germany), where his main research interests lie in political geography, border regimes, internal borders of the EU and post-structuralist discourse theory.

Theresa Dietl studied Cultural Studies (B.A.) at the universities of Frankfurt/Oder (Germany), Miami (USA) and Coimbra (Portugal) as well as European Urban Studies (M.Sc.) in Weimar (Germany). She worked as a research assistant at Bauhaus University Weimar, in urban and community development offices in Berlin and Dublin (Ireland) and co-founded an organization for community development through participatory processes in Berlin. She is currently working on her PhD on the potentials of rural development through participatory artistic projects as a scholarship holder of the national foundation Cusanuswerk.

Werner Dietl studied Geography at the University of Vienna (Austria). He finished his doctorate in Geography in 2011. His research activities focus mainly on evaluation research in the context of spatial planning and spatial research in rural areas. He is currently employed as a teacher for Geography at AHS Heustadelgasse (Vienna) and lecturer at the Department of Geography and Regional Research, University of Vienna.

Andreas Friedwagner, Benedikt Hahn and Thomas Langthaler are working for the Austrian consulting company Verracon. Their main fields of expertise are transportation and regional planning. They carried out numerous studies aiming at improving mobility services for different target groups in different Austrian and European regions.

Martin Heintel studied Geography and Philosophy at the University of Vienna (Austria). He holds both a PhD (1996) and a postdoctoral degree (Habilitation) (2004) in Human Geography and is currently a professor at Vienna University. In 2006/2007 he served as Marshall Plan Chair for Economics and Regional Economic Development at the University of New Orleans (US). He has been a visiting professor to several universities, e.g., German University of Technology in Muscat (Oman), Babeş-Bolyai University Cluj-Napoca (Romania) and University of Salzburg (Austria). His main fields of interest include spatial and regional development, cross-border cooperation, regional management and mega-cities in South East Asia.

Ulrike Grabski-Kieron studied Applied Geography at the Ruhr-University Bochum (Germany). She holds a PhD (Dr.) in Geography (1985) and completed her postdoctoral studies in 1993 (Habilitation). In these works, her research focused on issues of rural development and environmental management. Later, she worked as a managerial engineer at a state agency of rural and urban development (1986–96). Since 1996 she is professor of local and regional development and spatial planning at Münster University (Germany). Her research interests relate to European rural development, environmental planning, planning methodology and governance research. She is member of the German Academy of Spatial Research and Regional Planning (ARL).

Marta Jacuniak-Suda studied Environmental Sciences, Geography and German Philology at the universities of Zielona Góra (Poland), Vechta (Germany) and Staffordshire (England). In 2015 she completed her PhD (Dr.) on regional governance in the context of integrated rural development and social entrepreneurship in Poland and Scotland at Oldenburg University (Germany). As a research fellow at the HafenCity University Hamburg, she is currently working on potentials and restrictions of urban-rural cooperation in metropolitan areas in Europe.

Leena Karrasch studied Biology, Water and Coastal Management and Environmental Infrastructure Planning at the universities of Münster, Oldenburg (Germany) and Groningen (The Netherlands). She is working in the field of participatory adaptive land use planning and sustainable management of coastal areas at COAST — Centre for Environmental and Sustainability research, University of Oldenburg.

Thomas Klenke graduated in Geology at the University of Münster (Germany) in 1984 and received a PhD (Dr.) in Geosciences from the same university in 1988. He started his professional career at the Geological State Service of Lower Saxony working on groundwater protection schemes in 1987. In 1988, he joined the Institute for Chemistry and Biology of the Marine Environment (ICBM) of the University of Oldenburg as PostDoc. He became managing coordinator of the Centre for Environmental and Sustainability Research (COAST) of the University of Oldenburg in 2008. His research interests range from environmental geology to transdisciplinary international research on coastal management, climate adaptation and bioenergy.

Stefan Kordel studied Cultural Geography and German Studies at the Friedrich-Alexander-University Erlangen-Nürnberg (Germany). He completed his PhD (Dr.) in Geography in 2015, titled “Striving for the ‘good life’ — home-making among senior citizens on the move. An analysis of German (pre)retirees in Spain and Germany in a continuum of tourism and migration”. Since then, he is a post-doctoral researcher, focusing on social geographic research within the realms of tourism and migration studies, e.g., lifestyle migration. His research is located in rural areas of Europe, e.g., coastal areas and the hinterland of the Iberian Peninsula and the Bavarian Forest, Germany.

Henry Lewerentz studied Geography, Geology and Political Sciences at the Ernst Moritz Arndt University Greifswald (Germany) and the University of Aberdeen (Scotland). In 2015 he completed his PhD. His work focuses on the economic, ecological and social possibilities of the usage of alternative engine technologies for motorized private transport in rural areas. Currently he works at a state planning agency in Schwerin (Germany) in the field of spatial, regional and urban land-use planning.

Suzannah-Lynn Billing studied Environmental Science and Sustainable Environmental Management at Plymouth University (England). In 2016 she completed her PhD on the role of agents for change in the development of the wave energy sector in Scotland at the University of Aberdeen and the University of the Highlands and Islands. As a postdoctoral researcher at the Scottish Association for Marine Science, her focus is currently on industry and governmental accountability and planning, social license to operate, and co-location within the aquaculture and marine renewable energy sectors.

Judith Miggelbrink studied Geography, Sociology, Spatial Planning and Administrative Law and Mathematics at the University of Münster (Germany). She received her PhD (Dr.) in Geography in 2001 with a study on recent discourses on space and region in geography. At the Leibniz Institute for Regional Geography in Leipzig she is coordinator of the research group "Spatial productions — society and space". Her main research interests lie in border studies including informal and illegal practices, land use conflicts in (post-)nomadic and indigenous societies, visual geographies, space as power technique as well as in methodological developments in geography and spatial sciences. Currently, she is inter alia engaged in a project on "Discourse and subject in shrinking regions".

Ingo Mose studied Geography, German Philology, Social Sciences and Pedagogics at Osnabrück and Oldenburg Universities (Germany). He holds both a PhD (Dr., 1987) and a postdoctoral degree (Habilitation) (1983) in Geography and is currently a professor at Oldenburg University, heading the Applied Geography and Environmental Planning Working Group. He holds a Jean Monnet Chair and has been a visiting lecturer to several universities across Europe, e.g., University of Vienna, University of Salzburg and Kingston University London. His work focuses on rural development, regional governance, protected areas and sustainable tourism in a European perspective.

Giorgio Osti studied Sociology at the University of Trento (Italy). He holds a PhD in Sociology and Social Research (1991) and he has been qualified (abilitazione) as full professor of Environmental and Territorial Sociology in 2013. He teaches sociology of migration and local development at the University of Trieste. His fields of research are rural development and environmental crisis; for the former he has studied Local Action Groups, mobility and welfare in fragile rural areas, for the latter he has made researches in ecological movements, waste management and energy transition.

Anja Reichert-Schick studied Geography and Political Sciences at Saarbrücken and Trier Universities (Germany). She finished her doctorate in Geography 2004 and pursued postdoctoral studies (Habilitation, 2014) on the subject "Exodus, shrinking and abandoned villages. History, determinants and implications caused by peripheralisation and regression processes in rural settlements." Since 2013 she has been a visiting professor at Trier University, heading the Economic and Social geography chair. Her research activities are focused on the spatial development of rural areas, border studies, demographic change as well as cultural landscape development and tourism geography.

Frank Rennie is Professor of Sustainable Rural Development at the University of the Highlands and Islands (Scotland) and is Assistant Principal at Lewis Castle College UHI. He is the Programme Leader for the online MSc in Sustainable Rural Development, and tutors mainly postgraduates and research students. His research interests lie in new approaches to online education and the benefits of networking for sustainable rural development. He has been involved in developing and delivering distributed learning (particularly networked digital solutions for rural areas) with a wide range of international partners and published a diverse range of resources related to rural issues, including over 30 books.

Peter Schaal studied Agriculture, Environmental Protection and Rural Development at Justus-Liebig-University Gießen (Germany) and Landscape Ecology and Landscape Development at the University of Natural Resources and Life Sciences in Vienna (Austria). He holds a PhD (1997) in Spatial Planning and works currently as a senior lecturer and researcher at Oldenburg University, at the Applied Geography and Environmental Planning Working Group. He also is head of the RaUm Consult GbR consultancy firm. His work focuses on spatial planning, natural resources, regenerative energies and climate change mitigation, rural development and GIS.

Rozanne Spijkerboer graduated from the Research Master Regional Studies at the University of Groningen (The Netherlands) in 2015. Currently, she is working as a PhD researcher at the Department of Spatial Planning and Environment, Faculty of Spatial Planning at the University of Groningen. Her research focuses on the opportunities for integration between the domains of energy, infrastructure and spatial planning, with a particular focus on the role of institutions and institutional change.

Annett Steinführer studied Sociology and Eastern European Studies at the universities of Leipzig (Germany), Glasgow (Scotland) and Brno (Czech Republic). She holds a PhD in sociology (2002). As a social scientist at the Johann Heinrich von Thünen Institute, the Federal Research Institute for Rural Areas, Forestry and Fisheries in Braunschweig (Germany), she focuses on the consequences of ageing in rural areas, governance processes with regard to basic welfare services and small-town research in Germany. Further interests relate to socio-spatial inequalities in regional and urban contexts.

Elen-Maarja Trell studied Environmental Management, Environment and Infrastructure Planning and Cultural Geography at the University of Tallinn (Estonia) and University of Groningen (The Netherlands). She holds a PhD (2013) in Spatial Sciences. Her research broadly focuses on the role of local level initiatives and public participation in creating more resilient and sustainable places. The themes she explores include: community resilience in declining (rural) areas; urban self-organization, sustainability and local food networks; community resilience to flooding; and governance of renewable energy initiatives.

Alice Wanner studied Social Anthropology and is currently pursuing a degree in Spatial Research and Planning at the University of Vienna (Austria). She focuses on transportation and mobility.

Tobias Weidinger studied Cultural Geography at the Friedrich-Alexander-University Erlangen-Nürnberg (Germany). Since 2016 he is working as a research assistant at the same university with a special emphasis on rural studies, particularly concerning tourism and migration. He is currently working on migration decisions of recognized asylum-seekers in rural Bavaria.

Norbert Weixlbaumer studied German and Italian Philology and Geography at the University of Salzburg (Austria). He finished his doctorate in Geography in 1986 and pursued postdoctoral studies (Habilitation, 1995). He is professor of Geography at the University of Vienna (Austria). His main research interests include protected areas and regional development in Europe, transboundary co-operations, behavioural geography, rural geography and geography of food.

Christian Zuidema is assistant professor in Spatial Planning at the Department of Spatial Planning and Environment at the University of Groningen. He has done extensive research on topics such as sustainable urban management, and governance renewal in environmental and energy planning. He holds a PhD in Spatial Sciences and authored various books on environmental policy and governance. In addition to his everyday commitments, he is organizer of the Groningen Summer School on Energy, member of the Energy Academy Europe and Board Member of the International Urban Planning and Environment Association.

Introduction

European rural peripheries revalued? Introduction to this volume

Annett Steinführer, Anja Reichert-Schick, Ingo Mose and Ulrike Grabski-Kieron

"Experience Life on the Edge" — with this self-confident slogan the Outer Hebrides Tourism Industry Association rebranded the Scottish Western Isles in 2012 in order to attract more visitors to a destination that had never been among the hotspots of European tourism. As seen on a standard European map, the chain of islands indeed forms an "edge" — both for the United Kingdom and for Europe with the next stop being Canada.

Marketing professionals applied further attributes related to "rareness" and "uniqueness", in addition to the less surprising draws "wilderness" and "nature". The Western Isles are, from a traditional perspective, a typical regional, national and European rural periphery — in terms of distance, difference, deprivation and, in many cases, dependency (Ferrão and Lopes 2004; Hechter 1975). But what exactly and how persistent peripheries are, has time and again been questioned by the academic community. In the 1990s, for example, Nolte (1991 and 1997) transferred Wallerstein's (1974) notion of global peripheries to European contexts and specified the concept of "inner peripheries" in a historical perspective. In the 2000s, by using the term "peripheralisation" (e.g., Keim 2001 and 2006) the academic discourse shifted to questions of the dynamics of how peripheries are created, maintained or overcome. We will return to the different notions of periphery and peripheralisation in the course of this introduction.

In geography, sociology and spatial planning, peripheries have always been perceived as a problematic socio-spatial category. Rural peripheries in particular are often considered as "residual spaces" of spatial development in Europe. Their structural disadvantages are seen as intertwined with their geographical location, an imbalanced economic structure, low population density and consumer demand, out-migration and ageing, but also with political and, not least, financial dependency on decisions in the urban centres of power. While this has been true for many peripheries for decades, and seems to be deteriorating further in the course of globalisation, there are also opposite developments. Some peripheral areas across Europe display signs of an economic renaissance and residential revitalisation, some have established innovative models of governance and have created new self-esteem. The Scottish Western Isles mentioned above are an impressive example for how far such a transformation of a formerly peripheral and, at the same time, marginalised region, can actually reach (Mose and Jacuniak-Suda 2011; Rennie and Billing in this volume; Schmied 2012).

These ambiguous patterns and developments of rural peripheries are the starting point of this volume. Based upon oral presentations and a vivid debate during the international conference "European rural peripheries caught between stagnation and take-off" at Oldenburg University (Germany) in November 2014, the papers ask: (1) how actors in certain rural areas identify and possibly revalue endogenous potentials and unlock new resources of their regions, (2) which role external actors, the local population and further stakeholders play in these processes, (3) which impact European, national and regional policies have with regard to the dissolution and, possibly, (re)-creation of rural peripheries and (4) which obstacles still exist to overcome structural economic, social and spatial disadvantages. The volume sheds light on (seemingly) peripheral areas and on processes of peripheralisation in Europe by combining theoretical and conceptual issues with empirical findings. Particular emphasis is given to questions of regional governance, to multiple actors of both peripheralisation and revitalisation, to economic transformations in the sectors of tourism and renewable energies as well as to challenges arising from the climate change and sustainability discourses.

This introductory chapter is structured as follows: The first section will unlock the topic surrounding rural peripherality, followed by a more dynamic perspective spawned by the academic discourse on peripheralisation. Subsequently, we discuss current transformation processes in European rural peripheries by paying particular attention to tourism, residential revitalisation and the energy sector, as well as to the relationship between endogenous and exogenous factors. In the final section, the main topics of the contributions in this volume will be presented, followed by some concluding remarks.

General notions of rural peripherality

Although fundamental to the social sciences dealing with spatial phenomena, the term "periphery" is conceptualised in rather different ways. First, it is a descriptive *locational term* identifying regions lying either at the edge or margin of something (Mayhew 2009, 376). This meaning at the same time comprises the notion of a certain distance to a centre. Even more important, though, "periphery" used in a locational sense often includes negative associations of backwardness or underdevelopment, whether implicit or explicit (Nitz 1997, 19–20; Hall et al. 2013, 75–76).

The second meaning of periphery is related to its use as a *functional-spatial term* (Nitz 1997, 20–21), thus in a relational sense. In applying this notion, periphery can solely be — and usually is — referred to from a centre (or "core") position (e.g., Friedmann 1973; Wallerstein 1974; Nolte 1991 and 1997; Kreckel 2004, 41–51; Mayhew 2009, 106). The core or centre is conceptualised as a node in a network of economic, transport and symbolic interactions, power relations and material flows while periphery is considered as its opposite. In this understanding, periphery and centre are complementary spatial categories with a dialectical relationship to each other (Heintel 1998, 17; Göler 2005, 35). Surprisingly, core or centre is, to date, framed in a highly normative way. Even a recent dictionary of geography referred to "centre" as per se prosperous (Mayhew 2009, 106).

Centre-periphery relationships have been most prominently studied on the global scale. This is reflected in the works of dependency, world-systems and polarisation theories (e.g., Amin 1976; Senghaas 1974; Wallerstein 1974). Also, in light of growing socio-spatial disparities within certain world regions, whether Europe, Asia or Africa, in the course of globalisation it is, however, necessary to look more closely at the notion of "periphery" at a regional and even local scale. This was one of the starting points for the academic and, in the first instance, historic discourse on "inner peripheries" in the late 1980s and early 1990s. Inner peripheries were defined by Nolte (1991) as regions within a state "where conditions are organised to the advantage of people living in another region which we call the centre" (ibid., 1). This conceptualisation thus comprises a strong claim, namely, that peripheries within states are systematically discriminated by the centre and its actors. We will see below that certain arguments contradicting this claim can be found as specific policies directed towards disadvantaged regions (often additionally called "peripheral") are in place across the European Union. Today, the idea of "inner peripheries" is well understood in academia, but is rarely more than a metaphor. As Peréz-Soba et al. (2012) pointed out, only "scarce and patchy information" is available which is "insufficient to achieve a comprehensive understanding of the concept" (ibid., 5). In their conceptualisation, the main features of inner peripheries are a specific location on the edge of, or between core areas, a decline of the main economic activity, demographic decline and a loss of local identity (ibid., 33).

The traditional core-periphery distinction has another major implication with regard to the spatial category in the focus of this volume: From the perspective of the centre, rural areas are systematically placed in the periphery. Therefore, Kühn (2016) is right in critically recalling the idea of defining urban centres by way of "excess importance" (Christaller 1933/2006, 26–27). In light of Christaller's Central Place Theory, rural areas are thus structurally characterised by a deficit in importance (ibid., 17, 29; similarly: already Heintel 1998, 32) — or, to put it differently, they are peripheral per se. Rural studies made strong efforts in the past decades to go beyond the urban-rural dichotomy (e.g., Champion and Graeme 2004; see already Miner 1952). But even a more sophisticated typology usually ranges from global cities, metropolises and agglomerations, on the one hand, to some kind of distant, remote and/or "very" rural areas, on the other. This latter pole is almost exclusively considered as areas with a relatively low population density, an absence of urban and economic centres and a low offer of qualified jobs as well as generally limited opportunities for economic development. These *rural peripheries*, as we call them, are further characterised by poor transport facilities or, in other words, as a rule, "they are located at the end of transport networks" (Hall et al. 2013, 76). At the same time, this spatial pole is perceived as areas with a higher-than-average share of agriculture and/or forestry and consequently large areas of open space. With the onset of industrialisation, and increasingly accompanied by a long-term rationalisation in agriculture, these features resulted in locational disadvantages both for the population and for businesses. Throughout the 20th century these locational disadvantages could not be compensated for by positive assets such as a relatively untouched, often ecologically valuable natural landscape (Brunotte et al. 2002, 35; Henkel 2004, 293). Until today, rural peripheries are regarded as areas requiring specific attention and associated funding, often reflected in the adoption of specific development programmes (such as the European Regional Development Fund/ERDF of the European Union or specific national programmes).

Rural peripheries are a well-studied subject of rural studies. Linguistically, this "socio-economic territorial specificity" (Peréz-Soba et al. 2012) is framed in very different ways, for example:

- "residual spaces" (Glatz and Scheer 1981 exemplified for Austrian regions),
- "most rural space" as a phrase from the German debate (Weiß 2002),
- "lagging regions" in contrast with "leading regions" in an analysis on EU member states (Terluin 2003),
- "rural spaces with grave development problems" as a further term used in Germany (Glaser et al. 2007, 74),
- "fragile rural areas" in France and Italy (Depraz and Osti in this volume),
- "structurally weak regions" in the German and Austrian discourse (Reichert-Schick and W. Dietl et al. in this volume)
- "remote rural areas" for example in Scotland and Italy (Rennie and Billing and Osti in this volume),
- "ultra-peripheral" in contrast with "peripheral" areas which are both considered subtypes of "inner areas" as part of a recent typology in Italy (Barca et al. 2014, 26; Osti in this volume),
- "hyper-rural space" as a spatial type recently invented in France (Bertrand 2014; Depraz in this volume).

There are, of course, certain differences in the underlying semantics and national discourses as well as in the socio-economic and demographic indicators applied. More importantly, from the list — which is certainly not exhaustive — it becomes obvious that most of the terms are connected with a certain stigmatisation. Interestingly, while globalisation and metropolisation have mostly produced expressions with positive connotations such as "global city", "metropolitan regions", "urban poles" (for example in France) or even "urban poles of attraction" (as in Italy; Barca et al. 2014, 26), rural peripheries are characterised by terms that often have clear negative connotations.

By way of a summary and in order to systematise the research literature, six characteristic features can be identified that are usually ascribed to (the problems of) rural peripheries. By way of a four-dimensional conceptualisation, Ferrão and Lopes (2004) had already highlighted *distance, dependence, difference* and *discourse* (called "comprehensive 4D approach"; ibid., 31). These four features will be taken as a starting point and expanded by *disadvantage* and *(relative) deprivation* to outline our understanding (Table 1).

Table 1: Characteristic features of rural peripheries

Feature	Description
Distance	The more difficult it is to overcome the distance to the periphery, the more peripheral its location to the centre becomes (Nitz 1997, 19). According to Heintel (1998, 17), the distance component appears to be the "most common" scale of investigation since it is quantifiable and linearly measurable. However, distance is relative because although it designates the peripherality of a location, it only does so to a certain extent. Rather, the decisive factors are the actual effort, including transport costs and the time, needed to overcome distances (Ferrão and Lopes 2004, 34; Göler 2005, 35).
Difference	By applying the concept of difference, Ferrão and Lopes (2004, 37–41) highlight the internal dynamics of rural peripheries, starting from the question why similar regions cope so differently with the same challenges and why a number of European rural peripheries experienced economic success in the past decades. They point to regional path dependencies and thus specific development factors as well as to endogenous milieus "containing institutions, collective learning processes, innovation practices and forms of governance" (ibid., 39). In light of emerging polycentric networks and modes of inclusion due to new information and communication technologies, the authors plead with readers "to put aside any evolutionary determinism" both in a positive and negative sense (ibid., 41).
Disadvantage	Originally, rural peripheries used to be described as structurally disadvantaged in terms of location. These disadvantages had already been cemented in the 19th or early 20th centuries by the absence of a railroad connection. Today it is rather the position of certain regions within physical, social, economic, institutional and cultural networks as well as the networks' configurations which create and recreate peripherality (Peréz-Soba et al. 2012, 5, 25). After having analysed a wide range of socio-economic, infrastructural, governance and natural characteristics, Terluin (2003), in her comparison of 18 European regions, highlights that low values for both the capacity of local actors and the strength of internal and external networks coincide with stagnation in non-agricultural employment growth as a measure of regional disadvantage in economic terms.

(Relative) Deprivation	In contrast to disadvantage, this feature does not focus on structural deprivation from labour, capital and capacities (as, for example, in the approach of Terluin 2003). Rather, relative deprivation relates — at least in the global North — to the relative share of the wealth available and to subjective, though collectively shared understandings of social discrimination and justice, respectively (Kreckel 2004, 335). Relative deprivation comes about by reflecting on one's own social status in comparison to that of another person which is defined as higher. To a certain degree this can also be transferred to individual perceptions of living in a disadvantaged region in comparison with other areas.
Dependency	In European welfare states with their redistribution mechanisms and political efforts to balance grave regional disparities, rural peripheries were made and have become dependent upon both external funding and decisions in the (metropolitan) centres of power (Kühn 2015, 372–373). This leads to ambiguous outcomes: on the one hand, financial means and modern infrastructure are made available to disadvantaged regions and deprived social groups. On the other hand, urban or, at least external, standards are set and applied, respectively, without paying sufficient attention to local and regional contexts (Beetz et al. 2008, 296, 304–307).
Discourse	To a great extent, rural peripheries are created discursively, by media, research and politics. Framing a region as peripheral is rarely meant positively but rather carries with it all the negative connotations in terms of economic, social, financial and demographic problems already mentioned (Bürk 2013). However, Ferrão and Lopes (2004, 41–42) from an economic point of view also point to success stories of rural peripheries and their narratives, be it with regard to environmental characteristics (e.g., wilderness or nature; Hall et al. 2013, 73–74), historical legacies (heritage; Ferrão and Lopes 2004, 42) or economic developments (Labrianidis 2004).

Source: Authors' considerations (building upon and expanding Ferrão and Lopes 2004)

The term peripherality is certainly important in socio-spatial research in both its locational and relational facets. Yet, there are two major shortcomings of the academic debate on rural peripheries:

1) While the socio-economic core-periphery discourse is important to point to growing regional disparities within Europe and national states, it strongly neglects *intra-regional differences*. These also seem to develop increasingly at a very small scale and thereby result in a spatial patchwork with shrinkage and growth, prosperity and decline in sometimes close neighbourhood. Thus also peripheries are subject to an inner hierarchisation. To quote Fassmann (1990): "Centres and peripheries

[may be] hierarchically organised [...] meaning also that peripheral regions have local centres. There is a ranking order of centres and peripheries with different weights. From that follows that there can be no clear-cut definition of centre and periphery but only a relative delimitation of the terms"[1] (ibid., 6).

2) The notion of peripherality is highly static. While location disadvantages and path dependencies certainly matter, economic and demographic (particularly migration) patterns can change within a short period of time. This problem was addressed in recent years by the evolving discourse on peripheralisation. Therefore, we will shed light on the *dynamics of creating, recreating and overcoming peripherality* in the next two sections.

Beyond a static understanding of peripherality: the concept of peripheralisation

The academic debate on peripheralisation developed in the early 21st century in Germany against the background of strongly increased regional disparities in the enlarged country (Keim 2001 and 2006). About ten years after reunification, the socio-economic transition in eastern Germany had resulted in cemented long-term unemployment of a considerable share of the population and continuously high rates of out-migration from rural areas (and most cities), predominantly to western Germany. State-subsidised modernisation of the technical infrastructure contrasted with losses in rural social services. The higher willingness of the young and better qualified, and of women, to leave their regions of origin, reinforced by historical legacies and the strong work-orientation of East German women, led to strong social and age-related imbalances. With regard to gender, Leibert (2016) underlines that "rural East Germany is the place with the most unbalanced sex ratios in Europe" (ibid., 278; for earlier research on this phenomenon in the Scandinavian context see, for example, Dahlström 1996). At the same time, both researchers and many politicians had lost their faith in a short-term adjustment of the level of socio-economic well-being in the two parts of Germany. In this vein, "new" regions entered the status as peripheral areas, while "traditional" ones (e.g., in the northwest of Germany) had lost their status, leading to the research question of how peripheries come about and how they might overcome this status.

---

1 This quote, as well as all subsequent literal translations from a German original, are by the authors.

In light of this profound socio-spatial restructuring of the regional mosaic in Germany, Keim (2006) conceptualised "peripheralisation" by highlighting the process characterised by negative economic, social and demographic developments as well as the decoupling of regions from the prevalent economic dynamics of the urban agglomerations and the financial dependency of the peripheralised regions on these centres. He considered peripheralisation as complementary to the term "centralisation" which describes both the socio-economic development of a region and the freedom to manoeuvre granted to social actors and institutions. Just like in the centre-periphery discourse, peripheralisation and centralisation are also mutually interlinked. In the 2000s, positive impacts of economic globalisation seemed to be restricted to the major urban agglomerations which almost exclusively benefitted from economies of scale and node functions within international networks. Even with agriculture and forestry becoming increasingly involved in supra-regional or even global markets, their local impact was usually very low. Beetz et al. (2008) state that rural peripheralisation thus also includes realising "only marginal welfare gains" from such global value chains (ibid., 305). These authors understand peripheralisation as "a category to underline the spatial aspect of regional socio-economic decline and demographic shrinking" (ibid., 305). In this light, peripheralised regions are characterised by a steady decline in public infrastructure and private services, lopsided dependency on economic and political centres and a decoupling from positive impulses for development. Thus, "peripheralised" is a distinct concept to depict spatial as well as economic and social deprivation. The term thus goes beyond marginalisation which is more focused on social inequalities (Bock 2016; Osti in this volume). The process of peripheralisation leads to areas being deprived of room to manoeuvre and the capacity to adequately ensure the quality of life of the region's population. In most cases peripheralisation causes various functional sectors to contract leading to a concentration of relatively few resource uses and economic activities: "This results in the reduction of a profile which previously had been more diverse in functional, economic, social and cultural terms. Peripheralisations lead to de-differentiation and fragmentation" (Keim 2006, 5).

Empirical research on rural peripheralisation particularly highlights the simultaneousness and overlapping of negative social, economic, demographic and infrastructural processes of decline (Naumann and Reichert-Schick 2013). Some research was also devoted to the discursive dimension of peripheralisation, be it stigmatisation from external actors (such as the media) or negative self-perceptions of the local population (Bürk 2013; Steinführer and Kabisch 2007).

However, in a critical perspective on the academic discourse on peripheralisation of about one decade, one has to state that it generated only little empirical research and is mostly predictaed on desk-based considerations (with the notable exception of investigations in a number of medium-sized towns in Germany: Bernt and Liebmann 2013). It even contributed to a renewed stereotyped urban-rural dichotomy without paying sufficient attention to intra-rural centralisations or peripheralisation processes within agglomerations or at the urban fringe (as considered, e.g., by Förste 2013 and Kühn 2016). While time and again the process character of a spatial and socio-economic decoupling of regions is highlighted, contrary developments of de-peripheralisation are rarely taken into account. But the "older" periphery debate had already pointed to the necessity to overcome static categories: "Peripheries [...] are not in principle tied to a location but can be subject to fluctuation, that is, to changes following recent economic policy processes and decisions. [...] A progressing market-oriented world system dissolves peripheries by integrating them into the world market — new peripheries are emerging and existing ones are being consolidated by decoupling them largely from the world market; above all because they are regarded as economically 'uninteresting' or 'politically' difficult to integrate" (Heintel 1998, 19–20). Thus, a certain ambiguity of the peripherialisation debate has to be underlined: on the one hand, in Germany (and partly beyond: Fischer-Tahir and Naumann 2013, Kühn 2015; Lang et al. 2015) it contributed to renewed attention to regional disparities that had seemingly been overcome. On the other hand, the discourse carried with it a kind of fatalism. Usually based on data of the official statistics and on information concerning quantitative service availability, peripheralisation was ascribed to a number of regions in line with a certain degree of inevitability of the negative and interlinked socio-economic processes considered. Yet, subjective perceptions of the local population and the decision-makers in these peripheralised regions or their adaptation strategies and new emerging forms of governance were only rarely taken into account (e.g., Steinführer et al. 2014; Naumann and Reichert-Schick 2013; Mose et al. 2014; Wiest 2016).

Set against this background, this volume wants to go a step further in paying more explicit attention to the emergence of rural peripheries, their production, re-production and, possibly, processes of de-peripheralisation. In the following section, some preliminary thoughts in this regard will be formulated based upon previous research.

Between dependence and revaluation: rural peripherality and (de-)peripheralisation in present Europe

As peripheries can no longer be looked upon as static, greater attention needs to be paid to processes of transformation. While in the past peripheries, and particularly rural peripheries, were usually perceived as highly dependent regions, in economic, social, cultural or political terms, even resulting in concepts of "internal colonialism" (Hechter 1975), today a much more varied picture of different types of rural peripheries has become apparent. This also includes a renaissance of formerly peripheralised and dependent regions that are about to overcome disadvantages and experience economic revival and thereby gain new self-esteem. At the same time other peripheries remain somehow "frozen" in their state of dependency. Schmied (2005) has characterised this as a process of "winning and losing", while Copus (2001) stressed the need for a rather "aspatial" approach to peripherality.

For a better understanding of the processes of transformation and redevelopment taking place, a closer look at the spatial functions of rural peripheries and the associated rural economies is indispensable. In some places these are currently undergoing rather dramatic changes. As primarily agricultural areas, rural peripheries have long been perceived as economically single-sided. With the *economic marginalisation of the agricultural sector*, the lack of alternative sources of income and subsequent outmigration, many rural areas have indeed experienced a process of peripheralisation as described in the previous section. This was (and still is) the result of the ongoing processes of intensification and industrialisation in agriculture often condemning agriculture in rural peripheries to losers in the global agricultural competition. Many of the high mountain landscapes of the Alps give evidence on the subsequent agricultural decline. Although the described observation is true for a considerable number of peripheries throughout Europe, it does not apply to all of them. It is worth mentioning that rural peripheries in the Scandinavian north are much more associated with forestry while agriculture hardly plays a role.

The same is true for many coastal peripheries (either in northern or southern Europe) where fisheries and other sources of income have played a more important role, thus providing a multitude of pillars forming the rural economy.

In effect, a more differentiated perspective is required when looking at the functions provided by rural peripheries (and rural areas at large). In contrast with the traditional agro-forestry picture, many rural peripheries have long been embedded in a much more multifunctional setting, providing jobs and income from a wider range of economic activities than just agriculture. With the increasing pressures agriculture is facing across Europe through regional and global competition, *multifunctionality* has gained conceptual and strategic relevance since the mid-1990s. Given the fact, that rural areas (and peripheries) have always provided — more or less naturally — a multitude of non-agricultural functions, such as fresh air, water, room for housing or infrastructure, to name but a few, multifunctionality is seen more and more as an "alternative" for the future development of these areas. With agriculture further losing its former importance as an economic sector — which is not paralleled by its significance with regard to land use — new functions, and accordingly forms of income, emerge, among them leisure and tourism, industrial activities, room for waste disposal as well as nature conservation, service provision and tertiary sector.

Interestingly, it was Christaller himself who was among the first scholars to pay attention to economic developments in non-central places: Three decades after the publication of the Central Place Theory, he identified tourism as the branch of the economy "that avoids central places and the agglomeration of industry" (Christaller 1964, 95). In his paper, originally given as a speech at the 12th Congress of the Regional Science Association in Lund in 1963, he explicitly referred to peripheries as "seasonal central points" (ibid., 96) stating: "Tourism is drawn to the periphery of settlement districts as it searches for a position on the highest mountain, in the most lonely woods, along the remotest beaches" (ibid., 95). Consequently, tourism studies add a further notion to the concept of peripherality: amenity landscapes and wilderness enabling nature-based and, more recently, ecotourism (Hall et al. 2013, 73–74). In many former rural peripheries across Europe, *tourism* is certainly a major factor in diversifying the local economy. Depending upon weather conditions, but also on amenities and infrastructure available and developed, in certain areas, such as the Alps and other mountainous areas, as well as in coastal regions, tourism contributes at least temporarily to the generation of local income (Reichert-Schick 2013; see also the papers by Reichert-Schick and W. Dietl et al. in this volume).

Reinvented rurality (called "musealisation" by an interviewee cited in the paper of Kordel and Weidinger in this volume) or artistic initiatives of external actors (T. Dietl in this volume) are further strategies to make rural peripheries more attractive to a predominantly metropolitan target group. Paradoxically, it corresponds to the very nature of tourism that its actors transform rural peripheries by way of "consuming" them and incorporating them in global networks and exploitation logics (Hall et al. 2013). Already Christaller (1964) hesitated to name certain regions that in his understanding were peripheral in order to avoid "the guilt to making these known" (ibid., 105). Still, in large parts, his text today reads like a travel brochure across Europe, although his claim that tourists look for "the silence and perceptibility of a rural milieu" which he contrasted with the "stony, narrow, dark and noisy towns" (ibid., 103) is today — in the heyday of urban tourism — only partly true.

To some extent, tourism overlaps with another revitalisation tendency to be observed in some rural peripheries in Europe: their (re)discovery as places for *secondary homes or destinations for retirement migration* (see the papers by Weidinger and Kordel and Depraz in this volume). But also permanent (Hoggart 1997) or job-related in-migration (Schmied 2012) have long been overlooked as trends in rural peripheries. These practices are a significant result of the freedom of movement, i.e., the far-reaching abandonment of barriers to labour and residential mobility within the EU. By analysing the case of Andalucía in the 1990s, Hoggart (1997) pointed to high rates of population turnover and in-migration even in the case of continuously high out-migration rates. Thus, methodologically, migration balances are insufficient measures as they rather cover than unlock small-scale population dynamics. For small towns in peripheral areas in Denmark, Fertner et al. (2015) pointed to the fact that between 2000 and 2013 this type of settlement experienced a positive, or at least stable, population. Both in central and in peripheral regions, this was mainly due to an increased immigration of elderly people to service towns. The authors interpret this trend as a sign for what they call "residential urbanism" which is decoupled from "job-created urbanism" (ibid., 124, 130). Furthermore, this analysis points to centralisation tendencies within rural peripheries that fundamentally question the large-scale core-periphery dichotomy as discussed above.

Finally, *energy production* represents another major branch of the economy which is of increasing importance in many rural peripheries across Europe (see also the papers by Spijkerboer et al., Rennie and Billing, Schaal and Lewerentz in this volume). A wide range of factors contribute to this development: national and European climate mitigation policies, the search for alternative energy concepts and sources of income, as well as the ongoing process of agricultural restructuring have supported the energy sector in becoming a major player in rural peripheries and rural areas as such. However, different observations regarding the development of energy production can be made in European perspective. Making use once more of the example of the Western Isles of Scotland, this example shows how particularly the use of renewable energies such as wind or wave power plays a significant role in the search for alternative development approaches (Mose and Jacuniak-Suda 2011; Rennie and Billing in this volume). In the context of the islands' present process of regeneration, questions regarding the extent of development as well as control and ownership have thereby become of crucial importance. Especially the local population's concerns about safeguarding the natural and cultural assets (as typical unique selling positions of the region) in favour of a sustainable development perspective have caused considerable friction and conflicts. As a result, participation and acceptance became issues of major concern regarding further development of energy production. In contrast, observations from the North Sea coast of Lower Saxony, Germany, illustrate problems of a well-developed "energy periphery" (Mose et al. 2010). For several decades the coastal region between the Elbe estuary and the German-Dutch border, traditionally known as a peripheral region, has developed into one of the most important areas of energy production in the whole country. This includes conventional fossil energies (coal, oil, gas) as well as various forms of renewables (wind, solar, biogas) that are either produced, traded (via several harbours) or stored (e.g., gas caverns) in the area. While energy production has thereby become a major driver of economic development in the area, it is also regarded a source of serious land-use conflicts. Current concerns are, for example, about the rather uncontrolled growth of maize production (*Vermaisung* in German) for the purpose of biogas production which is contributing to severe losses in biodiversity as well as damage to the scenic beauty of the landscape with negative impacts on recreation and tourism (Mose and Schaal 2012). Against this background it becomes obvious that energy production may well contribute to the redevelopment of rural peripheries, but there seems to be an urgent need for political settlements about clear development goals in order to avoid land-use conflicts and lacking acceptance among the local populations in Germany and elsewhere (see also Karrasch and Klenke and Spijkerboer et al. in this volume).

The example of energy as a (re)new(ed) source of income and economic development in rural areas, but also the above-mentioned cases of labour and residential mobility within the EU lead to the central question of which factors are decisive for the transformation of (former) rural peripheral and peripheralised areas. There has been a long academic debate on the role of exogenous, in contrast to endogenous, factors (a short overview on these issues is provided by Bock 2016, 29). To begin with, there is a widespread consensus that endogenous factors like social capital, a strong feeling of identity, immaterial capacities and material resources to enable collective actions are crucial preconditions for a revitalisation of rural peripheries (Reichert-Schick 2013, 284–285). However, with the concept of "neo-endogenous rural development", Ray (2006) pointed to the fact that rural development in the European Union is strongly influenced also by "extra-local" actors, as manifested in the various policy approaches of the EU since the early 1990s.

Indeed, EU policies addressing regional, and particularly rural, development have been of decisive importance in the restructuring of rural peripheries towards more multifunctional spaces and economies. Due to a more differentiated perception of problems of rural areas, a new emphasis laid on spatial multifunctionality, the sustainability discourse of the 1990s as well as changing planning cultures towards multi-level governance, gradually a conceptual and instrumental reorientation emerged. In 1996, the Cork Declaration "A Living Countryside" (EU COM 1996) made this political change public. Rural development in terms of agriculture, forestry, fisheries and food business, environmental protection and conservation, sustainable resource use, preservation of the cultural landscape and social life was ascribed a major role to play not only for rural areas but also for the territorial cohesion in Europe as such. At the same time, principles of subsidiarity were enhanced while rules of procedures for funding schemes became modified according to the newly set imperatives (Fischler 2000). This paradigm shift in the European policy for rural areas consisted of three major approaches: First, all regional actors in the spheres of economic, social and ecological developments were to be addressed and included in local activities. Second, the regions were conceived as spaces of co-operation by also neglecting administrative borders. Third, spatial development was to be realised as a task of public-private partnerships. Thus from the mid-1990s onwards, particularly the agrarian sector was to a much greater degree perceived as an actor in rural areas with regional, landscape and societal implications, thus going beyond its mere economic and financial functions (Sánchez and Fock 2008; Segert and Zierke 2007). European funding was and still is complemented by national as well as sector-related funding schemes which make the rural landscape of funding highly diverse and differentiated.

Rural peripheries were in particular affected by the fact that the LEADER instrument (*Liaisons entre actions de développement de l'économie rurale*), which had been installed in 1991 with its focus on endogenous development and innovative project as well as communication-oriented strategies, was confirmed. In the years to follow, it became an indispensable instrument within the European Agricultural Fund for Rural Development (EAFRD) (Grabski-Kieron and Krajewski 2007; Dax et al. 2016; Cawley and Jacuniak-Suda in this volume). LEADER thus developed to an essential impulse for realising the concept of an integrated rural development. In contrast with earlier policy concepts that had predominantly addressed agriculture and its structural problems, integrated rural development aims at combining regional and participatory approaches with strategic planning as well as a realisation of projects within a limited time period. It is intended as an integrated, multi-sectoral approach with a precise problem-orientation and instruments to be applied at the regional and local level (EU COM 2006; Grabski-Kieron 2012). LEADER prominently tackles this paradigm of integrated rural development.

Yet, while the principle intention and the notion of innovation inherent to LEADER is widely supported also by academic observers, evaluation studies are, however, somewhat disillusioning: Dax et al. (2016) in their comparative analysis between the implementation of LEADER in Austria and Ireland in the 2007–13 funding period argue that the initiative falls far behind its potential to impact rural regions in a positive way. Centralisation, a reduction in regional targeting and implementation difficulties are but some of the problems addressed by the authors. Notwithstanding this critique, regionalised and local rural development concepts remain crucial instruments both for a strategic as well as project-oriented rural development that also set impulses for rural peripheries (Friedel and Spindler 2009).

Finally, it needs to be highlighted that it is not only the structural and investment funds of the EU (such as ERDF/European Regional Development Fund, ESF/European Social Fund, CF/Cohesion Fund or, as just described, EAFRD) but also further policies which contribute to both peripheralisation and de-peripheralisation processes, among them EU border and environmental policies, respectively (Beurskens et al. and Grabski-Kieron in this volume). While social and regional policy instruments intend to balance structural locational disadvantages in European welfare states, in times of the states' fiscal crisis public services retreat particularly from rural peripheries, leaving it up to a sporadic local mix of welfare schemes and regional actors attempting to moderate these processes, as is the case, for example, in Italy (Osti in this volume).

Current transformations and revaluation processes in European rural peripheries: the contributions in this volume

Set against the background outlined so far, this volume provides an overview of research on current and former peripheral areas as well as on processes of peripheralisation and de-peripheralisation in Europe. The geographical scope of the papers covers rural peripheries and border regions in Austria, France, Germany, Ireland, Italy, The Netherlands, Poland, Scotland and Spain. The papers are structured into four parts with altogether sixteen contributions.

In the *first section*, different conceptualisations and types of rural peripheries are described by taking the examples of Scotland, France and Italy.

The paper by *Frank Rennie and Suzannah-Lynn Billing* on community land ownership and rural resilience focuses on a, in the Scottish terminology, "remote rural area" which was already referred to at the beginning of this introduction: the Western Isles or Outer Hebrides. The authors discuss impacts of the Scottish land reform on community welfare. Renewable energies in form of windfarms are introduced as an opportunity for rural communities to capitalise on the natural resources of their region and, consequently, to become independent from far-away centres.

Samuel Depraz provides thorough insights in various definitions, the neglect and redefinition of rural areas in France. With metropolisation and peri-urban growth being major processes of socio-spatial restructuring in France, he argues that rural peripheries (in the French debate sometimes called "hyper-rural" areas) partly benefit from a stronger residential base along with a demographic reversal. Yet the author also discusses ambiguities of this development by indicating, for example, a lack of internal means and, thus, a dependence on a circulation of resources on the national scale.

A further term in the language of European rural peripheries is introduced by *Giorgio Osti* who devotes his attention to "fragile areas" in rural Italy. He is particularly interested in the question of territorial disparities and operationalises them with a wide range of indicators of social inequality, such as health or life experience. Evidence is provided for the great variety of rural trajectories while the author at the same time points to the problem of "unbalanced" rural welfare which is not calibrated to the varying needs of distinct regions.

Different European policies and their impacts on regional governance, participation and daily life are at the focus of the *second part* of the volume.

By going back to as early as 1991, *Mary Cawley* applies a long-term perspective on the implementation and governance problems of LEADER in Ireland, and discusses the changing relationship of this instrument to national and regional policies. While the LEADER initiative in Ireland contributed in its early years to a redistribution of responsibility for local development from the central to the local scale, the author in recent years observes a recentralisation of rural development issues along with an undermining of the principle of subsidiarity.

An interesting counterpoint to this long-term perspective is provided by *Marta Jacuniak-Suda*. In her paper, she deals with the theory and practice of LEADER in Poland, a comparatively young EU member state. She shows that participants of the public sector dominate the local action groups established in the course of implementing LEADER which contradicts the very idea of multi-actor and public-private projects. However, the author also argues that in spite of all critique, new forms of rural governance are about to evolve.

Ulrike Grabski-Kieron sheds light on governance processes in the field of nature conservation by taking the example of cross-border biotope networks in Germany and the Netherlands. While the former peripheral character of these regions contributed to their biodiversity, which is today considered as valuable and worthy of protection, the region is also subject to de-peripheralisation processes in the course of the European integration. Therefore, regional stakeholders claim a necessity of biodiversity conservation. The author particularly underlines the crucial role of informal instruments and processes.

Also the final contribution in this section is devoted to internal borderlands of the European Union. *Kristine Beurskens, Judith Miggelbrink and Christoph Creutziger* question the mainstream story of open borders unequivocally contributing to European integration. They argue that EU-internal borders might be regarded by borderland populations as places of crime and insecurity. Therefore, peripherality here relates to the perception of these regions as peripheral to state control. Securitisation, then, is a local strategy implemented via community-based initiatives and technical devices in order to rectify the perceived failure of the state.

The papers of the *third section* discuss exogenous and endogenous impulses to rural revitalisation through tourism and in-migration.

The specific importance of tourism to initiate de-peripheralisation processes but also the factors impeding such efforts is the focus of *Anja Reichert-Schick*'s contribution. She provides evidence on two hotel villages in Italy and contrasts them with a similar concept realised on a former manor in northeast Germany. Hotel villages are discussed as a means of an at least temporary revitalisation of depopulated villages in non-tourist regions. It becomes obvious, however, that no standard solution concerning finances and actors exists.

Private tourist entrepreneurs, hotel managers and other actors producing rurality are of particular interest to *Stefan Kordel and Tobias Weidinger*. They present findings from a case study in the Bavarian Forest, a traditional rural periphery at the German-Czech border the position of which has not decisively changed since the EU enlargement in 2004. The authors show how tourist entrepreneurs implement different notions of rurality, particularly that of the "rural idyll" which is linked to positive notions of peripherality, such as distance from modernity and stressful urban lifestyles — while at the same time providing accommodation facilities with technical achievements of the 21st century.

The same two authors, *Tobias Weidinger and Stefan Kordel,* in another common paper, investigate a topic of intra-European migration: the housing mobility of German (pre)retirees to rural Spain. While rooted in the long-term debate on amenity migration, the authors' perspective is specific in focusing on migrants' decisions and post-hoc rationalisations of intra-European moves in times of the financial and real estate crisis in Spain.

Finally, *Theresa Dietl* takes up a topic which is only rarely considered in rural studies: the role of public and participatory art in village development. By describing and analysing one out of three projects initiating contemporary art in villages in rural Germany in 2013, she shows how the local population participated in the project and reflects on the implications of such informal involvement for future village planning.

The *fourth and final part* highlights new perspectives in the fields of renewable energies, climate change and sustainable mobility. It starts with three papers relating to recent developments at the coasts of the North Sea from Germany and The Netherlands, respectively.

Leena Karrasch and Thomas Klenke introduce a model of participatory adaptive land use planning and assessment (PALUPA) for improving ecosystem-based management in northwest Germany. Within the framework of an inter- and transdisciplinary research project, local adaptive land use management strategies were developed in a joint effort with regional decision-makers and knowledge-brokers. The paper provides insights into the regional challenges to combine climate change adaptation and rural development in peripheral areas.

The contribution by *Rozanne Spijkerboer, Elen-Maarja Trell and Christian Zuidema* considers the potential of renewable energies for local development by taking the example of the Eemsdelta region in the Netherlands. This region has a strategic orientation towards a leading position in a bio-based economy. The authors frame the related policies by applying the concept of social-ecological resilience. It becomes obvious that there is still a large gap between rhetorical claims and the actual realisation of how benefits from renewable energies might be materialised to counteract rural peripheralisation.

Renewable energies are also a major issue in the paper by *Peter Schaal* who is interested in the implications of the German energy transition (*Energiewende*) for local policies in rural areas in northwest Germany. While rural peripheries are a hotspot of the production of renewable energies in Germany, climate change mitigation is no major issue for local administrations, as mirrored by missing integrated concepts to enable climate mitigation. Yet, their active participation is needed in order to achieve the ambitious German climate goals.

Werner Dietl, Andreas Friedwagner, Benedikt Hahn, Martin Heintel, Thomas Langthaler, Alice Wanner and Norbert Weixlbaumer discuss a major issue in the development of rural areas, let alone rural peripheries: mobility for daily activities. Based on empirical research on the use of public transport for shopping in Burgenland (Austria), a "structurally weak peripheral rural region", as it is called by the authors, they provide policy recommendations to municipalities and local suppliers to improve the mobility of the customers.

Mobility is also the focus of the final paper by *Henry Lewerentz*. He argues that costs for individual mobility in rural peripheries are increasingly transferred from the public sector to private households. This is particularly due to prolonged distances in the course of centralisation of public and private services. Based on empirical data from northeast Germany, the author concludes that alternative engine technologies, in particular electromobility, could create economic, ecological and social advantages for the households and rural peripheries in the future.

Conclusion

"Periphery" and "core" (or "centre") seem to be strong and persistent concepts in regional studies, similar to the well-known rural-urban dichotomy. Yet, the often normative and usually stigmatising use of the term "periphery," as still present, for example, in handbooks of geography (e.g., Mayhew 2009, 106–107) needs to be overcome. In 2001, the economic geographer Andrew Copus already fundamentally questioned the applicability of a spatial understanding of peripherality (Copus 2001), while Frank Rennie and Suzannah-Lynn Billing (in the first paper of this volume) go even one step further: they suggest that the periphery is considered by many rural dwellers as the centre whereas the city is experienced as remote and peripheral. Such a paradigm shift — once more applied from the perspective of the Western Isles at the northwest edge of Europe — could be a fruitful starting point for future research and policies relating to European peripheral(ised) regions.

References

Amin, S. (1976). L'impérialisme et le développement inégal. Paris: Minuit.

Barca, F., Casavola, P. & Lucatelli, S. (2014) (Eds). A Strategy for Inner Areas in Italy: Definition, Objectives, Tools and Governance. Materiali UVAL 31. Rome.

Beetz, S., Huning, S. & Plieninger, T. (2008). Landscapes of Peripherization in North-Eastern Germany's Countryside: New Challenges for Planning Theory and Practice. International Planning Studies 13 (4), 295–310.

Bernt, M. & Liebmann, H. (Eds.) (2013). Peripherisierung, Stigmatisierung, Abhängigkeit? Deutsche Mittelstädte und ihr Umgang mit Peripherisierungsprozessen. Wiesbaden: Springer VS.

Bertrand A. (2014). Hyper-ruralité. Un pacte national en 6 mesures et 4 recommandations pour "restaurer l'égalité républicaine". www.logement.gouv.fr/IMG/pdf/rapport_a_bertrand_vdef.pdf. Accessed 13 May 2016.

Bock, B. B. (2016). Rural marginalisation and the role of social innovation; a turn towards nexogenous development and rural reconnection. Sociologia Ruralis, online first: DOI 10.1111/soru.12119.

Brunotte, E., Gebhardt, H., Meurer, M., Meusburger, P., Nipper, J. & Martin, C. (2002). Lexikon der Geographie in vier Bänden. Dritter Band: Ökos bis Wald. Heidelberg and Berlin: Spektrum, Akademischer Verlag.

Bürk, T. (2013). Voices from the margin. The Stigmatization Process as an effect of socio-spatial Peripheralization in small-town Germany. In Fischer-Tahir, A. & Naumann, M. (Eds.), Peripheralization: the making of spatial dependencies and social injustice (pp. 168–186). Wiesbaden: Springer VS.

Champion, T. & Graeme, H. (Eds.) (2004). New Forms of Urbanization. Beyond the Urban-Rural Dichotomy. Aldershot, Burlington: Ashgate.

Christaller, W. (1933/2006). Die zentralen Orte in Süddeutschland. Eine ökonomisch-geographische Untersuchung über die Gesetzmäßigkeit der Verbreitung und Entwicklung der Siedlungen mit städtischen Funktionen. Darmstadt: Wissenschaftliche Buchgesellschaft.

Christaller, W. (1964). Some Considerations of Tourism Location in Europe: The Peripheral Regions — Underdeveloped Countries — Recreation Areas. Papers in Regional Science 12 (1), 95–105.

Copus, A. K. (2001). From Core-periphery to Polycentric Development: Concepts of Spatial and Aspatial Peripherality. European Planning Studies 9 (4), 539–552.

Dahlström, M. (1996). Young women in a male periphery — experiences from the Scandinavian North. Journal of Rural Studies 12, 259–271.

Dax, T., Strahl, W., Kirwan, J. & Maye, D. (2016). The Leader programme 2007–2013: Enabling or disabling social innovation and neo-endogenous development? Insights from Austria and Ireland. European Urban and Regional Studies 23 (1), 56–68.

European Commission (EU COM) (1996). The Cork Declaration — A Living Countryside. European Conference on Rural Development, November 1996, Cork, Ireland. ec.europa.eu/agriculture/rur/cork_en.htm. Accessed 5 May 2016.

European Commission (EU COM) (2006). The LEADER approach: A basic guide. Luxembourg: Office for Official Publications of the European Communities. ec.europa.eu/agriculture/rur/leaderplus/pdf/factsheet_en.pdf. Accessed 20 May 2016.

Fassmann, H. (1990). Arbeitsmärkte in der Peripherie. In Fassmann, H. & Pröll, U. (Eds.), Standort Burgenland. Probleme und Entwicklungschancen der Peripherie (pp. 6–22). Eisenstadt: Prugg.

Ferrão, J. & Lopes, R. (2004). Understanding peripheral rural areas as contexts for economic development. In Labrianidis, L. (Ed.), The future of Europe's rural peripheries (pp. 31–61). Aldershot: Ashgate.

Fertner, C., Groth, N. B., Herslund, L. & Carstensen, T. A. (2015). Small towns resisting urban decay through residential attractiveness. Findings from Denmark. Geografisk Tidsskrift — Danish Journal of Geography 115 (2), 119–132.

Fischer-Tahir, A. & Naumann, M. (Eds.) (2013). Peripheralization: the making of spatial dependencies and social injustice. Wiesbaden: Springer VS.

Fischler, F. (2000). Ländliche Entwicklung im 21. Jahrhundert. Zeitschrift für Kulturtechnik und Landentwicklung 41 (4), 112–120.

Förste, D. (2013). Innere Peripherien in großstädtischen Kontexten — das Beispiel Berlin. In Bernt, M. & Liebmann, H. (Eds.), Peripherisierung, Stigmatisierung, Abhängigkeit? Deutsche Mittelstädte und ihr Umgang mit Peripherisierungsprozessen (pp. 193–215). Wiesbaden: Springer VS.

Friedel, R. & Spindler, E. E. (Eds.) (2009). Nachhaltige Entwicklung ländlicher Räume. Chancenverbesserung durch Innovation und Traditionspflege. Wiesbaden: VS Verlag für Sozialwissenschaften.

Friedmann, J. (1973). A Theory of Polarized Development. In Friedmann, J., Urbanization, Planning, and National Development (pp. 41–64). Beverly Hills and London: Sage.

Glaser, R., Gebhardt, H. & Schenk, W. (2007). Geographie Deutschlands. Darmstadt: Primus.

Glatz, H. & Scheer, G. (1981). Neue Entwicklungsstrategien für strukturschwache ländliche Regionen. Wien: Institut für Höhere Studien.

Göler, D. (2005). Rückzug aus der nördlichen Peripherie Russlands? Jüngere räumliche Entwicklungen im hohen Norden Ostsibiriens und des Fernen Ostens — ein Beitrag zur peripheren Transformationsforschung. Leipzig: Leibniz-Institut für Länderkunde.

Grabski-Kieron, U. (2012). Le développement rural intégré en Allemagne: une démarche de planification et d'action prise entre ambitions et réalités. In Laquement, G., Born, K. M. & v. Hirschhausen, B. (Eds.), Réinventer les campagnes en Allemagne (pp. 185–198). Lyon: ENS Éditions.

Grabski-Kieron, U. & Krajewski, C. (2007). Ländliche Raumentwicklung in der erweiterten EU — Chancen und Probleme. Geographische Rundschau 59 (3), 12–20.

Hall, C. M., Harrison, D., Weaver, D. & Wall, G. (2013). Vanishing Peripheries: Does Tourism Consume Places? Tourism Recreation Research 38 (1), 71–92.

Hechter, M. (1975). Internal colonialism. The Celtic fringe in British national development, 1536–1966. Berkeley: University of California Press.

Heintel, M. (1998). Einmal Peripherie — immer Peripherie? Szenarien regionaler Entwicklung anhand ausgewählter Fallbeispiele. Wien: Universität Wien, Institut für Geographie und Regionalforschung.

Henkel, G. (2004). Der Ländliche Raum. 4th extended edition. Stuttgart and Berlin: Gebrüder Borntraeger.

Hoggart, K. (1997). Rural Migration and Counter-urbanization in the European Periphery: The Case of Andalucía. Sociologia Ruralis 37 (1), 134–153.

Keim, K.-D. (2001). „Aufbau Ost". Schrumpfende Städte und peripherisierte Regionen. Wanderungsbewegungen in den ostdeutschen Ländern. In Schader-Stiftung (Ed.), wohn:wandel. Szenarien, Prognosen, Optionen zur Zukunft des Wohnens (pp. 66–76). Darmstadt: Schader-Stiftung.

Keim, K.-D. (2006). Peripherisierung ländlicher Räume — Essay. Aus Politik und Zeitgeschichte 37 (6), 3–7.

Kreckel, R. (2004). Politische Soziologie der sozialen Ungleichheit. 3rd extended edition. Frankfurt/M. and New York: Campus.

Kühn, M. (2015). Peripheralization: Theoretical Concepts Explaining Socio-Spatial Inequalities. European Planning Studies 23 (2), 367–378.

Kühn, M. (2016). Peripherisierung und Stadt. Städtische Planungspolitiken gegen den Abstieg. Bielefeld: transcript.

Labrianidis, L. (Ed.) (2004). The future of Europe's rural peripheries. Aldershot: Ashgate.

Lang, T., Henn, S., Ehrlich, K. & Sbignev, W. (Eds.) (2015). Understanding Geographies of Polarization and Peripheralization. Perspectives from Central and Eastern Europe and Beyond. Basingstoke: Palgrave Macmillan.

Leibert, T. (2016). She leaves, he stays? Sex-selective migration in rural East Germany. Journal of Rural Studies 43, 267–279.

Mayhew, S. (2009). A Dictionary of Geography. Oxford and New York: Oxford University Press.

Miner, H. (1952). The folk-urban continuum. American Sociological Review 17, 529–537.

Mose, I. & Jacuniak-Suda, M. (2011). Designing sustainable pathways: the Highlands and Islands in transition. Findings from empirical case studies in the Western Isles. In Borsdorf, A., Stötter, J. & Veulliet, E. (Eds.), Managing Alpine Future II: Inspire and drive sustainable mountain regions (pp. 58–70). Wien: Verlag der Österreichischen Akademie der Wissenschaften.

Mose, I., Jacuniak-Suda, M. & Fiedler, G. (2014). Regional Goverance-Stile in Europa. Eine vergleichende Analyse von Steuerungsstilen ausgewählter LEADER-Netzwerke in Extremadura (Spanien), Warminsko-Mazurskie (Polen) und Western Isles (Schottland). Raumforschung und Raumordnung 72 (1), 3–20.

Mose, I., Oberdörffer, J. & Scheele, U. (2010). Küste unter Strom. Die deutsche Nordseeküste als Energiestandort. Praxis Geographie 40 (3), 10–15.

Mose, I. & Schaal, P. (2012). Probleme der Intensivtierhaltung im Oldenburger Münsterland. Lösungsstrategien im Widerstreit konkurrierender Interessen. Neues Archiv für Niedersachsen no. 2, 50–69.

Naumann, M. & Reichert-Schick, A. (2013). Infrastructure and Peripheralization: Empirical Evidence from North-Eastern Germany. In Fischer-Tahir, A. & Naumann, M. (Eds.), Peripheralization. The making of spatial dependencies and social injustice (pp. 145–167). Wiesbaden: Springer VS.

Nitz, H.-J. (1997). Der Beitrag der historischen Geographie zur Erforschung von Peripherien. In Nolte, H.H. (Ed.), Europäische innere Peripherien im 20. Jahrhundert (pp. 17–36). Stuttgart: Steiner.

Nolte, H.-H. (1991). Internal Peripheries. A Definition and a Note. In Nolte, H.-H. (Ed.), Internal Peripheries in European History (pp. 1–28). Zur Kritik der Geschichtsschreibung 6. Göttingen and Zürich: Muster-Schmidt.

Nolte, H.-H. (Ed.) (1997). Europäische Innere Peripherien im 20. Jahrhundert. Stuttgart: Steiner.

Pérez-Soba, M. et al. (2012). GEOSPECS: Inner Peripheries: a socio-economic territorial specificity. Luxemburg : ESPON Coordination Unit.

Ray, C. (2006). Neo-endogenous rural development in the EU. In Cloke, P., Marsden, T., Mooney, P. H. (Eds.), Handbook of Rural Studies (pp. 278–291). London et al.: Sage.

Reichert-Schick, A. (2013). Abwanderung — Entleerung — Wüstung? Verlauf, Determinanten und Implikationen von Peripherisierungs-, Regressions- und Wüstungsprozessen in ländlichen Siedlungen gestern und heute (unpublished postdoctoral thesis/Habilitationsschrift). Universität Trier.

Sánchez, J. M. T. & Fock, T. (2008). Die EU-Agrarförderung im regionalen Kontext — am Beispiel von Mecklenburg-Vorpommern und Galicien (Nordspanien). Ländlicher Raum 59 (5), 21–24.

Schmied, D. (2005). Winning and losing. The changing geography of Europe's rural areas. Aldershot: Ashgate.

Schmied, D. (2012). Zuwanderung von Arbeitsmigranten als Rettung für die europäische Peripherie? Das Beispiel Schottland. In Schmied, D. (Ed.), Peripherien in Europa zwischen Rückbau und Umbau (pp. 97–124). RURAL 6. Göttingen: Cuvillier.

Segert, A. & Zierke, I. (2007). Regionalinitiativen. Entwicklungsressource strukturschwacher ländlicher Räume. Raumforschung und Raumordnung 65 (5), 421–434.

Senghaas, D. (1974). Peripherer Kapitalismus: Analysen über Abhängigkeit und Unterentwicklung. Frankfurt/M.: Suhrkamp.

Steinführer, A. & Kabisch, S. (2007). Binnen- und Außenimage von Johanngeorgenstadt. In Wirth, P. & Bose, M. (Eds.), Schrumpfung an der Peripherie. Ein Modellvorhaben — und was Kommunen daraus lernen können (pp. 107–123). München: oekom.

Steinführer, A., Küpper, P. & Tautz, A. (2014). Adapt and Cope: Strategies for Safeguarding the Quality of Life in a Shrinking Ageing Region. Comparative Population Studies 39 (2), 345–370.

Terluin, I. J. (2003). Differences in economic development in rural regions of advanced countries: an overview and critical analysis of theories. Journal of Rural Studies 19 (3), 327–344.

Wallerstein, I. (1974). The modern world-system. New York: Academic Press.

Weiß, W. (2002). Der Ländlichste Raum — Regional-demographische Begründung einer Raumkategorie. Raumforschung und Raumordnung 60 (3/4), 248–254.

Wiest, K. (2016). Migration and everyday discourses: Peripheralisation in rural Saxony-Anhalt from a gender perspective. Journal of Rural Studies 43, 280–290.

1 At the edge?
Rural peripheries of and in Europe

Community land ownership and rural resilience: A case study from the Highlands and Islands of Scotland

Frank Rennie and Suzannah-Lynn Billing

Abstract

Over recent years, the Scottish Parliament has prioritised land reform as a means of encouraging rural development and the promotion of greater social inclusiveness. As a result, the formation of local community-owned land trusts as a mechanism for greater local participation in the control of local improvement has become a particularly significant aspect of sustainable rural development in the Highlands and Islands of Scotland. To illustrate this, a case study of the Galson Estate Trust, Isle of Lewis, Scotland, is presented. Community ownership of land is presented as being instrumental in the shift from short-term, economically-marginal, externally-dependent activities towards solutions that are long-term, sustainable, and resilient to unwanted change. The process of local empowerment accompanying this shift in perceptions is regarded as having substantial benefits for social capital and skills acquisition at the community level. Early results indicate that ownership of the land, together with the capacity-building process of community participation in its management, has encouraged a sense of local empowerment, together with providing employment, confidence, and other social improvements.

Key issues and challenges faced by very rural communities

Over the past three decades, much of the attention on strategies of survival for rural communities has been focussed on improving our understanding of the concept of "sustainable development". To some extent, the nature of "sustainability" relates to the extent of the contrast between very rural and very urban areas. In this context, the terms "remote" and indeed "peripheral" are both descriptive and emotive, because they have a direct implication that they are further away from the "centre" of things, i.e., from large urban populations. For many rural dwellers, however, their own society is at the centre, and it is the city which is remote and peripheral, so the terms peri-urban, rural, and very rural are less value-laden terms which can be used to describe the characteristics of the urban-rural continuum.

Generally speaking, attitudes towards development in rural areas have moved from a post WW2 "food security" perspective, with agricultural self-sufficiency being a key criterion of a sustainable rural society, to a view that encompasses broader definitions of the countryside in terms of land-use diversity, environment, and social development (Nicol 1998).

Concepts of sustainable development in the communities of the Highlands and Islands of Scotland have followed this broad trend in society attitudes, moving from subsistence agriculture as a key indicator, to a basket of factors including economic goals (Copus and Crabtree 1996), social measures (Parman 1990), and habitat management as ways of trying to understand the human interaction with the natural environment (Beaufoy et al. 1994). A number of in-depth studies (Dùthchas 2001) have attempted to evaluate specific communities to determine appropriate strategic approaches for sustainable development in areas that are economically fragile but internationally recognised for the quality of their natural environment (Rennie 2008). These studies have supported the need for a broad-based approach to community development as an essential component of sustainability (Bridger and Luloff 1999). Increasingly there has been a shift towards encouraging participation and decision-making of sustainable development issues at the local level of the community (rather than simply the responsible national agencies). It has been noted that *"definitions of sustainable community development stress the importance of striking a balance between environmental concerns and development objectives while simultaneously enhancing local social relationships"* (ibid., 381).

This period also produced a number of studies on the perceived quality of life and poverty issues in rural communities (Shucksmith et al. 1995) and this brought to the fore issues of social and economic contrast between rural and urban areas. Included in these discussions was the perception of rural advantage and disadvantage, particularly in terms of public services, housing, and cultural, social and economic opportunities. Much of the literature over the next two decades investigated these various factors as a contribution to the "quality of life" debate and related to the concept of social exclusion. This concept was often used as an inter-changeable aspect of poverty and in a study of the changing nature of how poverty and social exclusion are measured. Levitas et al. (2007, 25) suggested the following succinct definition:

> *"Social exclusion is a complex and multi-dimensional process. It involves the lack or denial of resources, rights, goods and services, and the inability to participate in the normal relationships and activities, available to the majority of people in a society, whether in economic, social, cultural or political arenas. It affects both the quality of life of individuals and the equity and cohesion of society as a whole".*

Partly as a result of the privatisation of public resources, and partly as a reaction against this, there has been a period of growth of the instances of public participation driving and managing development activities. This has come in several forms:

1) the creation of employment and economic opportunities through community businesses and other social enterprise initiatives (often based upon the non-distribution of profit and/or the re-circulation of profit within the community);
2) the acquisition of resources, such as buildings, forests, and landed estates;
3) community management of key services, often on a sub-contact to local or national government and agencies.

This citizen engagement led to the spread of the concept of Asset-based Community Development (ABCD) as a strategy for community-driven development (Mathie and Cunningham 2003). This approach was expounded as an alternative approach to build upon community strengths and assets, rather than to focus on deficiencies and negative aspects. Key points of ABCD are that it encourages participatory approaches to community inquiry and analysis, then builds upon this to strengthen civil society and social capital to build the capacities of individuals and community organisations. ABCD should evolve from the strengths and activities present *in* the community — it is not a process that is done *to* the community by external organisations. This approach to an understanding of sustainable community development, which does not simply consider economic growth versus environmental protection, but also emphasises the importance of social aspects of development, has led to a resurgence of interest in communities of place. In particular, it has built upon the ideas of the concept of social capital (Portes 1998) to investigate the effects of the local skills, organisations, networks, and reciprocal trust present within the community.

A common aspect of development activities in very rural communities is that they are often a response to situations of (actual or perceived) market failure. The challenging conditions of the very rural community — transport links, employment opportunities, population levels, demographic diversity, social services — have meant that solutions that have been used in urban communities are not necessarily socially or economically viable in the rural context. In fact, as classical economics demands the free mobility of both labour and capital, the counter demands of economics and community-based development can confuse the concept of "community" and can interfere with the free-flow of vital goods, labour, and other resources (Mathie and Cunningham 2003). On the other hand, *"Social capital, with its emphasis on trust and co-operation, is an attractive concept. After all, it holds out the possibility of developing voluntaristic solutions to problems that have not been solved through market mechanisms, government programs or legislation"* (Bridger and Luloff 2001, 467).

In recent years, the focus has shifted from identifying indicators of sustainable communities, to an analysis of the ability of communities to adapt to change. A contextual environment of continuing change (often due to external factors, beyond the control of the community) is recognised as a significant contributory factor in social sustainability and the ability to cope with these changes has been termed 'resilience'. Community resilience has been defined as:

> *"the existence, development, and engagement of community resources by community members to thrive in an environment characterised by change, uncertainty, unpredictability, and surprise. ...members of resilient communities intentionally develop personal and collective capacity that they engage to respond to and influence change, to sustain and renew the community, and to develop new trajectories for the communities' future" (Magis 2010, 402).*

An advantage of this approach is that it can be construed in such a manner as to include the assessment of sustainable development, social exclusion and social capital, all within a common framework of understanding of the ability of communities to cope with changing circumstances. It also utilises different disciplinary perspectives and different historical and environmental contexts to inform both policy and practice in the transformation of communities in response to changing circumstances. The concept has been used effectively, for example, in analysing both reactive and proactive community actions by drawing upon the recent experiences of community land purchases in Scotland as a means of stimulating community enterprise and local development (Skerratt 2011). This concept of community resilience challenges the idea that community action is only achieved as a response to external threats, and it highlights the proactive change being implemented by geographical communities in attempts to improve their conditions (Skerratt 2013).

Introduction to the region

A major issue in the pursuit of sustainable development for many rural communities has been the need to develop a measure of resilience to unwanted change, which is often externally driven. This resilience has been sought through utilising the human and natural resources of the area in a manner that sensitively exploits the ability of these resources to adapt to and benefit from change. This case study illustrates one such rural community and the conclusions that can be drawn from its responses to change.

Historically, the rural communities of the Highlands and Islands of Scotland have been closely associated with extensive agriculture (low input/low output) and a complementary high level of nature-conservation-value land use (Rennie 1991; SCU/RSPB 1992). Over the past two centuries an evolutionary trend can be traced in community perspectives relating both to attitudes towards the conservation of the natural environment (Smout 1990) and in the understanding of the meaning(s) of sustainable development (Rennie 2008) in relation to these rural communities.

One of the early actions of the re-formed Scottish Parliament was to address the issue of land reform, particularly with regards to the crofting areas of the Highlands and Islands, which have a historical legacy of contested and highly emotive land ownership issues (Wightman 2013). Crofts are very small units of tenanted agricultural holdings on which the tenants (crofters) have inherited security of tenure. The special laws governing crofting have their origins in the historical resistance to eviction and out-migration of the rural population of the Highlands and Islands. Both before and subsequent to the Land Reform Act of 2003, there has been an extensive contemporary dialogue on land reform in Scotland (Rennie 1995) placing this regional debate in a national (Hoffman 2013; LRRG 2014) and an international context (Bryden and Geisler 2007). In part, the debate has been centred on the political justification and process for enabling greater community responsibility in the ownership of land, on the mechanisms for funding the purchase of private land by communities, and on the ongoing management of this land for the common good. The result has been a dramatic growth in the ownership and management of crofting estates by the resident communities. Using both locally generated and external funds, over 50 communities throughout the region have now acquired the title to the land on which they reside, with a corresponding change in emphasis for management practices and development agendas (Hunter and MacLean 2012).

Where previously, many landed estates in the Highlands and Islands were viewed primarily as providing sport (shooting and fishing) for paying guests, or simply for collecting croft rents, the new community-owned estates have based their business plans on perceived benefits and improvements for the local, resident community. The combined membership of the organisation founded by these Land Trusts, Community Land Scotland, currently owns and manages around 500 000 acres (CLS 2014). In the Outer Hebrides, it is calculated that, from less than 10% pre-2003, more than 50% of all land is now community owned, with 75% of the population now living on community owned land. The main significance in legal terms is that the land-owning rights are transferred from a private owner to a non-profit-distributing company, owned by the community and managed by democratic election by the members of that community. Land reform, as currently practiced in Scotland, is an attempt to create a more equal distribution of land ownership, and in effect *"implies changes in the balance of power between the individual property owners, communities, and the state"* (Bryden and Hart 2000, 3). Under the 1976 crofting Reform Act, crofters were enabled to purchase their land, but the long-established right of security of tenure has meant that the vast majority of crofters have chosen to remain tenants, paying an annual rent to a landowner. The popularity of current community landownership initiatives, in contrast to the general lack of popularity of the crofters' pre-existing right to purchase their individual croft, can be seen as a confirmation of a belief in the collective power to influence effective local development, without introducing new limitations to the individual's management of their croft (a type of tenant subsistence agriculture).

In the early part of the 20th century, the whole of Lewis and Harris, the largest island in the Outer Hebrides, was owned by Lord Leverhume, and when he left the region in 1924 he sold the whole island in a number of smaller parcels which became separate estates. Most of these estates were crofting areas in which the crofters' security of tenure was well established and legally guaranteed. In practice, this means that landowners have limited management rights, but are able to collect the annual croft rents. The area in the north-west of Lewis, now known as Galson Estate Trust, was sold by Lord Leverhulme in 1924 to a private landlord for just £ 500 — less than £1 for every hundred acres acquired. During the decade following WW1, there were numerous incidents of civil disruption and land agitation in the Highlands and Islands (Leneman 1989) as returning war veterans and impoverished communities sought to realise government promises of access to land for agriculture and house-building. There were numerous land-raids and civil disruptions in Lewis, and there were several land re-settlement villages in the study-area of the Galson Estate. The Galson Estate subsequently passed into the hands of three families who were descendants of the new owner, and who lived on the Isle of Lewis but not on the estate.

Following the 2003 Land Reform Act of the Scottish Government, which specifically enabled crofting communities to initiate legal proceedings to purchase the land on which they were tenants, a number of estates in the Highlands and Islands took initial steps to acquire the land, including the villages of the Galson Estate. Initially the Galson Estate buy-out began as a so-called "hostile" buy-out (the owners had not intended to sell) but in the course of subsequent negotiations, this became an amicable settlement. In 2004, this community organised a vote of the electorate over 16-years-of-age on the land-area covered by the estate, in which there was a 72% turnout and an 85% majority vote in favour of a buy-out. A Trust — a company limited by guarantee, with charitable status — (officially known as Urras Oighreachd Ghabhsainn) was then set-up, and a working group was formed to steer the negotiations to secure community ownership (Galson Estate Trust 2014). Over the next few years, the working group negotiated grant and loan funding — usually through contractual agreements with public agencies and charities — towards the cost of purchasing the land, at a cost of £600 000. A second public ballot was then held and, with a 76% voter turnout, a Board of ten Trustees was elected from 30 nominees to manage the business of the company. In 2007 the 22 villages of the Galson Trust Estate, with a resident population of just over 2 000, were successful in purchasing the 55 800 acre (22 600 hectare) estate in the north-west of the Isle of Lewis.

From an early stage it was apparent that although the existing croft rents and other accrued income from the estate would enable the Galson Trust to remain solvent, additional income would be required to engage with new development in these communities. During the public meetings, consultations, and meetings of the Trust during this period, it was apparent very early-on that the members of this community had long-term aspirations for developing numerous social and economic activities on the estate, rather than just the short-term profits characterised by the majority of landowners, and this was reflected in some of the suggestions in the community consultation survey. Some of these activities which have a relevance to the concept of sustainable development have already been set in motion by the Trust; including a contract with the Local Authority to manage local domestic waste disposal; local environmental schemes; and a focus on the reduction of fuel poverty by achieving greater energy efficiency in local housing. Around 71% of householders in the islands (Scottish average is 27%) indicate that they spend more than 10% of their income on fuel, (TEAS 2014) 18% are in severe fuel poverty (at least 20% of their income); and 11% are in extreme fuel poverty, spending 30% of their income on fuel. The production and efficient management of energy locally is therefore of prime community concern.

The energy resource

During the same period of time (2004) that the community were organising their bid to purchase the Galson Estate, an application was made by a private company to create the Lewis Wind Farm — the largest wind farm proposal in Europe at that time. The aim was to construct 234 turbines, each a 90 m conical steel tower with 100 m diameter rotor and a 3MW generation capacity. Roughly half of these proposed wind turbines were to be located on the land of the Galson Estate in a deal with the syndicate who owned the land at that time. There was considerable concern in these communities about the impact of such a massive development locally, about the perceived loss of the ability to influence local development issues in general and that communities were not fully consulted and would realise minimum benefits from any developments (Wemyss 2011).

A well-documented formal public enquiry followed (ibid.) and in the end, amid considerable local opposition to the proposal, the wind-farm planning application was rejected. It might be expected that there would then be a residual apprehension — even hostility — towards any further plans for the erection of wind turbines. In actual fact, the initial 2004 feasibility study by the Trust into local community management options specifically identified wind and wave energy developments as possible sources of revenue funding for the estate. A subsequent community consultation (Handley 2013) on opportunities for local development indicated that 19% of the responses gave a 'high priority' to renewable energy generation options. When the subsequent (current) turbine site was identified as potentially suitable, a ballot of the township in this location was also strongly in favour of the development. This more positive view of community owned wind turbines over external commercial companies has also been identified in other localities (Warren and McFadyen 2010).

In 2008 the Trust established a wholly-owned subsidiary trading company called Galson Energy Ltd in order to progress the wind energy generating ambitions of the community and feed profits back into the main holding charity (the Trust). This resulted in full planning permission in 2009 for three wind turbines near the village of Ballantrushal, the first of which was successfully commissioned in early 2014. It is significant that of only four individual objections to the planning proposal, only one was from a person living on the estate (MacKinnon 2014) compared with very vociferous opposition (Wemyss 2011) and 10 924 letters of objection (Pasqualetti 2011, 211) to the Lewis Windpower proposal for turbines in the same geographical area.

The total cost of the first turbine phase was around £ 2.4 million (plus VAT) for the installation of a functioning 900 Kilowatt turbine, with two additional turbines planned for the same location. Start-up investment for the first turbine was through a cooperative bank loan, and a further £ 700 000 funding towards the other two turbines was raised through a share investors scheme open to the public. It is anticipated that when fully operational, each turbine will produce revenue of up to £ 150 000 per year, yielding as much as £ 450 000 annually for re-investment in the local community. This income will be deployed to three key areas; 1) repaying the investment costs of the turbines; 2) creating a reserve fund for the Galson Trust; 3) establishing a development fund to help facilitate selected community activities on the estate.

In addition, the west coast of the Galson Estate Trust has attracted attention from three wave energy developers since 2005 due to its high energy marine environment and relatively unobstructed coastline. In 2011 a 4 MW site and a 40 MW site were leased by the Crown Estate Commission to two different wave energy developers; Voith Hydro Wavegen Ltd. and Aquamarine Power Ltd. This resulted in a planning proposal for a long breakwater to be erected near a local slipway and 40–50 Oyster devices (harnessing kinetic energy from the waves) to be placed within 500m of the shoreline. In these negotiations, the community-owned estate emerged as a key agent of influence (Billing 2015).

In 2011, the Trust formalised the processes for developing renewable energy initiatives by hiring a full time Renewables Development Officer (Galson Estate Trust 2014), shifting the responsibility for negotiations from skilled volunteers to professionals who have been employed to fill specific roles. The Galson Estate Trust currently employs seven staff in various areas of development where no employment existed under previous private owners.

Discussion

A pragmatic conceptualisation of sustainable communities has been described (Bridger and Luloff 1999) where five inter-related targets are identified (Table 1). There is an explicit understanding that the success of sustainability as a practical concept is multi-factorial, cross-sectorial, and a combination of empirical measurement with the acknowledgment of less tangible influences. The Trust has initiated development activities in each of these five areas.

Table 1: Indicators of greater community sustainability applied to the case study

	Sustainability targets	Galson Trust achievements
1)	emphasis on increasing local economic diversity	Increased local employment
2)	the importance of self-reliance	Ongoing achievements in independent revenue accrual
3)	reduction in energy use coupled to the careful management of waste	Management of schemes for household energy efficiency and for the suitable management of domestic waste
4)	protection and enhancement of the biological diversity of natural resources;	Management activities on a local nature reserve
5)	a commitment to social justice housing, employment, access to public services, and local participation in local decision-making	Discussion of local initiatives for the creation of social housing and also the creation of a partnership project to address the relief of local fuel poverty

Source: Based upon Bridger and Luloff 1999

In some respects, this more 'rounded' approach towards what constitutes "sustainable development" in rural communities resonates with the concept of the *"shadow landscape"* (Bryant et al. 2011) in which multiple indicators are combined. This concept is perhaps less positive in its positioning of rural communities, but it attempts to make a comparative analysis of the multi-faceted perspectives of rural living in order to better understand the many inter-linked complexities. Supplementary approaches have emphasised the need to understand the actors (local and external) — their values, practices, and culture — in order to enhance our knowledge of the rural space (Madsen and Adriansen 2004).

This builds upon earlier work that focuses on the community as a source of social organisation and local action (Bridger and Luloff 2001). This approach takes a broad view of "sustainability" as something which does not diminish the economic argument (Macmillan 2000) but also encompasses both social equity and a healthy natural environment. It emphasises the recognition of local communities as a driver in promoting local action and the cultivation of social capital — a factor considered to be undermined by the historic legacy of concentration of private landownership in the Scottish Highlands and Islands (Bryden 1996). Specifically, the recent growth in community land ownership can be argued as a response through which *"in an increasingly fragmented and uncertain world, the search for a geographically based community becomes a means of exerting some control over at least a portion of one's life"* (Bridger and Luloff 2001, 460).

The movement to purchase the Galson Estate was stimulated directly by a sense of optimism generated by the Land Reform Act 2003, and the opportunity for perceived local benefits. This is consistent with a review of community resilience in the context of Scottish land reform (Skerratt 2013), showing that communities adopted deliberate proactive actions as an essential component to enhance their own sustainability, rather than simply reacting to external threats. This led to the perception of the external wind-farm development as an impediment to greater local participation and control. This was accompanied by a public awareness that collective action to acquire and manage the estate offered greater possibilities for long-term collective good versus simply short-term profits to an individual. This conflict is a feature of all of the recent community trust buy-outs. In the specific case of the Galson Estate Trust, it is clear (through the dialogue of community and public meetings) that the collective trust and collaboration associated in forming the Trust has led to a strengthening of the social capital of the 22 villages on the estate. This is evident in three main ways; firstly in the enhanced level of public engagement with the work of the Trust through business meetings and subsidiary activities; secondly through the mobilisation of skills and services from individuals and groups in the community (e.g., making use of volunteers and retirees); thirdly, in the increased activeity of clubs, societies and events within the community. The operational management of the Trust, run by a volunteer Board employing professional staff, is perceived by the community (through interviews and public meetings) to provide realistic alternatives to the regional failure of 'traditional' market mechanisms, the disinclination of private landowners to engage in certain issues, and other external forces of negative change in a very rural area.

This case study appears to be consistent with the reflections of Skerratt (2011) that several communities in the Highlands and Islands (of Scotland) have utilised the mechanism of the Community Land Trust to identify independent revenue streams (often from renewable energy generation) and to use these for local (re)investment. This measure of economic independence is associated with a growing community confidence and sense of self-identity (as measured by the initiation of new local economic and heritage activities) as well as a shift in attitudes towards long-term developmental aims and a decreased dependency upon external funding sources. Specifically, there is a growing realisation, as evidenced in local public meetings and in the regional media, that the association between community land ownership and new opportunities to engage in alternative energy generation can provide independent sources of revenue to enable the community to progress long-term developmental goals.

The revenue from the community wind turbine(s) is considered by the Trust Board to be a commitment to self-managing long-term community sustainability, a revenue stream to assist community resilience, and a reward for this shift in development perspective. The enabling legislation that has allowed the Galson Estate Trust (and other communities) to develop their natural assets for the benefit of the local community as a whole has helped to dispel *"the view routinely trotted out by Scotland's landed elite that it is land use rather than land ownership that matters"* (Macleod 2014, 17). Significantly, the Scottish Government has recently released the far-reaching final report of the Land Reform Review Group (LRRG 2014) which includes recommendations to extend the benefits such as have been experienced by the Galson Estate Trust, by the transfer of another one-million acres of Scottish land into community ownership. Implicit in the title of this report — *"The Land of Scotland and the Common Good"* — is the notion of the collective benefits of public social capital and local empowerment that can accrue following the process of successfully acquiring community land ownership.

Conclusion

In a study of the relationship between community resilience and community land ownership in Scotland (Skerratt 2011) it was noted that *"The Scottish Government is committed to establishing resilient communities.... Evidence from the research shows that community land trusts are delivering that brief"*. The report identifies five areas of significant change in the perspectives and capacities of communities such as the Galson Estate Trust that are coeval with the establishment of community ownership. These are:

1) A shift from short-term to long-term responsibilities in the management of the estate (exemplified by the commitment to energy generation and environmental enhancement initiatives);
2) The shift from a dependence on funding from external agencies towards a business model that creates revenue streams and seeks partnerships (using the land as security);
3) A growing sense of community identity, concomitant with community democracy being able to exercise greater control over local social and economic issues (e.g., the possible release of land for social housing or securing service contracts on behalf of Local Authorities).
4) The identification of revenue streams and local re-investment as a conscious step towards sustainability facilitates a measure of financial independence, and the use of local assets for development reduces the community's perception of being peripheral to decision-making that affects their lives (encouraging a more localised and more resilient approach to development).
5) The employment of paid staff rather than a continued reliance on volunteerism and in the cultivation of appropriate skills, both on the (voluntary) Board and for those employed within the community (e.g., energy generation management, improved domestic energy efficiency, local waste handling).

The recent socio-political changes associated with land reform in Scotland, especially in the Highlands and Islands, (such as rural housing, environmental engagement, sports management, and economic initiatives) have provided an important platform for greater public awareness of opportunities for the long-term sustainability of rural communities in this region. The combination of community land ownership and of new opportunities to capitalise on the natural resources of the region for the generation of energy (predominantly wind and marine sources) is perceived to have improved the level of resilience of these rural communities. In particular, the acquisition of community land ownership has provided a) a vehicle for widespread public engagement with a more democratic and consultative approach to local development issues; b) the physical asset of the community land, both as a source of income and as a base for further development; and c) the legal and physical ability to initiate revenue-raising initiatives (such as the generation of energy from local resources) which can be used to support social and economic benefits in the community. This association of the community owning the land assets and the ability to generate independent streams of community income is a catalyst for changing how these communities are beginning to re-conceive different ways of encouraging rural sustainability and community resilience.

References

Bailey, N., Flint, J., Goodlad, R., Shucksmith, M., Fitzpatrick, S & Pryce, G. (2003). Measuring Deprivation in Scotland: Developing a Long-Term Strategy ACRE Briefing Papers. Cirencester: ACRE. www.scotland.gov.uk/Resource/Doc/47176/0025569.pdf. Accessed 30 March 2015.

BBC News Scotland (2012). Wave power proposal for Shader in Lewis shelved. BBC News. www.bbc.co.uk/news/uk-scotland-highlands-islands-20796004. Accessed 6 January 2015.

Beaufoy, G., Baldock, D. & Clark, J. (1994). The Nature of Farming: Low intensity farming systems in nine European countries. London: Institute for European Environmental Policy.

Billing, S.-L. (2015). Agents for change in the sustainable development of renewable energy in the Highlands and Islands region of Scotland. PhD dissertation (unpublished). University of the Highlands and Islands.

Bridger, J. C. & Luloff, A. E. (1999). Toward an interactional approach to sustainable community development. Journal of Rural Studies 15 (4), 377–387.

Bridger, J. C. & Luloff, A. E. (2001). Building the sustainable community: Is social capital the answer? Sociological Inquiry 71 (4), 458–472.

Bryant, R. L., Paniagua, A. & Kizos, T. (2011). Conceptualising 'shadow landscape' in political ecology and rural studies. Land Use Policy 28 (3), 460–471.

Bryden, J. (1996). Land tenure and rural development. In 3rd Annual John McEwen Lecture. www.caledonia.org.uk/land/bryden.htm. Accessed 6 January 2015.

Bryden, J. & Geisler, C. (2007). Community-based land reform: Lessons from Scotland. Land Use Policy 24 (1), 24–34.

Bryden, J. & Hart, K. (2000). Land reform, planning and people: an issue of stewardship? In Paper given at the RSE/SNH Millennium Conference, The future for the environment in Scotland: Resetting the Agenda? Edinburgh. www.caledonia.org.uk. Accessed 6 January 2015.

CLS (Community Land Scotland) (2014). www.communitylandscotland.org.uk. Accessed 6 January 2015.

Copus, A. K. & Crabtree, J. R. (1996). Indicators of socio-economic sustainability: An application to remote rural Scotland. Journal of Rural Studies 12 (1), 41–54.

Dùthchas Project, (2001). Area sustainability strategies for peripheral rural areas January 1998–April 2001 Dùthchas — The Final Report. www.duthchas.org.uk/frameset1.html. Accessed 6 January 2015.

Galson Estate Trust (2014). www.galsontrust.com. Accessed 6 January 2015.

Handley, S. (2013). Community Consultation Survey: Summary Report 2013. Galson: Urras Oighreachd Ghabhsainn (Galson Trust Estate).

Hoffman, M. (2013). Why community ownership? Understanding land reform in Scotland. Land Use Policy 31, 289–297.

Hunter, J. & MacLean, C. (2012). From the low tide of the sea to the highest mountain tops: Community ownership of land in the Highlands and Islands of Scotland. Stornoway: The Islands Book Trust.

LRRG (Land Reform Review Group) (2014).The Land of Scotland and the Common Good. The Scottish Government. www.scotland.gov.uk/Resource/0045/00451087.pdf. Accessed 6 January 2015.

Leneman, L. (1989). Fit for Heroes? Land settlement in Scotland after World War I. Aberdeen: Aberdeen University Press.

Levitas, R., Pantazis, C., Fahmy, E., Gordon, D, Lloyd, E. & Patsios, D. (2007). The Multi-Dimensional Analysis of Social Exclusion. www.poverty.ac.uk/definitions-poverty/social-exclusion. Accessed 30 March 2015.

MacAskill, J. (1999). We have won the land. Stornoway: Acair.

MacKinnon, N. (2014). Consultation Background. Galson: Urras Oighreachd Ghabhsainn (Galson Estate Trust).

Macleod, C. (2014). Radical boost to land reform vision. West Highland Free Press, 30 May 2014, p. 17.

Macmillan, D. C. (2000). An economic case for land reform. Land Use Policy 17 (1), 49–57.

Madsen, L. M. & Adriansen, H. K. (2004). Understanding the use of rural space: the need for multi-methods. Journal of Rural Studies 20 (4), 485–497.

Magis, K. (2010). Community resilience: An indicator of social sustainability. Society and Natural Resources 23 (5), 401–416.

Mathie, A. & Cunningham, G. (2003). From clients to citizens: Asset-based Community Development as a strategy for community-driven development. Development in Practice 13 (5), 474–486.

Nicol, R. A. E. (1998). Indicators of sustainable development: A critical review of progress and a case study of the Orkney Islands, MSc thesis (unpublished). Edinburgh: University of Heriot Watt.

Parman, S. (1990). Scottish Crofters: A Historical Ethnography of a Celtic Village. Fort Worth: Holt, Rinehart and Winston.

Pasqualetti, M. J. (2011). Social barriers to renewable energy landscapes. The Geographical Review 101 (2), 201–223.

Personal Communications (2014). Interview with Neil Mackinnon, Urras Oighreachd Ghabhsainn. 2014: May 14th.

Portes, A. (1998). Social Capital: Its origins and applications in modern sociology. Annual Review of Sociology 24, 1–24.

Rennie, F. (1991). The way of crofting. In Magnusson, M. & White, G. (Eds.), The nature of Scotland: Landscape, Wildlife, and People (pp 107–116). Edinburgh: Canongate Press.

Rennie, F. (1995). The Dingwall Agenda. www.caledonia.org.uk/land/dingwall.htm. Accessed 06 January 2015.

Rennie, F. (2008). Human ecology and concepts of sustainable development in a crofting township in Folk Life. Journal of Ethnological Studies 46 (1), 39–57.

SCU/RSPB (Scottish Crofters Union and the Royal Society for the Protection of Birds) (1992). Crofting and the environment: A new approach. Inverness.

Shucksmith, M., Chapman, P. Clark, G. M., Black, S. & Conway, E. (1995). Rural poverty and disadvantage: How it is experienced and how can it be tackled? Aberdeen: University of Aberdeen.

Skerratt, S. (2011). Community land ownership and community resilience. Scottish Agricultural College, Rural Policy Centre Research Report. www.communitylandscotland.org.uk/wp-content/uploads/2014/06/commlandownerfulllowres.pdf. Accessed 6 January 2015.

Skerratt, S. (2013). Enhancing the analysis of rural community resilience: Evidence from community land ownership. Journal of Rural Studies 31, 36–46.

Smout, C. (1990). The Highlands and the roots of green consciousness 1750 – 1990. Scottish Natural Heritage, Occasional Paper No. 1. www.snh.org.uk/pdfs/publications/corporate/occaspapers/Occ%20Pap%201%20the%20highlands.pdf. Accessed 6 January 2015.

TEAS (The Energy Advisory Service) (2014). The Western Isles fuel poverty survey report. Stornoway.

Warren, C. R. & McFadyen, M. (2010). Does community ownership affect public attitudes to wind energy? A case study from south-west Scotland. Land Use Policy 27 (2), 204–213.

Wemyss, R. D. (2011). A major wind farm development in a valued natural environment: A thematic discourse analysis of public responses to a proposed wind farm on the Isle of Lewis, Scotland. PhD dissertation (unpublished). Open University and the University of the Highlands and Islands.

Wightman, A. (2013). The poor had no lawyers: Who owns Scotland and how they got it. Edinburgh: Birlinn Ltd.

Xodus Group (2012). Ltd. Siadar Wave Energy Project, pp. 16–18. www.scotland.gov.uk/Resource/0039/00399091.pdf. Accessed 6 January 2015.

The dynamics of hyper-rural areas in France: New economic bases for rural peripheries

Samuel Depraz

Abstract

In a context of strengthened metropolisation and urban growth, official French spatial categories have erased the references to rural areas as a space distinct from cities and their influence, so that the adjective "rural" appears now to be an out-dated socio-spatial scientific reference. However, this paper intends to demonstrate that rural peripheries in France are still to be considered as a research object per se, since recent changes into the most remote, scarcely populated areas in France have occurred. Based on three recent studies, this paper will show that "hyper-rural" peripheries are growing again, with positive allocations of revenues and new, non-productive economic bases — but with remaining structural weaknesses at the same time. That is why an adapted regional policy is still claimed for those "hyper-rural areas".

Rural issues in France are being more and more challenged by the overwhelming question of metropolisation. An official focus has been made on competitiveness and innovation, after the Lisbon Strategy of the European Union (2000), thus promoting the most productive regions around metropolitan areas with new, dedicated programs such as, for instance, the competitiveness poles and excellence research units. The claim for territorial equity and the request for specific measures for rural areas have now difficulties to be heard (Dubourg 2015). However, recent works about rural areas in France have advocated that new methods are necessary to define and study rural areas, so that our understanding of them will show new, specific trends. Indeed, in-depth analysis is suggesting that even the uppermost rural peripheries cannot be systematically considered as "lagging behind" anymore. Rural peripheries are actually benefiting from the reallocation of external resources thanks to a stronger residential base, and thus are gaining increasing interest in terms of socio-economic development. They are now questioning the ability of public policies to shift towards a new paradigm, not only focused on economic growth and productive functions — where rural areas remain effectively weak — but also on this residential base, which is a multi-dimensional economic system with different levers for local development.

In order to demonstrate these points, our contribution will proceed in four sections. Section 1 will analyse the way rural peripheries are defined, arguing that those definitions show how rural areas are considered by the current pro-urban official policies. Section 2 will introduce three alternative methods that have been recently promoted to identify "hyper-rural" areas in France, in order to focus on the issues of those very specific territories. Thus, section 3 shall develop the most striking trend of French rural peripheries — that is a demographic reversal with a new, positive migration balance. If service coverage remain low, the average wealth and the level of all-day life facilities is growing, so section 4 will discuss the dimensions of the resulting potential of economic growth that is actually developing in those rural peripheries.

French rural peripheries: how to define them?

The instability of French definitions for rural areas

Naming issues, defining target territories and choosing indicators is the first, if not the most important part of spatial public policies. Actually, the way problems are perceived by the institutions is influencing their methods greatly. The evolution of the definition of rural areas in France has shown such an influence for a long time (Mathieu 1998). Indeed, it is no coincidence if there are still difficulties to encompass rural areas properly in France. Today there is neither a satisfying, nor a stable definition for rural peripheries.

The one and only stable, available definition — which is also the oldest — remains the characterization of rural municipalities at local level, though it dates back to 1846. It is based on a famous, but discussed threshold of 2 000 agglomerated inhabitants per urban pole (*unité urbaine*). With this definition, rural areas in France today covers 14.2 million inhabitants, that is about 23% of the metropolitan population covering 78% of the national territory and approximately 80% of French municipalities (CGDD 2013). The resulting density is very low (33 inhabitants/km^2), when the national average itself remains lower than in most neighbouring countries (115 inhabitants/km^2).

However, this legacy of the official definition is neither sufficient to define the diverse panel of rural areas properly, nor useful to speak about rural peripheries. The missing criteria, which has proved to become more and more important with time, is the influence of the nearest urban agglomerations on rural life, considered in terms of services coverage, commuting working population, etc. In 1999, a new measurement of rural areas, based on a dynamic approach of local workers' migrations, has thus been implemented by the national statistical office, the INSEE. Urban poles of at least 5 000 working places have been defined, as well as their influence area, with a threshold of 40% of commuters among active population. The two elements formed the "mostly urban areas" (*aires à dominante urbaine*), also including urban poles (*pôles urbains*) and their close rural surroundings (*couronnes périurbaines*). As a result, rural municipalities with strong urban influence have been recognized as "periurban areas", distinguished from of a third, independent, "mostly rural areas" type, that is municipalities out of any urban influence.

The "mostly rural areas" were themselves diverse and included, on the one hand, small rural poles (*pôles ruraux*), with 1 500 up to 4 999 working places and their influence area, and, on the other hand, rural peripheries (*communes rurales isolées*), that is the municipalities that were remaining out of the influence of any rural pole. This whole category of "mostly rural areas" meant approximately 18 to 22% of national population (Poncet and Belot 2008, 12), among which rural peripheries (*communes rurales isolées*), out of the influence of any urban pole, counted for 4.7% of national population, after the 2009 census results.

The overriding influence of urban studies on regional analysis

However, this typology has been suppressed in 2011, arguing that rural areas cannot be only defined in terms of urban influence. As a consequence, there is now no specific tool to measure rural peripheries in the official definitions any longer. The sole influence of all urban poles, small or big, is still measured, but without any consideration for the rural/urban characteristics of local areas anymore, so that "95% of the population is now [supposedly] under urban influence" (INSEE 2011, our emphasis). If "rural" considerations have disappeared, urban ones are still included in that definition.

Such a turnaround could be seen, as well, as the result of the overwhelming weight of urban studies on regional analysis, since many researchers in the field of regional analysis are proclaiming that "urbanity is becoming, through an astonishing epistemological reversal, totally expositive of our contemporaneous societies and their transformations" (Bonnet et al. 2012, 5). This is the aftermath of and old, passionate debate in France that dates back to Jacques Lévy's 1993 famous paper, *Oser le désert* ("Let's dare rural deserts"), which denounced that the question of rural peripheries was over-considered: "Some areas, in the Massif central for instance, are becoming actual deserts. It is inconvenient for those who stay, and solidarity must prevail in their favour, but is it, basically, that important?" (Lévy 1993, 8). The same author is still claiming today, in very efficient shortcuts, that "rural populations are urban people that live in the countryside" (Levy 2013) or that "the French urbanization process is, out of some minimal exceptions, completed" (Lévy, in Cailly and Vanier 2010, 55); as a consequence, researchers should stop designing specific rural analysis and adopt a more general, urban-centred approach on regional development studies, if they want to reflect the general trend of socio-economic realities in France properly.

Three recent approaches of rural peripheries in France

However, some research institutions dedicated to spatial prospective (DATAR, CGDD) and rural areas (INRA Dijon, L'Œil laboratory — University of Grenoble), but also official reports by the Senate, have provided recent counter-proposals that give a sharper insight into French rural peripheries. I will henceforward refer to three updated, very precise studies about French rural peripheries, summarized in Table 1 as for the main statistical features of each definition.

Table 1: Three recent approaches of French rural peripheries

Definition	Goal	Criteria	Resulting population
1. Hyper-rural areas (INSEE 2012; Bertrand 2014)	Services coverage	Low density of "all-day life basins", non-aggregated population, time to basic services (>17 min)	5% (2009)
2. Type 4/Type 5 "remote rural areas" (CGDD 2013)	Housing conditions	Multivariate analysis (demographic balance, rate of construction) + resulting distance (> 15 km from any urban unit)	7.4% (2009)
3. Low-density areas (DATAR 2012)	Territorial development, prospective	Density (two thresholds: <30 inhab./km² ("low density areas") and <10 inhab./km² ("deserted areas")	7.3% (2009)
Isolated rural areas (INSEE 1999)*	*Measure of the urban influence*	*Commuters' migrations (<40% commuting to any employment pole of at least 1 500 workplaces)*	*4.7% (2006)*

* *This definition has been abandoned in 2011*

The first one has been implemented by the INSEE itself with the help, among others, of the INRA and the DATAR, and offers an alternative to the sole definition of urban influence through working places. The zoning in "all-day life basins" (*bassins de vie*) has tried to identify "the smallest territories in which inhabitants can access the rifest facilities and services" (INSEE 2012). Those 1 666 small areas are structured around a pole of services: that is an urban pole counting at least 50% of 31 intermediate equipment (supermarket, book-shop, high-school, swimming-pool, police-station, medical laboratory, etc.), as listed at national level, and attracting the nearest rural municipalities around it. This study gives a sharper insight into big urban areas since smaller, local areas are identified, on the margins of metropolised countryside, with their specific features. In a second step, the European definition of rural/urban areas has been applied, following a smart grid of 200 square meters that allows to identify areas with at least 300 inhabitants/km² each, and at least 5 000 inhabitants when aggregated, that will be qualified as urban. The rate of urban aggregated squares into each basin makes urban (>80% urban population), intermediate (50–80%) or rural basins (<50% urban population). The result is more interesting, since 75% of all-day life basins can be considered as rural, gathering 31% of the national population.

Taking in account this new statistical framework of the INSEE, an official report to the Senate (Bertrand Report 2014) defined a total of 250 "hyper-rural" basins, counting at least 75% of rural population and/or located beyond a 17 min threshold to the nearest urban pole. Hyper-rural areas concern 5% of the national population and 26% of national territory. Urgent measures could thus been advocated in the name of a spatial justice, such as a legal, official recognition of hyper-rurality, assorted with a minimum guaranteed coverage with public facilities and a better dotation in experts for territorial engineering. The report underlines the importance of defining an "abandon threshold", under which the decline of spatial equipment cannot be reverted anymore.

The second analysis was produced in December 2013 by the new *Commissariat général au développement durable* (CGDD 2013), which is in charge of a coordinated analysis of spatial inequities in the perspective of an improved sustainability. In this study, the CGDD tried to get a closer view of the evolution of rural areas and, mostly, of the evolution of housing conditions in rural municipalities. The CGDD created an interesting typology in 5 classes, based on (a) demographic changes, (b) the rate of housing constructions or renovations and (c) the socio-economic characteristics of rural households. Types 1 and 2 of this typology describe narrow suburban areas and coastal, border, connected areas, very close from urban poles. Type 3 describes touristic rural areas, a very dynamic countryside though far from any urban pole, and quite wealthy also. Types 4 and 5 describe then remote rural areas (15 km or more from any urban centre for type 5), with even 30 to 40% of them belonging to "isolated rural municipalities" after the older INSEE typology. Types 4 and 5 gather 7.4% of the national population on 38.8% of the territory. This typology confirms a global relationship between a peripheral position in urban-rural connexion and weaker socio-economic indicators.

The third analysis is the result of an expert panel conducted by the prospective work of the DATAR, *Territoires 2040*, about low-density areas in France. The word "rural" has been cautiously avoided, in order to skip the debate on words and the negative connotations associated to older categories such as "deep rural areas" (Béteille and Montagné-Villette 1995), "isolated rural areas" or recalling the "France of emptiness" of the 80's (Béteille 1981). Only the density criteria at municipal level (UAL 2) has been kept, with a threshold of 30 inhabitants/km² ("low density area") and even 10 inhabitants/km² ("deserted area"). The motto of the collective work was that the concept of "low density, often mixed up with the idea of a deprived area, shall not systematically be perceived as a challenged, if not declining area" (Barthe and Milian 2011, 144) and the experts insist on the fact that "international comparison goals lead to produce aggregated indicators at intermediate scales (NUTS 2 or 3)" where "contrasts are erased and specific features of low-density areas are combined with others, depending on other types of areas" so that the result is often undifferentiated. In contrary, isolating low-density areas at municipal level produces a sharper picture for the 4.5 million concerned inhabitants (7.3% of national population), among which 0.5 million live in deserted areas. Their basic characterization leads to speak about human and capital scarcity, and the dominance of agricultural, forestry and natural resources.

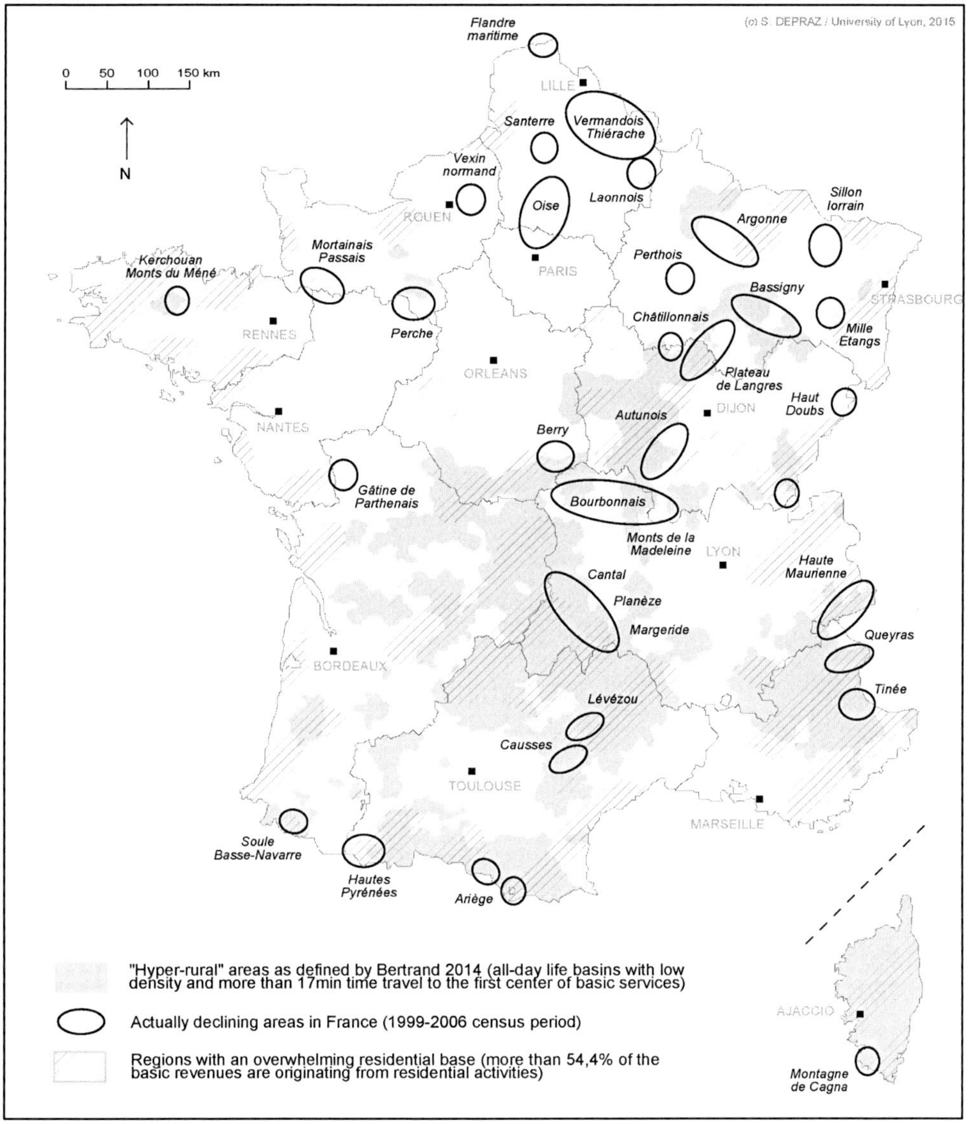

Figure 1: The compared extension of French rural peripheries
Source: Depraz / University of Lyon 2015

However, the crossing of the three approaches produces some interesting results about rural peripheries in France, not the least interesting one being the clear signs of a demographic turnout, with the revival of French rural peripheries — even though services and capital remain weak.

A demographic reversal: the revival of French rural peripheries

Global growth for rural areas

An important shift has been noticed in France since the turn of this century: rural areas are not declining anymore in France. In average, rural municipalities — considered after the local level definition, and taking into account the possible changes of municipalities from one category to another — have grown by 11.6% between 1999 and 2009, when urban ones only by 4.9%. The increase is now positive for 85% of rural municipalities (Dedeire et al. 2011; Lacquement and Chevalier 2011). The same global statement can be given when considering rural all-day life basins: they face a +0.9% yearly increase, when urban basins are only growing by 0.6% a year.

This growth is a combination of both natural increase (+2%) and migrations (+9.6%). But if we take into account the sole migration balance, 70 000 rural inhabitants become urban dwellers each year in France, but 100 000 urban ones become suburban or rural, so that recent studies have confirmed the existence of an "urban exodus" (Guérin 1983; Hervieu and Viard 1996; Merlin 2009), producing a de-densification of urban centres (Guieysse and Rebour 2012).

Truly enough, those global data include the massive growth of suburban areas, which has been the very first explanation given to the "rural rebirth" (Kayser 1990) since 1975 — the year when rural exodus started to stop in France. There is indeed a broader and broader influence of urban poles and commuting populations on their surroundings, seemingly obliterating the remaining of any rural peripheries in the long term.

Endogenous growth within rural peripheries at the same time

But if we consider rural peripheries more specifically, the statement is also positive. The 2013 CGDD typology asserted that, even in remote rural areas, the growth was clear: +0.8% increase per year (period 1999–2009). The same measure has been observed with the criteria of low density areas, where the growth was of +0.3% a year in the period 1999–2006. The report on hyper-rural areas stated similarly that a "slight rebound" (Bertrand 2014) had occurred in the demographic evolution of the most remote rural areas since the mid-2000 years.

Admittedly, natural growth is here negative, and rural populations remain quite older than national average (33% of the population is 65 or older, compared to 28.5% for all rural areas and 17% nationally), but this is not a sign for a weaker rural area since the migration balance has reverted the trend with new inhabitants coming from Paris area, or even from abroad: mostly young retired persons, coming back to their native village, and thus not depending very closely from employment poles or public services for education. The share of holiday homes is also important (>25%, national average: 10.3%), among which the share of foreign owners is high (up to 16%, national average 7.3%). Citizens coming from Great Britain and the Netherlands are the keenest investors in remote, non-touristic rural areas. Most dwellings (>50%) are the conversion of former holiday homes or vacant houses into permanent lodgements, so the dynamics of the building sector remain weak. All those indicators show a picture opposite to suburban growth, which is mostly driven by young, commuting people causing a strong dynamics of new constructions in the rural fringes of employment poles.

Last, but not least, this analysis about low-density areas has dismissed the old idea of an "empty diagonal" in France, since the south-western part of it is now continuously growing. Only Champagne, Lorraine and Aisne (Picardie) remain quite weak, and not only in remote rural areas, but also within old industrial cities, expressing actually a regional trend rather than a specific rural one.

Some remaining weaknesses regarding health and services coverage

Rural peripheries in France are thus at a turning point with visible changes, but still classical, inherited weaknesses. Situations of social vulnerability remain higher in remote rural areas, with 6% of dwellings without any access to bath or shower, and still 9.5% of empty dwellings (national average: 7.1%). The access to facilities in rural basins remains very variable. Upper-level equipment is scarce (only 25% of all possible facilities are available in rural basins, against 77% in urban basins). Travel distance for work is even twice higher than the average in hyper-rural areas.

Three sectors are more concerned than others with scarcity: education (1h20 for a return trip on average for secondary schools in rural basins), sports/leisure and health services (43–44 min both). New governmental plans for the rationalisation of public services for justice, education and health, such as the RGPP Act of 2007 (*revision générale des politiques publiques*) have led to reduce between 2008 and 2015 by 300 the sites of the national defence, to close 387 local courts on 1206 in a two year-period, and to shut down 20% of maternity services in France between 2002 and 2012 — with only one left per *département* in remote rural areas such as Creuse, Gers, Haute-Loire, Lozère, Pyrénées orientales, Haute-Saône, Tarn et Garonne, placing many remote rural municipalities beyond the security threshold of 45 min travel time (Vanlerenberge 2015).

As another consequence, stroke hazards are rated as 2 time more dangerous than the average in hyper rural areas, due to the higher vulnerability of population (15 per million and per year versus 8 in average) and the distance to stroke centres (16‰ lethal cases versus 11‰ in average) (Bertrand 2014). This notion of "threshold" has been the core of the Bertrand report, claiming for a maintained specific policy for hyper-rural areas.

As a consequence, hyper-rural areas are facing a contradictory situation, with lower facilities and a demographic rebound at the same time, as if public policies had not been able to consider rural areas in their most recent aspects. One of the reasons for this is that the analysis of rural development has been kept in a too narrow economic approach, and has missed the effective factors of the rebirth of rural peripheries.

The importance of new, non-productive economic bases in French rural peripheries

The relative decline of the productive base

The contradictory situation that rural peripheries are encountering can be described under the theory of the economic bases, after the recent works of Laurent Davezies (2008) and Magali Talandier (2007; 2009). This theory, inspired by the earlier analysis of Werner Sombart, Homer Hoyt, Douglass North and Charles Tiebout (Krumme 1968), is trying to understand which factors are really effective to foster local and regional development. It will distinguish between basic and non-basic activities, where basic activities contribute to export the produced added value towards external consumers, and import revenues and wealth at local scale, and thereby have a leading effect on local employment and domestic economy. The neo-classic theory has most of all insisted, among basic activities, on the role of the productive sector (agriculture, industry, commercial and private services), considered as the main source of revenues for territories. Many works have thus defined the impact of the productive base through a multiplier effect (Perroux 1955; Hoyt 1956).

Public policies have been very careful about maintaining as much productive activity as possible in remote rural areas: for the weakest rural areas, the ZRR program (*zones de revitalisation rurale*) allows tax reductions and social contributions' relief for new employers. This zoning, dating back to 1995 but confirmed in the 2005 Act on rural areas, is concerning 5 million people in low-density areas with demographic decline and/or active population decline and/or strong agricultural occupation (>30% of active population). The same 2005 act is also giving tax breaks for young farmers, promoting collective employment, releasing multi-jobs agreements and helping rental dwellings for young workers in rural areas: all sets of measures are concentrated on the productive sector. However, the task is difficult, since the departure of older students and young workers remains dominant in rural areas. A typology made by the Observatoire des territoires of the DATAR (Hilal et al. 2012) confirmed, through a multi-varied analysis, the remaining of "fragile countryside" with ageing workers' and farmers' rural areas. As such, the stated "weakness" or "crisis" of remote rural areas is solely based on the measurement of the economic growth, since all used criteria are only considering employment and productive activities.

The shift towards new economic bases

Actually, the population growth in French rural peripheries can only be explained if the question of regional development is decoupled from economic growth. Today, "the geography of growth has been disconnected little by little from the geography of local development" (Pecqueur and Talandier 2011, 123). Admittedly, Paris and the biggest metropolis are attracting always more economic value. The differences in regional BIP have grown, and Paris itself produces 29% of the French BIP (27% in 1999). In terms of economic growth, rural areas become weaker and weaker. However, the share of regional income concentrated in Paris and the biggest metropolitan areas is *declining* in favour of rural areas (22% instead of 25% in a 10-year period): there is a "catching up effect" in terms of regional income. If rural areas' income remains in average 20% lower than in urban areas, the difference is lowering since 1999 (Davezies, in Bertrand 2014). The explanation of this "scissor effect" can be found in the "circulation of wealth" (Davezies 2008) and the redistribution of incomes through the presence of new populations that have not generated the income on-site: tourists, holiday dwellers, pensioners, state officers, commuters, etc. So the new economic base for rural areas — including now rural peripheries — is not the productive base anymore (22%) but are, in a much more significant extent, non-productive bases such as the residential (45%), the public (10%) and the health and social (23%) bases (Figure 2). The residential base includes pensioners, tourists, holiday dwellers, whose wealth has been brought to the countryside, and not produced on-site. Public, health and social bases are developing on public transfers and subsidies in a large part. So "development processes are originating from the circulation of resources, which have been produced in distinct areas than where they are consumed" (Pecqueur and Talandier 2011, 132). Rural areas are definitely multi-functional, mixed areas (Perrier-Cornet 2002) depending on several forms of economic bases, among which the balance is evolving towards residential functions.

Basic activities				Non-basic activities
Productive base	Residential and touristic base	Public base	Sanitary and social base	Domestic, or "presential" economy
Creation of added value	Private transfer of added value	Public transfer of added value	Public transfer of added value	Local recycling of added value
22% of national income	45% of national income	10% of national income	23% of national income	—

→ Circulation of wealth (Davezies, 2008): towards other bases and domestic economy

Figure 2: The new economic bases of the national economy in France
Source: Based on Pecqueur and Talandier 2011

Therefore, "goals of local economic development is triple, and is based on (a) creating, (b) captating wealth and (c) allowing the circulation of wealth in order to (d) maximize the expenses at local scale" (Pecqueur and Talandier 2011, 123). The length of the stay of external population in rural areas, converted in equivalent-persons per year and added to the permanent population, has given birth to the *presential rate*: this indicator is now scrutinized with higher attention, since the residential base appears to be the most important, growing driving belt for the economy of rural peripheries.

New goals in public policies for rural peripheries

The 2005 Act for Rural Areas has partially considered the importance of fostering the image of rural areas, and the quality of services for incomers: attention to rural heritage, protection of natural and agricultural landscapes, maintenance of healthcare centres and retail shops through the public and social basis have been recognized, even in a context of shortage in public resources.

New goals, as identified by the prospective work *Territoires 2040* of the DATAR (2012), have been set for low-density areas:

- assuring a social cohesion between more heterogeneous populations and their needs, facing at the same time demands for mobility and sedentary lifestyles, for in-place economy and intermediate goods and services, etc.;
- fostering local commitment and cooperation in rural development projects thanks to national and European actors' networks (Pays — becoming *Pôle d'équilibre territorial et rural, Parcs naturels régionaux, Pôles d'excellence rurale*, LEADER program of the European Union), with a stronger inter-municipal cooperation that has been made compulsory since January 2014 for every municipality in France;
- but also encouraging informal cooperation (ex: participative and cooperative enterprises, citizen's initiatives);
- and — still — valuating the fixed assets of rural areas, that is "local resources whose valuation can only be obtained through the presence of the producer and/or the consumer on-site": classic productions (agro-industrial productions, quality food products and certified timber extraction), with a stress on higher added value, but also new emerging resources such as bioenergy in the framework of the new Act of Grenelle II (2010) on climate adaptation;
- compensating the disadvantaged areas and paying for the environmental quality and services thanks to public transfers and the new CAP 2014–2020.

As such, a shift towards a qualitative approach of rural areas has been suggested. Economic growth and production shall not be the first targets when addressing "lagging behind" rural areas, since they produce a negative, one-sided vision of the current processes that are occurring in French countryside. A new policy is suggested, considering residential resources as an amenity for rural development, and even a necessary, positive externality for economic production.

Conclusion

Rural peripheries in France, even difficult to define with precision, are concerning roughly 5 to 7% of the national population and over more than one quarter of the national territory. Those low-density, remote areas, sometimes described as "hyper-rural" areas, are still facing structural weaknesses in the field of productive activities and services coverage, but are characterized since the beginning of this century by a demographic reversal, so that most rural municipalities, even in peripheries, are now slightly growing. No "empty diagonal" can be observed anymore, but rather a new "urban exodus", though with regional exceptions in the north-eastern part of the country.

It would be wrong to despise this trend, seeing rural areas only as a picturesque framework for a dependent, non-basic local economy with the disappearance of classical productive sectors. Positive migratory trends induce an efficient reallocation of resources in favour of rural areas, even if those resources are more and more produced in metropolitan areas. Wealth transfers, both private and public, constitute an effective economic base — residential, touristic, administrative and social at the same time — that shall now be taken in account in the reshaping of public policies towards rural areas. Priority shall not only be given to the creation/defence of productive activities — anyway more and more limited to fixed assets. The catchment of external resources, driven by rural newcomers, bringing new potentials for endogenous initiatives, is becoming crucial for public policies in order to foster territorial projects. Rural geography has still a role to play in the identification of those new economic trends and in the monitoring of the population renewal of rural peripheries, hoping to anticipate and mitigate the local conflicts that will erupt in such a multi-functional rural scene.

References

Barthe, L. & Milian, J. (Eds.) (2011). Les espaces de la faible densité — état des lieux et problématiques. In DATAR, Des systèmes spatiaux en perspective. Revue d' études et de prospective n°3, coll. Territoires 2040 (pp. 141–160). Paris: La documentation française.

Bertrand A. (2014). Hyper-ruralité. Un pacte national en 6 mesures et 4 recommandations pour "restaurer l'égalité républicaine". www.logement.gouv.fr/IMG/pdf/rapport_a_bertrand_vdef.pdf. Accessed 13 May 2016.

Béteille, R. (1981). La France du vide, Paris: LITEC, coll. Géographie économique et sociale.

Béteille, R. & Montagné-Villette, S. (Eds.) (1995). Le rural profond français. Paris: SEDES.

Bonnet, F., Charbonneau, J-P., Conrad C., Dessus D., Donzelot J. et al. (2012). Pourquoi maintenant? Tous urbains n°0, Editorial, 5–7.

Cailly, L. & Vanier, M. (2010). La France. Une géographie urbaine. Paris: Armand Colin.

CGDD (2013). L'habitat rural entre 1999 et 2009: des évolutions contrastées. Le point sur n°179. www.developpement-durable.gouv.fr/IMG/pdf/LPS179.pdf. Accessed 6 June 2015.

DATAR (2012). Des systèmes spatiaux en prospective. Revue d'études et de prospective n°4, coll. Territoires 2040. Paris: La Documentation française.

Davezies, L. (2009). L'économie locale ' résidentielle'. Géographie Économie Société 11 (1), 47–53.

Davezies, L. (2008), La république et ses territoires. La circulation invisible des richesses. Paris: Seuil, coll. "La république des idées".

Davezies, L., Guilluy, C., Donzelot, J. & Béja, A. (2013). La France périphérique et marginalisée: les raisons du ressentiment. Esprit 3, 23–33.

Dedeire, M., Razafimahefa, L., Chevalier, P. & Hirczak, M. (2011). Dynamiques des espaces ruraux en France: vers un modèle des trajectoires démographiques? Espaces, population, sociétés 3, 512–537.

Dubourg, P. (2015). Is rurality a thing of the past?, Métropolitiques. www.metropolitiques.eu/Is-rurality-a-thing-of-the-past.html. Accessed 6 June 2015.

Guérin, J-P. (1983). L'exode urbain: nouvelles valeurs, nouvelles élites. Revue de géographie alpine 71 (3), 267–277.

Guieysse, J-A., Rebour, T. (2012). Territoires ruraux: déclin ou renaissance? Population et Avenir 710, 4–8.

Hervieu, B. & Viard, J. (1996). Au bonheur des campagnes. La Tour d'Aigues: Les éditions de l'aube.

Hilal, M., Barczak, A., Tourneux, F-P., Schaeffer, Y., Houdart, M. et al. (2012). Typologie des campagnes françaises et des espaces à enjeux spécifiques (littorales, montagne et DOM). Paris: Datar.

Hoyt, H. (1956). Homer Hoyt on Development of Economic Base Concept. Land Economics 30, 182–186.

INSEE (2011). Le nouveau zonage en aires urbaines de 2010. 95% de la population vit sous l'influence des villes. Insee première, 1374, www.insee.fr/fr/ffc/ipweb/ip1374/ip1374.pdf. Accessed 6 June 2015.

INSEE (2012). Le nouveau zonage en bassins de vie de 2012. Trois quarts des bassins de vie sont ruraux. Insee Première, 1425, www.insee.fr/fr/ffc/ipweb/ip1425/ip1425.pdf. Accessed 6 June 2015.

Kayser, B. (1990). La renaissance rurale. Sociologie des campagnes du monde occidental. Paris: Armand Colin.

Krumme, G. (1968). Werner Sombart and the Economic Base Concept. Land Economics 44 (1), 112–116.

Lacquement, G. & Chevalier, P. (2011). Ländliche Räume in Frankreich. Typen jüngerer sozioökonomischer Entwicklung. Geographische Rundschau 63 (2), 36–42.

Levy, J. (1994). Oser le désert? Des pays sans paysans. Sciences Humaines, hors-série n°4, 6–9.

Levy, J. (2013). Nos communes sont des fossiles. Interview in the magazine L'Express, 12/08/2013. www.lexpress.fr/actualite/politique/jacques-levy-nos-communes-sont-des-fossiles_1272957.html. Accessed 6 June 2015.

Mathieu, N. (1998). La notion de rural et les rapports ville-campagne en France Les années quatre-vingt-dix. Économie rurale 247 (1), 11–20.

Merlin, P. (2009). L'exode urbain, Paris: La documentation française.

Pecqueur, B. & Talandier, M. (2011). Les territoires à base économique résidentielle et touristique. In DATAR, Des systèmes spatiaux en perspective. Revue d' études et de prospective n°3, coll. Territoires 2040 (pp. 121–138). Paris: La documentation française.

Perrier-Cornet, P. (Ed.) (2002). Repenser les campagnes. La Tour d'Aigues: Les éditions de l'aube.

Perroux, F. (1955). Note sur la notion de pôle de croissance. Economie appliquée 8 (1/2), 307–320.

Poncet, J-F., Belot, C. (2008). Le nouvel espace rural français, Rapport au Sénat n°468. www.senat.fr/rap/r07-468/r07-4681.pdf. Accessed 6 June 2015.

Talandier, M. (2007). Un autre modèle de développement hors métropolisation. Le cas du monde rural, thèse de doctorat en urbanisme, aménagement et politiques urbaines, University Paris XII. www.talandier.com/thesetalandier.pdf. Accessed 6 June 2015.

Talandier, M. (2009). Une autre géographie du développement rural: une approche par les revenus. Géocarrefour 83 (4), 259–267.

Vanlerenberge, J-M. (2015). La situation des maternités en France, Rapport au Sénat n° 243, Paris. www.senat.fr/rap/r14-243/r14-2431.pdf. Accessed 6 June 2015.

The Unbalanced Welfare of Italian Fragile Rural Areas

Giorgio Osti

Abstract

The aim of this paper is to describe welfare in fragile rural areas in Italy with especial attention to health conditions, considered as factors crucial for understanding people's wellbeing. The diversified development trajectories of Italian rural areas would require welfare services calibrated on specific disadvantages. Yet Italian rural welfare is 'unbalanced' for several reasons: it has followed regional disparities instead of compensating for them; it is retreating from peripheral areas because of the state's fiscal crisis; it is too closely centred on cash benefits rather than interpersonal services. There are sporadic local welfare mix schemes attempting to moderate these processes, but the situation remains all in all spatially and socially unfair. The paper is structured as follows: first, it places Italy in an international scenario concerning the links between economic development and wellbeing; second, it analyses the disparities internal to Italy, paying especial attention to the rural-urban dimension; third, it comments on the welfare policies necessary to reduce Italy's territorial disparities.

Introduction

The aim of this paper is to describe welfare in fragile rural areas in Italy with especial attention to health conditions, these being considered factors crucial for achieving people's wellbeing. The starting point is *local development* (Osti 2013), a multidimensional process intertwined with welfare policies which causes social and territorial differences.

Three premises are necessary before entering into details on the paper's topic:

- The rural/urban dichotomy is indeed a very problematic notion, but it is still an efficient and synthetic way to represent the spatial dimension of social processes (Schaeffer et al. 2013). It includes low population density, peripheral location with respect to important services, the predominance of primary activities. When these features are accentuated, rural areas become 'fragile' that means at risk of demographic and environmental collapse (Devereux 2001).
- Wellbeing is a broad condition of *psychosocial life* (Martikainen et al. 2002) which has, beside health, education, and income parameters (Human Development Index), some social dimensions like cohesion and territorial identity. In general, wellbeing is considered to be linked to self-realisation — a typical individual goal — and to peaceful relations with other people (*civicness*). Wellbeing is both an aim and a factor able to improve economic activities, the community's self-esteem, and political participation.

- Welfare consists of the socio-political activities deliberately undertaken to enhance the wellbeing of a population. Traditionally, public bodies have delivered welfare, but more recent sources include non-profit organisations, households (welfare mix), and even single individuals, and corporations, in accordance with a broader notion of *social investment* (Hemerijck 2015). Basic welfare policies concern health, education, employment and social assistance, which includes housing and antipoverty measures. A special role is given to insurance funds against sickness, work accidents, and ageing (pensions). A broader conception of welfare which includes culture and sport will not be considered in this paper.

Welfare measures have both a reparatory purpose — to compensate the disparities caused by development — and an ideal purpose — to provide a universal service for better human wellbeing. Italian rural areas have followed different development trajectories. They therefore need not only standard services delivered in accordance with a principle of equity (welfare state), but also services calibrated to their specific advantages and disadvantages (welfare community). This is a balance not easy to achieve. Indeed, Italian rural welfare is 'unbalanced' in several respects: it has followed regional disparities rather than compensate for them; it is retreating from peripheral areas because of the state's fiscal crisis; it is too closely centred on cash benefits rather than interpersonal services.

To show these unbalances, the paper first locates Italy in an international scenario concerning the main links between development and wellbeing. It then analyses the disparities internal to Italy with especial regard to the rural-urban dimension. Finally, it illustrates the welfare policies applied in fragile rural areas, highlighting some innovative practices.

Development, territorial disparities and health

Italian rural areas are fully placed within the world trend of contraposition to urbanisation:

> *"Urbanization is inextricably linked to economic development. It accompanies, and in many ways drives, the structural transformation from largely agricultural to increasingly manufacturing and services-based economies. As part of this process, 'leading' areas emerge, characterized by higher average standards of living than in 'lagging' areas. The spatial inequalities that emerge are often viewed as an inevitable part of the development process, rewarding higher productivity, economies of scale, and agglomeration effects in the more advanced sectors of the economy. While 'leading' and 'lagging' areas often correspond to urban-rural divides, the same spatial disparities may be replicated on larger or smaller geographical scales, for instance in the form of North-South divides or intra-urban inequalities" (Dudwick et al. 2011, 191).*

Intra-urban inequalities can be seen as class disparities and can be greater than those between countryside and towns, says Costa (2008), using a fundamental parameter of wellbeing like population health. Nevertheless, if we adopt a large-scale temporal perspective, like the *divergence-then-convergence* paradigm, the disparities in the living standards of rural and urban populations are still marked.

> *"Our cross-country findings [...] suggest that few low-and middle-income countries fit neatly into the WDR [World Development Report] 2009's paradigm that urban-rural welfare disparities first diverge and then converge once countries reach upper-middle income status" (Dudwick et al. 2011, 192).*

The issue is not only how much the territorial disparities converge in developing countries but also whether developed ones maintain their own social differences between cities and countryside. The suspect is that a moderate rural-urban divide still exists in richest countries, but masked by more important differences within the areas. Giovanni Costa reached this conclusion, although he did not mention the rural/urban divide.

> *"As said, for studies at macro geographical level (countries or regions), the correlation between (average) income and health should be direct under a certain minimal survival threshold [5000 dollars, author's note], while above that threshold the correlation with income disappears, becoming once again direct if income is substituted with the equity of income distribution. Vice versa, when smaller and sufficiently uniform geographical areas are considered, a direct correlation of health is observed with both income and its equal distribution" (Costa 2009, 259, my translation).*

The economy and wellbeing are connected: when development is low and uneven, the general level of health is low as well. This may explain also why countries with relatively high average incomes but great internal differences — like the USA or Russia — do not have proportionally good results in terms of life expectancy. Then, social disparities within rural areas of each country must be included in the analysis in order to prevent misinterpretations (Haynes and Gale 2000).

Life expectancy is generally high in European countries, with some internal differences. The best situation is in the Mediterranean countries, followed by the Nordic ones (Eurostat 2014a). The worst situation is in eastern European countries. Italy is among the best ones, although its income level is not the highest. Other western European countries are very close in the ranking. "With regard to life expectancy at birth, the main differences (up to 7–10 years in some cases) are across countries, *not between rural and urban regions within the same country*" (Bertolini et al. 2008, 93, italics added).

If we consider life expectancy as a robust indicator of wellbeing, the partiality of the economic explanation is evident. GDP is more valid for eastern European countries, quite appropriate for Northern ones, and lacking for southern European countries. According to Costa (2014), Italy's better position is due to three factors: less smoking among poor women, better diet, and the smaller presence of unmarried impoverished mothers. If we consider food preparation, which in Italy is still largely a female activity, we understand how wellbeing depends on local and critical family balances. But food and drink habits have their own importance: the Mediterranean diet is better in preventing illnesses typical of rich countries like diabetes and cardiovascular diseases (D'Alessandro 2013). Also, the custom of drinking less alcohol and only during meals is favourable for good health.

Epidemiologists maintain that also typical social variables, like intense relationships, are favourable for good health (Cohen et al. 2000). In the Mediterranean countries, especially in southern Italy, this is a source of criticism against Robert Putnam's (1994) idea of low social capital and corresponding low wellbeing (Piselli 2001). This is a further reason for studying wellbeing not just as the product of economic development and welfare policies.

Returning to the specific features of rural Europe, to be noted is that

> *"the difference is more strongly expressed for males, who have a higher mortality rate in rural areas than in urban areas, although the trend in females is the same. One of the few national studies available, carried out in Northern Ireland in 2007, indicated in rural populations aged 25–74 years a lower mortality due to circulatory, respiratory and malignant diseases and a higher mortality due to external causes compared to urban populations (...) It is, therefore likely that, in the rural setting, citizens may have significant advantages, as compared to urban residents, for quite a number of health effects, but, probably, with a higher severity of consequences of accidents and other concerns" (Eugloreh 2008).*

Living in a rural area is not necessarily a handicap; there are specific compensations or trade-offs, a negative process has positive side effects and vice versa. For example, rural people suffer rarity of healthy services, but they enjoy a less dense built environment, preventing the heat islands (Gabriel and Endlicher 2011). The mentioned idea of higher severity accidents is related to car mobility. In rural areas there are fewer accidents but they are more severe because most traffic occurs on routes crossing residential areas. The absence of motorways or bypasses makes car and lorry movements more dangerous (Piano 2009; Eurostat 2014b). The same is the case of farming, the typical rural activity (WHO Regional Office for Europe 2010, 16). The low standardisation of operations, the use of dangerous agri-machines, and a mentality little used to prevention provoke numerous severe accidents among farmers. It is likely that a superficial attitude towards risk prevention concerns all rural people and matches objective gaps of health structure (WHO Regional Office for Europe 2010, 23).

Advantages result from better food, less polluted air, low density of activities; generally, contact with green spaces has therapeutic effects (Eugloreh 2008; Van den Berg et al. 2016). Nevertheless, it is dangerous to generalise because the environmental factors are very specific, and they are sometimes contrary to common sense. Pollution sources are generally much localised. Frequent in rural areas are factories with high environmental impact and low opposition by local people. In regions with intense agro-industrial activities, using many types of chemicals, the supposedly better quality of rural life may be the reverse.

The case of Brittany, a region with many polluting agriculture activities, shows the difficulty of setting a health divide between urban and rural: "the amalgamation of all causes of standardised mortality ratios (SMR) show only a moderately increased risk in the rural areas compared with the overall regional level (+4% in men, +5–7% in women). The analysis of cause specific SMRs display higher rural mortality for cardiovascular diseases and external causes of death, road traffic accidents in particular (+24% in men). Among all specific causes investigated, only lung cancer mortality risk appears to be higher in urban areas" (Grimaud et al. 2004, 499).

Nevertheless, the overall picture is positive for European rural areas if we consider the incidence of mental disease, depression and suicide. The evidence is weakened by the different compositions of the population in urban and rural contexts. However, "most published studies claim that there is a higher prevalence of mental health problems, or at least of depression, in urban areas" and "odds ratios for psychiatric morbidity adjusted for socio-demographic variables supported the idea of rurality as a protective factor" (European Communities 2004, 49). The underlying reasons are framed within communitarian factors that may have already disappeared in the European countryside.

> *"(There is) a higher risk of depression in urban areas than rural areas because of the decline in community relationships and social isolation in the city; greater stresses with housing, work, marriage, childrearing and with security in urban environments, in interaction with the resources available to cope with the stress of urban life and high levels of hostility; poor social integration and social withdrawal and socio-cultural disintegration, including family and marital disintegration which limits social networks" (European Communities 2004, 50–51)*

Possibly, Durkheim's thesis on suicide as due to rapid economic changes and therefore less widespread in the quieter countryside is still valid. Slower economic development produces less mental stress. However, observers note a *convergence*: that is, suicide rates are becoming similar between rural and urban areas, even if for different reasons (Scramaglia 1994; Scocco 2009).

By contrast, rural areas maintain the gap (*divergence*) in the economic domain. According to the European Network for Rural Development, "some 14% of the population in the EU's predominantly rural regions suffers from employment rates of less than half the EU average, and there are areas of low per-capita GDP" (Vashchyk and Siudek 2015, 8). It is confirmed that wealth per person is lower in rural Europe because of the lower incomes of farmers and farm workers (Bertolini et al. 2008). However, the largest differences are along the west/east Europe axis. Rural poverty is especially a problem in the former Soviet countries, where rural poverty is provoked by the restructuring of former publicly-owned farms and by the total absence of industrial activities in rural districts. Industrialisation — a process which also concerns agriculture activities — is a key factor in the differentiation of rural Europe. This is especially evident in Italy.

The diversities in rural Italy

In this section, the effort is to demonstrate that the economic development of the Italian countryside has been differentiated, and that the economic trends of rural areas do not fit perfectly with indicators of wellbeing. Consequently, welfare policies for rural areas (which are considered in the next section) are unbalanced because they follow uneven economic trends rather than compensate for them.

In Italy, two features interact with each other: first, since national unification in the nineteenth century, northern and southern Italy have maintained a wide economic gap, despite numerous interventions by national governments and the European Union. Second, in the 1960s Italy's so-called 'economic boom' followed different trajectories that are still visible. Bearing the North/South divide in mind, and considering rural areas in particular, three main pathways are apparent:

1) Local industrialisation: industrial districts or more modest agglomerations of small factories are widespread in most of northern-central rural Italy; the process has been caused partly by the decentralisation of big industrial complexes, and partly by the endogenous development of local crafts traditions.
2) Tourism development: so-called winter-summer tourism has reversed the misfortunes of many mountain areas, especially high Alpine valleys. Double season brought many customers hosted in new hotels, increasing a permanent tourist industry. In central Italy and in some upland areas, cultural tourism linked to mediaeval towns, monasteries, and typical wine-food products attract visitors in every season of the year.
3) Some rural areas have developed entire food *filiere*, that is, specific crops transformed and marketed locally. The examples abound: the apple *filiere* in South Tyrol, the fruit juice industry in Romagna, the rice district between Lombardy and Piedmont, the beef cycle in Tuscany, flowers on the Liguria Riviera, and wine districts in every region.

These patterns have overlapped in some rural areas, producing high incomes and resident's pride. They are not absent in southern Italy, but more rare and less robust, despite the abundance of natural and artistic beauties. The causes of the South's backwardness are numerous, and they have been debated since national unification (Giunta and Martinelli 1995). Contrary to large depressed areas in other European countries, like eastern Germany and Andalusia, the general economic conditions of southern Italy have deepened the divide with northern Italy.

A less marked divide is exhibited by inner areas. 'Inner or internal area' is an expression typically used by scholars of the '*Mezzogiorno* question', who have divided the South into *flesh and bone* areas (Pazzagli 2015). Southern Italy developed along the coasts and in concentrated towns, leaving the inner zones depopulated and dominated by poor farming. This dependency on the flesh-bone pattern is still high and can be limitedly extended to Italy as a whole.

In 2014, government has launched the *National Strategy for Inner Areas*. This is not only a structured development policy; it is also a renewed way to recognise rural deprivation (Barca et al. 2014). The notion of inner areas (*aree interne*) is a framework especially important for understanding unbalanced welfare. It is inspired by place-based development (Barca 2011), but more importantly it focuses on marginality in terms of distance from healthcare and education structures (Higgs and White 1997). Moreover, it considers not only pure distance in kilometres or minutes from an urban centre providing those services, but also accessibility to a railway station. The role of railways in the redemption of rural areas throughout Europe is evident. In Italy, the arrival of the train initially represented the end of isolation and recently the possibility to commute to distant places. Of course, the capacity of railways to support rural development has been affected by the modernisation of lines and trains. In Italy, despite the presence of a widespread network developed since the end of the nineteenth century, branch lines have not been modernised, as has instead happened in nearby Switzerland.

Because railways allow commuting over long distances, they create the suburbanisation of even remote areas; rural students can thus commute daily even to attend secondary school. The same applies to the employees of large urban companies. For other services, not even a high-speed train is enough to ensure a rapid arrive at destination. In this case, access to the main services is assured by private means of transport; non-possession of a car or a driving licence immediately becomes a source of marginalisation.

The map of inner Italy is consistent with what was said above about rural diversification: not all rural areas are marginalised (Figure 1). Those included in the urban networks can be considered in some respects urban hinterlands or even suburban areas. An appropriate term for these latter is *peri-urban* areas, recently associated with the idea of *urban farming*. These areas are markedly different from inner ones; their populations have constantly increased, and they have taken advantage of Italy's polycentric urban structure. The price they have paid is the location within their borders of heavy infrastructures, such as motorway interchanges, waste incinerators, airports, shopping malls, fuel storage facilities, and so on. But the location of these infrastructures is sporadic, impeding any general labelling of peri-urban areas as disadvantaged. Inner rural areas, by contrast, have a more marked phenomenology: constant loss of population, ageing, lower levels of education and income, and distance from a high level hospital and a set of secondary schools, measured in more than 40 minutes with a private means of transport.

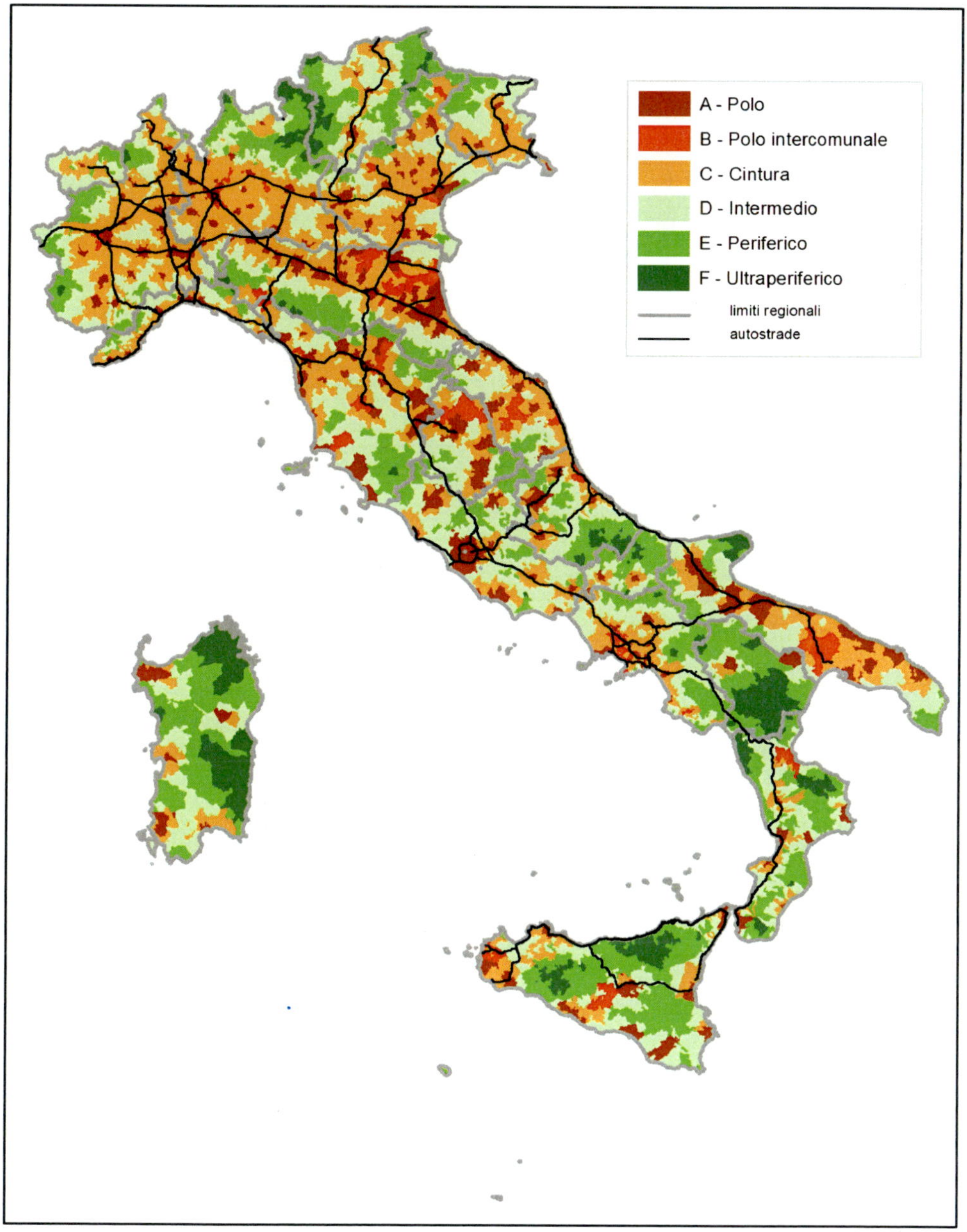

Figure 1: Map of Italian inner areas

Remark: In the continuum from poles (A) to ultra-periphery (F), areas D (intermediary), E (peripheral) and F are considered 'inner'. B refers to intercommunal poles and C to peri-urban areas.

Source: L'Agenzia per la coesione territoriale 2014

It should be easy to conclude that rural Italy is divided between peri-urban and remote areas. The problem is that further factors have intervened to change the geography of the Italian countryside. Besides the economic processes outlined above, also policies and the capacity for self-organisation matter. These two aspects are intertwined because the pressure on political bodies for favourable measures depends also on local mobilisation and cultural resources, like a distinct language, heritages, libraries, institutions for local commons.

Political actions can be classified into two types: inter-governmental and local development policies. The former are summarised in *regionalisation* (Mazzoleni 2009). This process started in 1970s and culminated in 2001 with the reform of the Fifth Title of the Constitution, which gave greater legislative autonomy to the twenty Italian Regions. Also competences on welfare and local development were devolved to the regions, with two important exceptions: the pensions (social security), including disability benefits, and education. Moreover, the fiscal systems remained largely centralised at national level. Consequently, the regions administered most welfare services, but depended on funds from the Ministry of Finance. There ensued long and tiresome negotiations, if not open conflicts, between national and regional governments (Fargion 2005). Ability or a lack of it in these negotiations produced many different situations at local level, even within the same region.

Local development policies are more complex because the actors involved are numerous and operate at four levels: local (municipalities and their consortia, provincial administrations), regional, national, and finally the EU level. As in the case of regional reform, the negotiations are very intricate. It is difficult to summarise such a dense inter-governmental network. Two features appear comparable to the rest of Europe: the emphasis on local action (*subsidiarity*), and the intent to promote inter-sectoral development. For the Italian countryside, this means matching agriculture, tourism and artisanship. The European Community's LEADER initiative can be cited as a symbol of pro-rural policies in Italy (Osti 2015). National policies, like *Territorial Pacts* and *Negotiated Planning* (Governa and Salone 2005), have been based on largely the same pattern: bottom-up participation, cooperation among sectors, co-funding. All of these policies concentrate more resources in the South, which is permanently included in the EU Objective 1 area, and they pay little attention to welfare measures. For the countryside, the latter have been left to the national pensions system, especially with the provision of disability benefits, and to ordinary regional policies, with the result, as said, of the close dependence of rural areas on the negotiation capacity of their administrators.

Wellbeing indicators in rural Italy

Given these different paths of development, one would expect wellbeing indicators of rural Italy to be diversified in the same way. Unfortunately, the main indicators are not available at the appropriate territorial scale. Life expectancy and suicide rates are not available at municipal level. For life expectancy at birth, one notes a gradient from North-East to North-West, Central Italy, and then the South and Islands. Nevertheless, the differences in life expectancy among the five subnational units are very small, more or less one year (Table 1). This seems to contrast with the sharper differences that we have seen for economic development.

Table 1: Life expectancy at birth in the Italian statistical units (2014), in years

	North-East	*North-West*	*Centre*	*South*	*Islands*
Males	80.8	80.6	80.5	79.5	79.6
Females	85.3	85.5	85.2	84.2	84.2

Source: Italian National Institute of Statistics (Istat 2015)

Inspection of the differences among provinces (NUTS 3) of the rich North shows a lower level of life expectancy in two more rural areas of Veneto: Rovigo, a typical flat agriculture zone (male 79.3 and female 84.5 years) and Belluno, an entirely mountainous province (male 79.7 and female 84.6 years). It is noteworthy that other almost completely upland provinces in the Alps (Sondrio, Verbanio-Cusio-Ossola, Valle d'Aosta) constantly have values below the average, with the sole exception of South Tyrol. Also for the completely flat provinces of northern Italy (Ferrara, Mantua, Cremona, plus the just-mentioned Rovigo), life expectancy is below the average at least within their region.

The strange situation of the Alps is reproduced at length for suicide rates (Scramaglia 1994). The same provinces mentioned above for lower life expectancy present higher levels of suicide, again with the exception of South Tyrol (CCIAA Sondrio 2013, 32). The latter, however, is not more virtuous of entire Italy but has rates close to the national average (and in the past the rate has been even higher in the province with the German-speaking minority). The flatter and more agricultural provinces of the North do not confirm the similarity with mountain areas, because they have variable suicide rates. Finally, on moving towards southern Italy, the rate decreases, with the exception of Sardinia. If we consider southern Italy as a more rural area, the general trend — suicide is an urban phenomenon — is confirmed. The Alps are more rural as well, but for them *ad hoc* explanations are necessary (Arnoldi 2009; Kim et al. 2011; Bonomi 2011).

Research shows that the formal education level is lower in Italian rural areas (Centro Nazionale PISA-INVALSI 2012). There used to be an unfavourable tradition in the countryside whereby rural families pushed their offspring to leave school early for rapid inclusion in farm work. Today, so few young people are involved in farming or forestry that they do not change the statistics. Nevertheless, the education divide is still present, being also caused by a 'brain drain' from the countryside. If we consider that higher human capital has a linear correlation with income, employment and health, it is easy to conclude that rural areas suffer generally from a material disadvantage.

Monetary wealth is an important component of wellbeing. However, the gap between rural and urban areas is more evident for consumption than for income, and according to some researchers (Salvio et al. 2009, 74), it is diminishing. Moreover, in rural areas real estate capital, like buildings, fields and woods is more important. In northern and Central Italy, the assets (*capitale netto* in Italian) of families living in more extreme rural areas are higher than those living in intermediate ones. Per-capita income has the expected ranking: it progressively decreases from urban, to peri-urban, to fully rural areas. Income and assets also diminish from North to South. "The most critical datum is the sharp difference of net capital value between North and South, with a ratio of more than 2 to 1, that greatly reduces the chances of integrating family income with profits coming from the assets" (Della Chiara et al. 2014, 37).

According to a multidimensional index of household poverty that includes income, assets, employment, education, one-parent family, and full time mother's job (lack of time indicator), the same study (Della Chiara et al. 2014) shows that Italy's peripheral rural areas are generally poorer. There are, however, a certain number of regions in which the classic ranking from urban to remote rural is not respected. For example, in Umbria, Toscana and Marche — upper-central Italy — the highest percentage of families above the threshold of multidimensional poverty live in the peri-urban zones. It is difficult to find an explanation of this regional and rural-urban interrelation. The same combined index can work differently in such variable territorial contexts. In any case, some points are clear: there is further proof that the main divide is along the

North/South axis; in Italian rural areas, people are poorer and less educated; and presumably they also have lower life expectancies. From a dynamic point of view, the differences in wellbeing between rural and urban areas have become less sharp, contrary to the North/South divide, which appears to be permanent.

The unbalances of Italian rural welfare

The task of a modern welfare system is not easy in such a variable situation. In Italy, the system has worked both following the trends and contrasting them. The measures more dependent on national policies, like pensions and education, have provided some kind of redistribution by delivering minimal resources to sparsely populated areas. For a period in Italy, the softening depopulation trend of remote rural areas was assigned to retired people who lived or had moved there. Unfortunately, that period has finished: the reform of pensions, which are now based on effective contributions made during the working life (*contributory* method, Agudo and García 2011), is reducing a certain revenue for rural families, especially those with an agricultural background. The schools are caught in a spiral of young population decrease and cutbacks in grants for all rural public services. In order to avoid closing, the national government has reduced the minimum number of children necessary to keep open a primary school in mountain areas. The secondary schools suffer more, and it is difficult to avoid their centralisation. Negotiations between local authorities and the peripheral units of the Education Ministry (MIUR) give flexibility to the application of student/school standards, but they create great variability among internal areas (Carrosio *in press*).

The welfare measures dependent on regional and municipal authorities have followed economic trends to a greater extent. Consequently, health and social services in rural areas reflect the degree of development of the entire district. The most innovative solution for low density ageing areas has consisted in domiciliary care services. These allow old or sick people to remain at home as part of the community, and to receive periodic services concerning both the house and the person. For services of this kind, the differences at regional level reproduce the North/South divide (Istat 2015, 286–287). Furthermore, there is a methodological problem: not only are there no data according to the rural/urban dimension but the indexes must be interpreted according to the local situation. For example, the low number of old people assisted with domiciliary services in southern Italy can be a real deficiency, while the same low number in South Tyrol may depend on the better integration of old people into the family and the community. The 'Mezzogiorno' is internally diversified, showing a welfare system in the Adriatic regions better than those overlooking the Tyrrhenian Sea (Bertin and Carradore 2016).

Hence, in rural areas welfare depends closely on a local mix of public services, local non-profit organisations and the internal resources of families. Those resources are not only grants or pensions, as said, but also organisational and professional ones. Unfortunately, the tendency is to privilege monetary benefits instead of real services (Bouget et al. 2015, 24). In Italy, the network of early childhood education and care services is very weak (Bouget et al. 2015, 13), presumably weaker in rural areas: a factor which dissuades "rural women from participating in the labour market" (Lucatelli et al. 2006, 8).

The perceptions of rural people on the quality of services provide new insights (Table 2). According to a national survey of more than one thousands interviewees (Ventura, Milone and Van der Ploeg 2008), it emerges rural people centre their attention on the means to obtain service, not on their presence nearby. In other words, rural people want better communication for the centres where best services are placed. The example of maternity wards is paradigmatic. Rural people prefer a relatively distant but large hospital in which to give birth to a small 'birth point' near their homes. Of course, this preference is affected by car property, roads accessibility and capacity to drive, so-called motility (Osti 2010, 298). Thus, the battle to keep small rural hospitals open is linked more to job losses and community prestige than to real expected services.

Table 2: Services/infrastructure that would improve the quality of life in own residential area (in per cent)

Leisure and sport activities for young people	31.4
Enlargement of road network	23.5
Public transport Improvement	18.4
Restoration of rural tracks and roads	12.6
Scepticism on quality of life improvement	8.0
Increase in meeting places	6.1
Total	*100.0*

Source: Ventura, Milone and Van der Ploeg 2008, 115

The results of this survey are useful also for a more general evaluation of rural welfare. The needs of young people are more frequently indicated than transport means. This not only legitimates a so-called *secondary welfare* (Maino and Ferrera 2013), less focused on older generations, but also highlights the main problem of remote rural areas: their low attractiveness for young people. Apart from problems of income and services, remote rural areas seem destined to be left when teenagers finish school and look for jobs. Of course, ageing is the biggest quantitative issue; but it can be tackled with higher-scale services like residential units located in rural areas near the former home, and organized with internal assistance services. In these co-housing units, some tasks, like cooking, cleaning, or gardening, are managed by healthier old persons; other tasks are carried out with the constant presence of specialised personnel.

It is much more difficult to organise an *activating* welfare for younger people. The kind of needs that they have cannot be simply satisfied by offering a service, like to equip a fitness/sport centre. They evidently need an *atmosphere* that rural areas totally focused on images of greenness and calmness cannot give. Secondary welfare centred on job creation can help to create a more lively atmosphere. Generally, unemployment is higher in Italian rural areas than in urban areas. This is a legacy of farming affected by seasonal and precarious work. Small-scale industrialisation blocked the Italian rural exodus of the 1960s and 1970s. *Countryside industrialisation*, as it was called (Fuà 1988), happened in a very diversified way, leaving the most remote areas and southern Italy untouched. Rural industrial areas, including those developing *agricultural product quality*, were able to organise a quite good network of social services (Di Iacovo 2004).

This process continued until the 2008 crisis, which heavily affected private consumption and public expenditure destinated to rural areas. Since then, average income has diminished and territorial disparities have increased (Ricolfi and Cima 2015). It is unlikely that the Italian government will have the resources for a massive redistribution, even if at the beginning of 2016 the Minister of Welfare announced a one billion euro plan for poor families (Redazione 2016). According to the government, this will be a permanent measure.

For rural areas based exclusively on primary activities and located far from urban centres, this measure will certainly be insufficient. The micro-imbalances of rural Italy will dramatically increase. However, there are signals of change. One very important sign is represented by migrants. In more attractive areas, these are rich Western-European citizens; in all rural areas, they are people from poor countries employed in domestic services and primary sector activities. Especially women from Eastern European countries have become an important component of local welfare systems because they assist numerous old and disabled people. Their presence will be stable as long as their remuneration is low and paid 'off the books'.

There is great attention to 'community' cooperatives of rural areas (Fazzi 2010). These are cooperatives legally defined as 'social' (National Law 381/1991), but specialised in services for a restricted territorial community. Indeed, they are working mainly for cultural services to tourists rather than for disadvantaged local people. In regard to multifunctional tasks, such community cooperatives could enlarge their action to include social services. At present, some family farms have become more flexible by developing various green care services. Nevertheless, the limitation of these schemes is their sporadic distribution and economic precariousness. They often arise from a special fund, a single project, or the winning of an EU tender.

The third signal of change is the functional and spatial aggregation of services for small rural municipalities, like the family and community nurse units (Obbia 2014). It is the packaging of several services to an intermediate level, in between the urban centre and the rural municipality (district scale). This would allow public money savings and a more constant presence of trained personnel. In Italy there are already municipal consortia for social services; in some regions it is the District Health Agency (*Azienda Sanitaria Locale*) that delivers the basic welfare services. Such consortia encounter the parochialism of local communities, an attitude that hampers the beneficial effects deriving from a limited rescaling and centralisation of some services (e.g., a nurse unit serving an entire valley). For their part, the District Health Agencies often impose either too large territorial scales or the subordination of social assistance to health services. This happens because the Agency must keep the prestige of its urban hospital high. Moreover, because health care is provided universally, it is valued more than social care by authorities constantly in search of consensus. Then, a dilemma internal to welfare – social versus health services — emerges. Its solution is linked to the negotiations on what means and how delivering 'essential levels of care' in areas so different for topography, income and political capabilities (Torbica and Fattore 2005).

Rural welfare in the hands of advocacy coalitions?

The paper conclusion has a *political flavour*. Italian rural welfares are unbalanced not only because of objective physical disadvantages, not only because of weak industrialisation, but also because of their differing capacities for advocacy and the creation of extra-local links. In a country like Italy, where eminent local political figures dominate (Tarrow 1977) and centre/periphery relationships are highly discretionary (Page 1991; Bobbio 2007) or marked by meso-corporatism (Lehmbruch and Schmitter 1982), rural welfare depends greatly on advocacy coalitions. They are heterogeneous and temporary networks often focused on a single issue. More importantly they cross the traditional territorial borders. For some issues they become even international. In the past a strong advocacy coalition was working in favour of rural development; nowadays its political strength is blurring and two new topics are emerging; one is to privilege rural *inner* areas and the other is to enrich social services for younger generations living in the countryside. Thus, the intertwining of *older* local development paths *with new* forms of rural welfare is a promising line of inquiry which here is only outlined.

References

L'Agenzia per la coesione territoriale (2014). Mappe di riferimento: classificazione Aree, quota popolazione over 65 nelle Aree Interne e nei Centri, variazioni demografiche 1971–2011 nelle Aree Interne e nei Centri. www.agenziacoesione.gov.it/it/arint/Cosa_sono/index.html. Accessed 21 December 2015.

Agudo, L.F. & García, M.A. (2011). Pension reform in Italy: Description and evaluation. Pensions: An International Journal 16 (2), 96–106.

Arnoldi, C. (2009). Tristi montagne. Guida ai malesseri alpini. Turin: Priuli&Verlucca.

Barca, F. (2011). Alternative Approaches to Development Policy: Intersections and Divergences. OECD Regional Outlook 2011, Paris: OECD.

Barca, F., Casavola, P. & Lucatelli, S. (Eds.) (2014). A Strategy for Inner Areas in Italy: Definition, Objectives, Tools and Governance. Materiali UVAL 31. Rome.

Bertin, G. & Carradore, M. (2016). Differentiation of welfare regimes: The case of Italy. International Journal of Social Welfare 25 (2), 149–160.

Bertolini, P. Montanari, M. & Peragine, V. (2008). Poverty and Social Exclusion in Rural Areas. Brussels: European Commission, Directorate-General for Employment, Social Affairs and Equal Opportunities.

Bobbio, L. (2007). I governi locali nelle democrazie contemporanee. Roma-Bari: Laterza.

Bonomi, A. (2011). La Malaombra. Il perturbante caso dei suicidi in una vallata alpina. Turin: Codice edizioni.

Bouget, D., Frazer, H., Marlier, E., Sabato, S. & Vanhercke, B. (2015). Social Investment in Europe. A study of national policies. European Commission, European Social Policy Network (ESPN).

Carrosio, G. (in press). Welfare services and Inner Peripheries of Italy. Dipartimento di Scienze Politiche e Sociali, Università di Trieste.

CCIAA Sondrio (2013). Benchmarking Report 2013. Sondrio.

Centro nazionale PISA — INVALSI (2012). Le grandi città rappresentano per l'istruzione un vantaggio o uno svantaggio? PISA IN FOCUS 2012/06 (June), OECD, 1–4.

Cohen, S., Gottlieb, B.H. & Underwood, L.G. (2000). Social relationship and health. In Cohen, S. Underwood, L. G. & Gottlieb, B. H. (Eds.), Social support measurement and intervention: a guide for health and social scientists (pp. 3–25). New York: Oxford University Press.

Costa, G. (2008). Geografia della salute in contesti urbani. In Nuvolati, G. & Tognetti Bordogna, M. (Eds.), Salute ambiente e qualità della vita nel contesto urbano (pp. 97–150). Milano: Angeli.

Costa, G. (2009). Le disuguaglianze di salute: una sfida per le discipline che si occupano di valutazione delle politiche. In Brandolini, A., Saraceno, C. & Schizzerotto, A. (Eds.), Dimensioni della disuguaglianza in Italia: povertà, salute, abitazione (pp. 251–270). Bologna: Il mulino.

Costa, G. (2014). Promuovere equità di salute e di sanità in Italia. In Costa, G., Bassi, M., Gensini, G.F. et al. (Eds.), L'equità in salute in Italia. Secondo rapporto sulle disuguaglianze sociali in sanità (pp. 13–42). Milan: FrancoAngeli.

D'Alessandro, A. (2013). La dieta mediterranea. Le evidenze scientifiche del suo ruolo protettivo nei confronti dell'aterosclerosi coronarica e delle malattie dismetaboliche. Bari: Cacucci.

Della Chiara, E., Montresor, E., Pecci, F. & Perali, C.F. (2014). La distribuzione del benessere in Italia: diversità tra famiglie urbane e rurali. Agriregionieuropa 10 (36), 34–39.

Devereux, S. (2001). Livelihood Insecurity and Social Protection: A Re-emerging Issue in Rural Development. Development Policy Review 19 (4), 507–519.

Dudwick, N. et al. (2011). From Farm to Firm. Rural-urban Transition in Developing Countries. Washington D.C.: World Bank Publications.

EUGLOREH (2008). The status of Health in the European Union: towards a Healthier Europe, 7. Health Determinants and Other Factors Involved in Disease, 7.3. Living and working environment (executive summary), October 2008. www.intratext.com/ixt/_ExeS-Eng/_PL.HTM. Accessed 21 January 2016.

European Communities (2004). The State of Mental Health in the European Union. Brussels: Directorate General for Health and Consumer Protection.

Eurostat (2014a). Life expectancy at birth, 1980–2013. ec.europa.eu/eurostat/statistics-explained/index.php/ Mortality_and_life_expectancy_statistics. Accessed 4 February 2016.

Eurostat (2014b). Road safety statistics at regional level. ec.europa.eu/eurostat/statistics-explained/index.php/ Road_safety_statistics_at_regional_level. Accessed 21 December 2015.

Fargion, V. (2005). From the Southern to the Northern Question. Territorial and Social Politics in Italy. In McEwen, N. & Moreno, L. (Eds.), The territorial politics of welfare (pp. 127–147). London: Routledge.

Fazzi, L. (2011). Social Co-operatives and Social Farming in Italy. Sociologia Ruralis 51 (2), 119–136.

Fuà, G. (1988). Small-scale Industry in Rural Areas: the Italian Experience. In Arrow, K. J. (Ed.), The Balance between Industry and Agriculture in Economic Development (pp. 259–283). London: Palgrave Macmillan.

Gabriel, K.M.A. & Endlicher, W.R. (2011). Urban and rural mortality rates during heat waves in Berlin and Brandenburg, Germany. Environmental Pollution 159 (8–9), 2044–2050.

Giunta, A. & Martinelli, F. (1995). The Impact of Post-Fordist Corporate Restructuring in a Peripheral Region: the Mezzogiorno of Italy. In Amin, A. & Tomaney, J. (Eds.), Behind the Myth of European Union: prospects for cohesion (pp. 221–262). London and New York: Routledge.

Governa, F. & Salone, C. (2005). Italy and European Spatial Policies: Polycentrism, Urban Networks and Local Innovation Practices. European Planning Studies 13 (2), 265–283.

Grimaud, O., Bayat, S. & Chaperon, J. (2004). Mortalité urbaine et rurale en Bretagne. Santé Publique 16 (3), 499–508.

Hanlon, N. & Kearns, R. (2016). Health in Rural Places. In Shucksmith, M. & Brown, D. L. (Eds.), Routledge International Handbook of Rural Studies (pp. 62–70). Abingdon and New York: Routledge.

Haynes, R. & Gale, S. (2000). Deprivation and poor health in rural areas: inequalities hidden by averages. Health & Place 6 (4), 275–285.

Hemerijck, A. (2015). The Quiet Paradigm Revolution of Social Investment. Social Politics, 22 (2), 242–256.

Higgs, G. & White, S. D. (1997). Changes in service provision in rural areas. Part 1: The use of GIS in analysing accessibility to services in rural deprivation research. Journal of Rural Studies 13 (4), 441–450.

Iref (2009). I servizi nelle aree rurali ed il ruolo dell'agricoltura. Roma, 18 June.

Istat (2015). Rapporto BES 2015. Il benessere equo e sostenibile in Italia. Roma.

Kim, N. et al. (2011). Altitude, Gun Ownership, Rural Areas, and Suicide. American Journal of Psychiatry 168 (1), 49–54.

Lucatelli, S., Savastano, S. & Coccia, M. (2006). Health and social services in rural Umbria. Roma, Ministry of Economy and Finance Public Investment Evaluation Unit. www.dps.mef.gov.it/documentazione/uval/materiali_uval/Muval12_Sviluppo_rurale_inglese.pdf. Accessed 8 February 2016.

Maino, F. & Ferrera, M. (Eds.) (2013). First Report on Second Welfare in Italy. Turin: Centro di Ricerca e Documentazione L. Einuadi.

Martikainen, P., Bartley, M. & Lahelma, E. (2002). Psychosocial determinants of health in social epidemiology. International Journal of Epidemiology 31 (6), 1091–1093.

Mazzoleni, M. (2009). The Italian regionalisation: a story of partisan logics. Modern Italy 14 (2), 135–150.

Di Iacovo, F. (2004). Multifunzionalità in agricoltura e sviluppo delle reti di protezione sociale: alcune ipotesi di lavoro. In Di Iacovo, F. (Ed.), Lo sviluppo sociale nelle aree rurali. Metodologie, percorsi di lavoro e promozione della multifunzionalità in agricoltura nella Toscana rurale (pp. 141–156). Milan: Franco Angeli.

Obbia, P. (2014). Introducing the family health nurse in Italy. International Journal of Integrated Care 14 (6), doi:10.5334/ijic.1610.

Lehmbruch, G. & Schmitter, P. C. (Eds.) (1982). Patterns of corporatist policy-making. London: Sage.

Osti, G. (2010). Mobility Demands and Participation in Remote Rural Areas. Sociologia Ruralis, 50 (3), 296–310.

Osti, G. (2013). The moral basis of a forward society: Relations and forms of localism in Italy. Local Economy 28 (3), 291–303.

Osti, G. (2015). LEADER LAGs: neocorporatist local regimes or examples of economic democracy? In Granberg, L., Andersson, K. & Kovách, I. (Eds.), Evaluating the European Approach to Rural Development. Grass-roots Experiences of the LEADER Programme (pp. 149–164). Aldershot: Ashgate.

Page, E.C. (1991). Localism and Centralism in Europe. The Political and Legal Bases of Local Self-Government. Oxford: Oxford University Press.

Pazzagli, P. (2015). Bone’s Territories: Territorial Heritage and Local Autonomy in Italian Inner Areas. TAFTERJOURNAL N. 84 – Settembre Ottobre, 1–8.

Piano, L. (2009). La sicurezza passiva degli autoveicoli. Criteri di progettazione e sperimentazione. Milan: Hoepli.

Piselli, F. (2001). Capitale sociale: un concetto situazionale e dinamico. In Bagnasco, A., Piselli, F., Pizzorno, A. & Trigilia, C. (Eds.), Il capitale sociale. Istruzioni per l’uso (pp. 47–55). Bologna: Il mulino.

Putnam, R. (1994). Making Democracy Work: Civic Traditions in Modern Italy (with R Leonardi and R Nanetti). Princeton: Princeton University Press.

Redazione (2016). Povertà, l'Alleanza: “Piano del governo deludente”, Vita. www.vita.it/it/article/2016/02/01/poverta-lalleanza-piano-del-governo-deludente/138122/. Accessed 1 February 2016.

Ricolfi, L. & Cima, R. (Eds.) (2015). Disuguaglianza economica in Italia e nel mondo. Dossier, Fondazione DAVID HUME per Il Sole 24 ORE, I.

Salvio, C., Sciulli, D. & Aiello, C. (2009). Verso la parità economica. In Barberis, C. (Ed.). Ruritalia. La rivincita delle campagne (pp. 59–81). Roma: Donzelli.

Schaeffer, P.V., Kahsai, M.S. & Jackson, R.W. (2013). Beyond the Rural–Urban Dichotomy: Essay in Honor of Professor A. M. Isserman. International Regional Science Review 36 (1), 81–96.

Scocco, P. (2009). Epidemiologia del suicidio nelle aree urbane. 13° Congresso SOPSI, "Psichiatria 2009: Clinica, Ricerca e Impegno Sociale", Roma, 10–14 February.

Scramaglia, R. (1994). Analisi degli studi successivi a Durkheim. In Durkheim, E. Il suicidio. Studio di sociologia (pp. 63–158). Milan: BUR.

Tarrow, S. (1977). Between Center and Periphery: Grassroots Politicians in Italy and France. New Haven: Yale University Press.

Torbica, A. & Fattore, G (2005). The "Essential Levels of Care" in Italy: when being explicit serves the devolution of powers. European Journal of Health Economics 6(Suppl 1), 46–52.

Van den Berg, M. et al. (2016). Visiting green space is associated with mental health and vitality: A cross-sectional study in four European cities. Health & Place 38, 8–15, doi:10.1016/j.healthplace.2016.01.003.

Vashchyk, M. & Siudek, T. (2015). Social Development of Rural Areas in the European Union Member States in 2000–2012. Roczniki naukowe ekonomiii rolnictwa i rozwoju obszarów wieskich 102 (1), 7–17.

Ventura F., Milone P. & Van der Ploeg, J.D. (2008). La vita fuori della città. Perugia: AMP Edizioni.

WHO Regional Office for Europe (2010). Rural poverty and health systems in the WHO European Region. Copenhagen.

2 European policies and their impacts on regional governance and participation

Changing relationships between local governance and local government: an historical perspective from the Leader programme in Ireland

Mary Cawley

Abstract

This contribution discusses the European Union (EU) Leader programme in promoting local governance. Its goals are to illustrate why and how relationships between governance and local government in rural areas can change over time, using Ireland as an example. Traditionally, Irish rural development was organised through central government and its agencies rather than through local government. New local area partnerships were formed to meet the Leader programme's requirement for subsidiarity in 1991. Since the late 1990s, the actions of the partnerships have come increasingly under local government influence. The contribution documents the international, national and local factors that have contributed to this process. A critical evaluation method is followed, based on analysis of official documentary evidence, reviews of the programme and discussion with representatives of four long-established partnerships. The results illustrate emerging threats to the subsidiarity conferred by Leader in Ireland.

Introduction

This contribution discusses the EU Leader programme approach (*Liaison entre actions de développement de l'économie rurale*, to be translated as "links between actions for the development of the rural economy") in promoting local governance. The geographical focus is the Republic of Ireland (referred to as Ireland) where many rural areas were classified as economically as well as geographically peripheral, when the programme was first introduced in 1991 (NESC 1994). The Leader programme identified broad developmental goals to be pursued by local action groups (LAGs), involving local people in the development actions, and provided support funding. A partnership approach introduced a new type of governance that differed from Irish local government (county councils). This chapter adopts an historical perspective and focuses on the ways in which the relationships between local governance and local government evolved in Ireland since 1991. It is hoped that the Irish evidence will contribute to a better understanding of these relationships in the context of rural development more generally.

The text is structured as follows: (i) governance and partnership are defined with particular reference to how they relate to rural development; (ii) the overall objectives of the EU Leader programme and its evolution over time are introduced; (iii) Leader as an innovative form of governance in Ireland is discussed; (iv) changes in the relationships between Leader governance and local government are traced from 1991 to 2013; (v) recent local government reforms in Ireland and their potential implications for the implementation of the Leader element of the Rural Development Programme (RDP) 2014–2020 are assessed; and (vi) conclusions are drawn relating to the evolution of the relationships between local government and local governance through partnerships and the implications for the ethos of subsidiarity in the Leader programme in Ireland.

Governance and partnerships

Local governance as a concept and structure attracted growing attention in research on urban and rural development during the 1990s (Goodwin 1998; Jessop 1995). Goodwin (1998, 6) has described governance as being broader than government and drawing attention "to the ways in which governmental and non-governmental organisations work together and to the ways in which political power is distributed, both internal and external to the state". Two particular sets of circumstances have been identified as contributing to the growing role of local governance in rural contexts for more than two decades. The first is associated with states reducing their long-established responsibilities for certain types of service delivery during a period of widespread economic recession, in the 1980s, when private and voluntary actors were encouraged to meet the deficits that resulted (Geddes and Benington 2001; Geddes 2006). For example, voluntary organisations became involved in delivering meals and home care services to the elderly, where small numbers of people are widely distributed throughout rural areas (Cawley and Nguyen 2008). The second set of circumstances that was conducive to promoting local governance in the EU, in the 1990s, was the increased attention being given to promoting greater community representation and inclusion in development-oriented actions (Ray 2000). Partnership between the public, the private and the voluntary/community sectors was advocated to promote rural development (Turock 2001). Partnership is defined by the OECD (1990, 18) as involving "systems of formalised co-operation, grounded in legally binding arrangements or in formal undertakings, co-operative working relationships and mutually adopted plans among a number of institutions".

Partnership was originally advocated in the early 1980s as a method of attracting investment from private investors to support public sector urban renewal projects (Peck and Tickell 1994). In the later 1980s, it became part of the then European Community's (ECs) adoption of rural 'subsidiarity' as a method of promoting involvement in policy making at the level at which actions are implemented (Ray 2000). An objective was to 'empower' local people through their closer participation in developments that affected their social and economic welfare (Benington and Geddes 2001). Several commentators express more critical views on the devolution of influence to local levels. Healey (2003) and Geddes (2006), for example, suggest that the concept of empowering local action groups provided an excuse for neo-liberal states to withdraw from their welfare obligations at a time when national budgets were under pressure. The power and responsibilities allocated to local actors were presented as positive features but they often required revenue to be raised locally to support whatever funds were transferred by the state. Extensive contributions of labour by volunteers and co-funding by the recipients of grant aid were also usually required (Moseley 2003). Secondly, Swyngedouw (2005, 1991) referred to concerns about the empowerment of local actors leading to "a substantial democratic deficit", vis-à-vis political representatives elected by universal franchise. He also questioned the legitimacy of public funds being made available to organisations formed predominantly of representatives, instead of officially elected members. The allocation of developmental responsibilities to local partnerships has also been viewed as leading to a 'hollowing out' of the legitimate role of the central state (Keating 1998), although this was not the case in Ireland, as discussed later.

The Leader programme

There are seven key features of the Leader approach to local development: it is area-based; is pursued in a 'bottom-up' instead of a 'top-down' way (from the local level instead of being designed and applied by central government, as was often the case in the past); actions are implemented by local public-private-voluntary partnerships; an integrated multi-sectoral approach is followed instead of a sectoral approach; the actions should be innovative with true potential for creating additional employment and introducing new economic and social actions; co-operation and shared learning take place among the businesses and other actors; and networking takes place between partnerships in order to share good practice (CEC 2006, 8–14). In these ways, Leader provides a mechanism for a transfer of influence and power from the central state to local levels. This transfer involves partnerships being responsible for devising and implementing strategies that are jointly-funded in substantial ways by the EU and national governments. Leader has strong spatial dimensions in that it is area-based and responds to the needs of the populations of geographical territories.

Five phases of the EU Leader programme have been implemented since 1991. The programmatic periods and the development objectives changed over time (Table 1). The first programming period was for two years, the second for five years and the following three for six year periods. The programme objectives were modified in response to new needs over time (Table 1). Emphasis was placed on economic and socio-cultural development, the development of a territorial identity and external networking in order to transfer good practice, in Leader 1 (1991–1993) (Ray 2000); innovation was assigned special importance in Leader 2 (1994–1999), as a method of creating local employment and contributing more generally to economic and social development (Government of Ireland 1995); LEADER Plus (2000–2006) encouraged thinking about the longer term potential of local areas, by encouraging the implementation of integrated, high-quality, original strategies for sustainable development (CEC 2006). During the first three phases, Member States and regions had stand-alone Leader programmes with separate financing set aside at EU level. From 2007 onwards, the Leader approach was integrated ('mainstreamed') within overall EU rural development policy. This meant Leader was included in national and regional general rural development programmes supported by the EU, alongside a range of other rural development axes, under a new European Agricultural Fund for Rural Development (CEC 2006). Special emphasis was placed on supporting agriculture in the RDP 2007–2013 (Dax et al. 2016). The RPD 2014–2020 aims to foster the competitiveness of rural businesses through investment, ensure the sustainable management of natural resources and climate action, and achieve a balanced territorial development of rural economies and communities, for example, through the rollout of fast and ultra-fast broadband in rural areas (CEC 2015).

Table 1: The EU Leader programme: evolution of key objectives over time

Time period	*Programme*	*Key objectives*
1991–1993	Leader 1	Economic and social development, territorial identity, external networking
1994–1999	Leader 2	Special emphasis on innovation
2000–2006	Leader Plus	Emphasis on longer-term sustainable development
2007–2013	Rural Development Programme	Leader became an axis in the rural development programme; agriculture was assigned particular attention
2014–2020	Rural Development Programme	Competitiveness of rural businesses, sustainable management of natural resources, climate action, balanced territorial development

Sources: CEC 2006, 2015

The implementation of the Leader programme differed in detail between countries. However, certain broad features were characteristic (Moseley 2003; CEC 2006). Thus, local partnerships in each country were invited to form and respond to an advertised call for proposals from the EC, under defined headings. The member states selected the successful LAGs for funding from EU sources combined with national funds. The selected groups appointed staff and called for proposals for projects under defined headings within their territorial areas. The submissions were subject to appraisal by the responsible national body (the rural development division of the Department of Agriculture in Ireland) under designated criteria agreed with the EC. Joint-funding by the proposers was required. Technical and other support was provided by the professional staff of the partnerships to implement the projects. Expenditure was subject to audit at both national and EU levels. The national programmes in their entirety were also subject to evaluation by independent evaluators.

Leader in Ireland: innovative governance and new territorial structures

A three year pilot programme for integrated rural development, based on partnerships in twelve less-developed western areas, preceded the introduction of the Leader programme in Ireland (O'Malley 1992). Irish rural development policy was, however, primarily sector-based, led by the government departments for agriculture, forestry, fisheries, industry and tourism and little intersectoral collaboration took place. Policy was decided at a national level by central government and was implemented through state agencies and delivered by their local personnel. Local government, which consisted of the county councils (and city councils), at NUTS4 (Nomenclature of Territorial Units of Statistics) level, in the twenty-six counties of the state (Figure 1), had limited involvement in rural social and economic development as an integrated activity. Their main responsibilities related to physical planning and public service supply in the form of housing, roads, water provision, sewerage facilities, technical education and libraries. Their main sources of income came from rates levied on commercial properties and from a registration tax on road vehicles, supplemented by transfers of funds from central government. Private residential property tax was abolished in 1977, which removed one of the main sources of councils' income. In an effort to supplement their income, following recession in 2008, when public expenditure was cut significantly, a charge on second homes was introduced by government in 2009. A private residential property tax (including second homes) was reintroduced in 2013. The latter tax is collected by the Revenue Commissioners but is to support local services. Water rates were introduced in 2015 but the funds are collected by a newly-constituted semi-state company, Irish Water, which is responsible for the operation of public water services, a function that previously resided with the county councils. For more than two decades, therefore, Irish local government has become increasingly dependent on central government for funding to support the delivery of public services. In the early 1990s, the remit relating to local enterprise development consisted in providing modest funding support to enterprises that employed less than ten people. Non-agricultural employment and enterprise support more generally at a local level were the responsibility of the regional offices of a state agency, Enterprise Ireland. It should be noted also that the NUTS3 regional authorities had limited involvement with the Leader programme in Ireland (Cawley 2009).

Figure 1: Irish counties

Source: Ordnance Survey of Ireland, mapping held under licence

In addition to Irish county councils having limited remits for integrated rural development, the population size and territorial areas of most counties were not ideal for implementing the Leader programme. Many of their populations exceeded the 100 000 upper threshold limit, associated with the partnerships envisaged to deliver the programme. Their territories were also larger than was considered optimum for integrated area-based actions. For these reasons, the emphasis on subsidiarity that was central to the ethos of the Leader approach brought about two changes in Ireland, in an organic, 'bottom-up' way: (i) new partnerships were formed in response to the call for expressions of interest in co-ordinating development actions in Leader 1 (1991–1993); and (ii) new territorial structures were formed within which the actions were to be implemented (Cawley 2009).

Ireland is viewed as being particularly successful in the formation of Leader partnerships which included the public, private and voluntary sectors (McDonagh 2001; Moseley 2003; O'Keeffe 2009, 2014). Because of a long tradition of voluntary rural action, organisations existed that were able to respond relatively quickly to the first call for proposals for funding (Sabel 1996). Representatives of state agencies and organisations with experience in supporting agriculture, employment, enterprise and tourism became members of the Leader LAGs (Moseley et al. 2001). Retail banks and farmers' co-operatives also joined. Research by Moseley et al. (2001) found that where local government was involved, it was usually as an ordinary member of a partnership, represented by an official on a similar basis to other partners. Because the territories adopted for the implementation of the programme were based on the formation of the partnerships, there was considerable variation in the areas involved (Figure 2). They included some counties, for example Wexford (no. 15) and Clare (no. 8), where sub-structures existed or liaison took place with subgroups to ensure representation across the county (NESC 1994). There were counties which contained several Leader areas (for example, County Mayo – nos. 5, 6, 7) and some partnership areas crossed county boundaries (for example no. 3 which included parts of counties Sligo, Leitrim and Roscommon). Therefore, many of the geographical territories reflected common interests, instead of being determined by county boundaries, and this was an innovative outcome of the Leader approach.

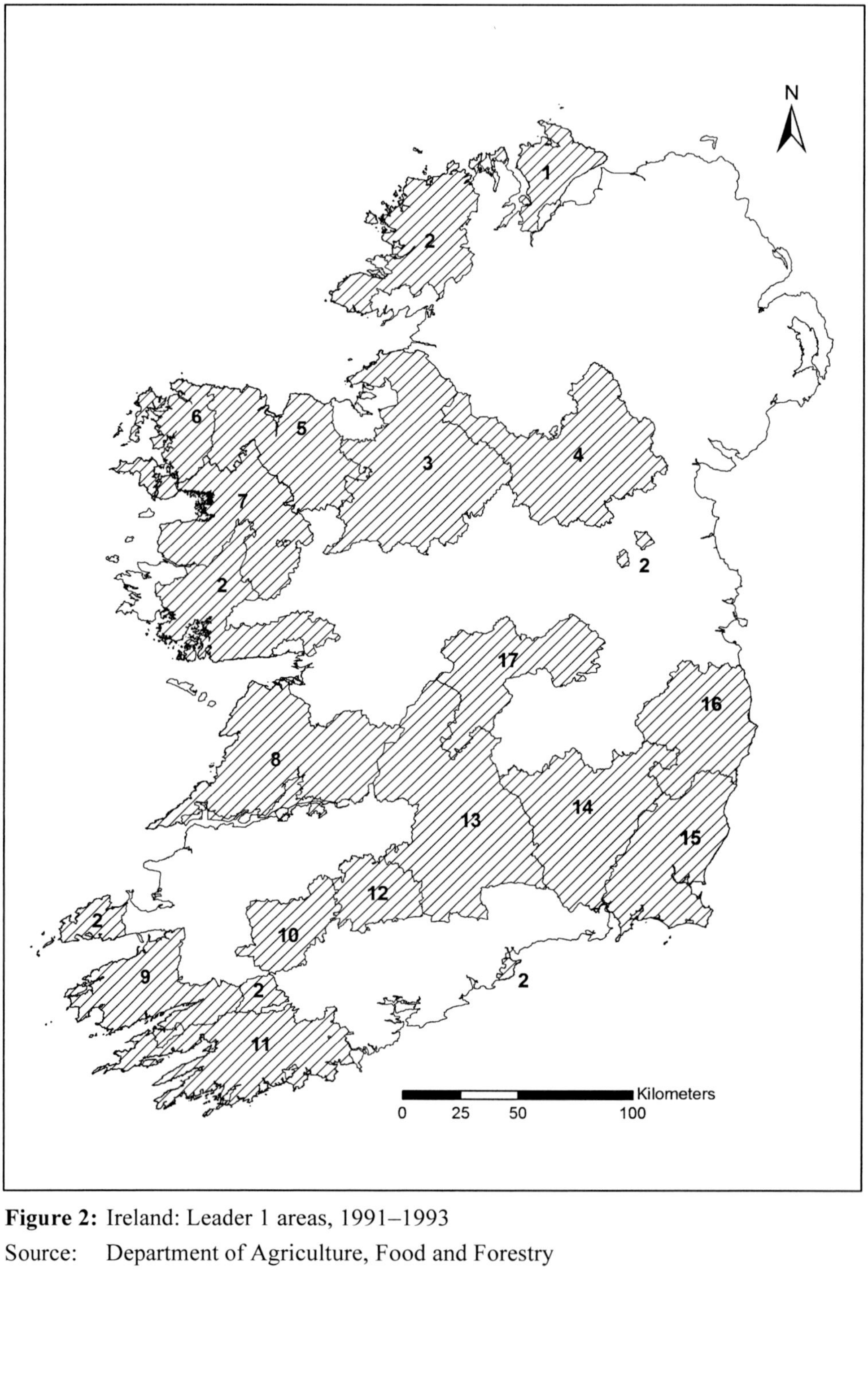

Figure 2: Ireland: Leader 1 areas, 1991–1993

Source: Department of Agriculture, Food and Forestry

This paper does not deal with the practical outcomes of the investment that took place through the Leader programme in Ireland, in terms of business formation, job creation, training in skills and local area enhancement. Information relating to these dimensions is available from several other sources (e.g., Storey 1999; Moseley et al. 2001; Dax et al. 2016). The priorities for funding changed over time but small scale manufacturing businesses and tourism enterprises were of particular importance in the early phases of Leader. The outcomes in Ireland are judged to have been most successful in promoting social empowerment, through the engagement of local volunteers in the LAGs, and local business and employment creation (O'Keeffe 2014). Weaknesses associated with Leader 1 (1991–1993) and Leader 2 (1994–1999) related to 'deadweight' (the funding of projects that might have taken place in any case) and the displacement of existing employment (Moseley et al. 2001). An evaluation of Leader Plus (2000–2006), however, noted that the cost per job was, in fact, less than that imputed originally and that deadweight was reduced considerably (Fitzpatrick Associates 2005). The European Court of Auditors (2010, 31) more recently commended Ireland for having checks in place to avoid deadweight in the RDP 2006–2013, although the associated documentation was not considered to be adequate.

Leader's changing relationships with local government in Ireland, 1991–2013

The Leader programme was subject to central control through the responsible Irish government department (usually, but not always as discussed below, the department of agriculture), but the local partnerships enjoyed considerable independence from local government. In fact, it could be said that the Leader partnerships compensated for the absence of governmental structures below the level of the county. Gradually, however, measures were introduced to bring them within the oversight of local government, beginning in the late 1990s. These changes are now discussed.

Ireland has taken part in all five phases of the Leader programme to date, Leader 1, Leader 2, Leader Plus, the RDP 2007–2013 and the RDP 2014–2020[1]. Leader was introduced as a fourth axis included with agricultural competitiveness, the maintenance of environmental quality, the quality of life in rural areas and diversification of the economy, as part of rural development more broadly

[1] The details of the respective roles of local government and the partnerships within counties, in the latter, was unclear in January 2016 when the paper was finalised

defined in the RDP 2007–2013 (DAFF 2007). The amounts of funding granted and the numbers of LAGs and their territories changed over time (Table 2).

Table 2: Leader programme in Ireland and relationships with local government

Time period	*Programme*	*N of Leader partnerships (total funding)*	*Relationship with local government*
1991–1993	Leader 1	16 (€44.8 million)	Local government representatives were ordinary members of some LAG Boards
1994–1999	Leader 2	34 (€95.58 million) Plus 2 tourism sector LAPs and 1 community LAP	From 1999, elected members of local government appointed to LAG Boards Requirement for reporting to new County Development Boards (CDBs)
2000–2006	Leader Plus	22 jointly EU-Ireland funded 16 funded by Irish government (€98.23 million)	On average, two elected members of local government in a county sat on each LAG Board
2007–2013	Rural Development Programme	37 (€425 million)	Four places were allocated for elected members on each LDC Board and one place for a nominee of the County Manager Endorsement of Annual Plans by CDBs required
2014–2020	Rural Development Programme	28 (€250 million) (respective roles of county councils and LDCs unclear in January 2016)	Local Government Reform Act 2014 brought LDCs within the remit of a new county Local Community Development Committee

Sources: Storey 1999, Kearney & Associates 2000, Fitzpatrick Associates 2005, DECLG 2014

Sixteen LAGs were funded under Leader 1 (1991–1993) and covered approximately 60% of the land area of the Irish state (Kearney and Associates 1997). Thirty seven LAGs (34 partnerships, two tourism entities and one community development entity) were recognised and funded under Leader 2 (1994–1999) and covered all counties in the state, excluding cities, reflecting the success of local groups in preparing proposals that were considered worthy of funding (Table 2). Again some Leader territories coincided with the county areas, some counties contained more than one partnership and some cross-county partnerships existed (Figure 3). An evaluation of Leader 1 suggested that there was a possibility of projects applying for funding from both Leader and social partnerships (established to offset unemployment in 1991 and later having wider roles relating to social inclusion), although the total founding received did not exceed the permitted amount under the Leader regulations (Kearney and Associates 1997). To avoid dual applications, the government transferred responsibility for economic initiatives from the social to the Leader partnerships in 1994 (Cawley 2007). The desirability of integrating rural policy more effectively with regional policy was proposed in an official policy White Paper on Rural Development in 1999 (Government of Ireland 1999). The White Paper reflected the attention that was being given to this issue at a wider European scale in the emerging European Spatial Development Perspective (Faloudi 2001).

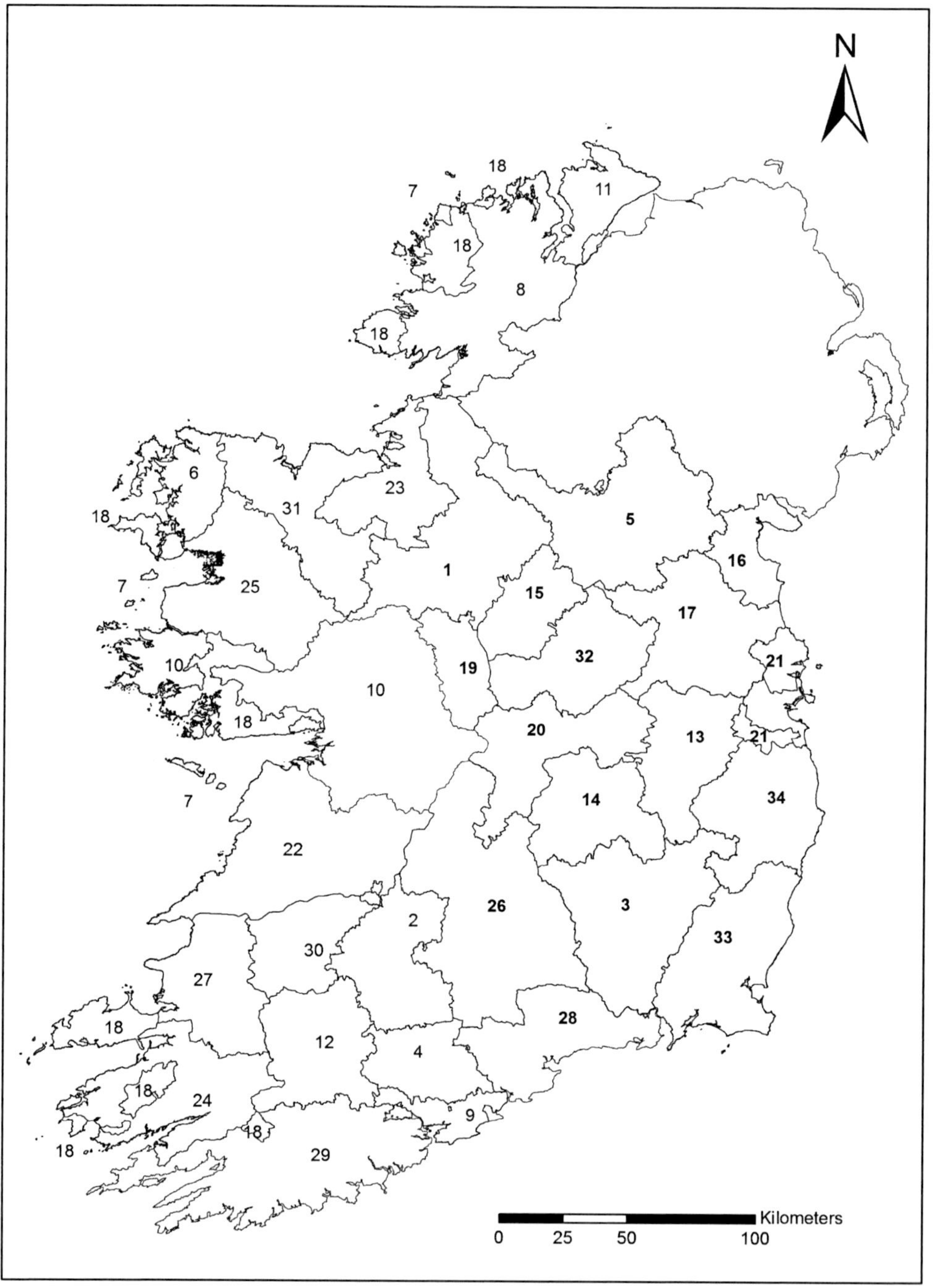

Figure 3: Ireland: Leader 2 areas, 1994–1999

Source: Department of Agriculture, Food and Forestry

By the late 1990s, as Leader 2 came towards an end, concerns were growing about a democratic deficit among the membership of the management boards of the LAGs (Walsh et al. 1998; Moseley et al. 2001), reflecting an issue identified by Sywngedouw (2005) as potentially problematic, in the case of such partnerships. Most of the members represented sectional interests of various kinds but, unlike county councillors, were not elected by universal franchise among the population. From 1999, in order to keep local government informed about the activities of the partnerships, central government required the LAGs to include elected representatives (county councillors) on their boards and to liaise with a County Development Board (Government of Ireland 1998). On average, two elected representatives were appointed (Kearney and Associates 2000).

The Leader Plus programme ran between 2000 and 2006. Twenty two LAGs were funded jointly by the EU and the national government and 16 were funded by the Irish government solely, reflecting the increased availability of funding as the economy grew strongly (Fitzpatrick Associates 2005) (Table 2). The National Spatial Strategy 2002–2020 (Government of Ireland 2002) linked rural development explicitly to regional development, in particular through the development of medium sized 'hub' towns as locations for the delivery of services to smaller towns and villages and surrounding rural populations. In 2002 also, the rural development function was moved from the Department of Agriculture to a new Department of Community, Rural and Gaeltacht Affairs (DCRGA) in order to highlight the integrated approach in Leader and to promote regional and social balance more effectively.

A review was initiated in 2003 to secure the most positive impacts for communities from expenditure through partnerships and to gain the most effective use of EU funding, in preparation for the EU RDP 2006–2013 (DCRGA 2003). Continuing concerns were expressed in the review about duplication of actions and investment between the social partnerships and the Leader LAGs, notwithstanding measures taken to avoid such duplication in 1994 (by allocating the economic functions to the Leader LAGs, as noted above). In August 2007, the Minister for Community, Rural and Gaeltacht Affairs announced the establishment of 25 new Integrated Local Development Companies (ILDCs) in rural areas across the state (DCRGA 2007). The title ILDC was used to highlight their role in integrating the activities of the Leader LAGs and the social partnerships and they were non-profit companies. Not surprisingly, this proposal evoked a negative response from some of the various partnerships because of involving a reallocation of responsibilities and mergers between existing groups. Following extensive discussion, the Minister agreed that 37 ILDCs would be established to retain more than one entity in the larger counties like Cork, Donegal and Mayo. By July 2009 (two years after the RDP was scheduled to commence), 25 ILDCs were in place, based on partnerships which already worked together in delivering social and economic programmes, and 12 additional companies were formed through mergers. In general, the links to the county structures were strengthened. These links were underlined by an allocation of four places on ILDC boards for elected representatives of the county council and one place for a nominee of the County Manager and a requirement to have annual plans endorsed by the County Development Boards. The latter requirement involved additional administration for both local government and the partnerships, especially in cases where an ILDC crossed a county boundary (a small number continued to do so). There were, however, many examples of partnerships and local authorities working productively together in village enhancement projects, for example (O'Keeffe 2009). In order to meet new EU strategic guidelines for rural development for 2007–2013 (CEC 2006), the ILDCs were also required to increase the membership from the community and voluntary sectors and from minority groups (youth, women, the Irish traveller community and immigrants) on their boards to seek to ensure that their needs were recognised and responded to.

An evaluation of the RDP 2007–2013 in Ireland identified several weaknesses which arose, in part at least, from changes in governance structures and procedures (Dax et al. 2016). Interviews conducted by the present author with representatives of four long-established Leader companies in western Ireland confirm these weaknesses. They related to delays in beginning the work of the programme, as a result of the delays with mergers. The need for additional budgetary reporting was an extra demand on staff time. It was found that agriculture-related projects took precedence over other projects, arising from Leader being integrated with agricultural policy. There was also little evidence of the 'new rural paradigm' based on local territory, as recommended by an OECD (2006) report. More far-reaching changes were to take place.

Local government reform: implications for the implementation of the RDP 2014–2020

A new Irish government was elected in 2011 and partnership governance became subject to review by a new Department of the Environment, Community and Local Government (DECLG). This was part of a quest for efficiency and effectiveness in delivering social and development measures, which echoed objectives of the EC for the RDP 2014–2020 (ENRD 2015), as well as budgetary concerns in Ireland. A review of the RDP 2007–2013 programme by Local Government/Local Development Alignment Steering Group (referred to hereafter for simplicity sake as the Steering Group), published in March 2012, identified the "limitations of the local development company model" (DECLG 2012a, 8). This is the model for integrating the various types of partnerships introduced in 2007 which are now referred to as Local Development Companies — LDCs. These were said to include: the considerable administrative burden carried by local development companies; the potential that existed for duplication and overlap between activities because of the many organisations involved in local development; the multiple arrangements for funding and reporting; and the demands and hidden costs associated with the requirement on various stakeholders to participate in multiple boards and structures at local level. It was further pointed out that, in general, local government had limited input into strategy or decision-making with respect to local development programmes and that their involvement should be increased.

The Steering Group recommended a sharing of office facilities and services between local government and the LDCs, as a method of gaining financial efficiencies, and oversight of local and community development programmes by a special committee at a county level, i.e., to integrate the activities of the LDCs with those of local government within the county territories (DECLG 2012a, 5). A national Inter-Departmental Group was also proposed to further oversee the management and delivery of local and community development programmes and interventions (ibid., 4), in order to enhance data capture across various programmes and avoid the possibility of duplication of expenditure. Thus, the various partnerships were to be brought within closer control of local government in a process of "territorial alignment" and of national government as part of a "whole government approach" (ibid., 13).

The Steering Group proposals were incorporated into an Action Plan for More Effective Local Government (DECLG 2012b) of October 2012. Finally, the Local Government Reform Act of 1 June 2014 (Government of Ireland 2014) gave effect to the increased influence being allocated to local government in economic and social development through the establishment of a Local Community Development Committee (LCDC), in each county, to replace the former County Development Board. As described in the Act, the role of the LCDC relates to "developing, co-ordinating and implementing a coherent and integrated approach to local and community development" (ibid., Chapter 2, 36[1], 49A). Specific guidelines for establishing the LCDCs were published in July 2014 (DECLG 2014). The membership represents a wide range of public, community and civic society interests but, where more than one partnership exists within a county, not all may be represented on the Committee at any one time. Under EC regulations, no one sector can hold more than 49% of the seats which serves to allay a fear that a combination of local government and the state could have overall domination. The LCDC is legally required to prepare a six-year Local Economic and Community Plan, implement, review, monitor and, if necessary, revise it. It has legal obligations also to "improve the coordination of public-funded local and community development programmes and reduce duplication" (DECLG 2014, 10) and to co-ordinate generally other local and community development programmes within its area of operation. There was an almost immediate public protest from members of the LDCs to the establishment of LCDCs and the wide-ranging powers allocated to them with respect to local development (Holland 2014). Several also expressed their concerns through the newsletter of their representative group, the Irish Local Development Network (ILDN News 2014a).

The results of the calls for expression of interest from LDCs for funding under the RDP 2014–2020 were unknown in January 2016. There was considerable uncertainty and concern being expressed publicly by the boards in relation to the potential role of local government in co-ordinating the programme. Another concern related to the reduction of 41% in the overall Leader budget between 2007–2013 and 2014–2020 and the lower government support for Leader than for other axes (particularly agriculture) of the proposed programme (ILDN News 2014a). An allocation of responsibility for Leader to local government was identified as potentially leading to the loss of expertise built up among LAG staff over many years and a loss of the voluntary inputs of members of the management boards (ILDN News 2014b). It was suggested also that the philanthropic funds which had been leveraged in the past, to supplement funding from the EU and the national government, would not be available to local government which is precluded from sourcing such funding (ILDN News 2014c).

Conclusion

An allocation of responsibility for local development and service delivery has been given to new locally based public-private-voluntary partnerships in both urban and rural areas in many countries for more than two decades. This has been viewed, including in Ireland, as having positive features in devolving greater influence to local communities over actions that affect their wellbeing and creating a sense of ownership and empowerment (OECD 2006). Negative features have been associated with undermining the role of elected local councillors and local government through the formation of boards whose membership comes predominantly from a range of state, private and voluntary organisations (Swyngedouw 2005).

The EU Leader programme assumed particular importance in promoting area-based local development through LAGs in the member states, since the early 1990s, where previously rural development was often dominated by 'top down' sectoral approaches designed by central government departments. The Leader approach was considered as being particularly innovative in Ireland which has a highly centralised system of government and limited financial resources that could be deployed for rural development at a local government level (NESC 1994; Turock 2001). In order to meet the Leader programme requirement for local territories within which actions would be devised and implemented, new territorial units emerged associated with the LAGs that were responsible for the implementation of Leader in Ireland. In many instances they diverged from the existing county structures and had considerable independence from local government, although often working with the latter on particular projects.

From 1999 concerns arose about this independence of partnerships, most of whose members lacked the democratic status of elected councillors, illustrating the issue of the 'democratic deficit' identified by Swyngedouw (2005). In the case of Leader in Ireland, measures were introduced gradually by central government, during each succeeding phase of the programme and its successor RDP programme, to increase the involvement of the elected councillors and local government in governance and to align the Leader territories with the county areas. A quest for greater efficiencies in public expenditure at EU and national level, following the recession of 2008, contributed to far-reaching reforms in Ireland which have served to reduce some of the remits of local government but assigned them instead a leading role in co-ordinating the delivery of the RDP Programme 2014–2020. Monitoring at a national level has been increased also. These developments appear to reduce the role of the local partnerships and, according to the latter, undermine the principle of subsidiarity on which Leader is based.

References

Benington, J. & Geddes, M. (2001). Social exclusion, partnership and local governance — new problems, new policy discourses in the European Union. In Geddes, M. & Benington, M. (Eds.), Local partnerships and social exclusion in the European Union (pp. 1–14). London: Routledge.

Cawley, M. (2007). Rural development in Ireland: a local governance perspective. In Hodgett, S., Johnson, D. & Royle, S. (Eds.), Doing development differently: regional development on the Atlantic periphery (pp. 246–266). Sydney, Nova Scotia: Cape Breton University Press.

Cawley, M. (2009). Local governance and sustainable rural development: Ireland's experience in an EU context. Revija za Geografijo 7 (1), 53–64.

Cawley, M. & Nguyen, G. (2008). Service delivery through partnerships in sparsely populated areas: evidence from France and Ireland. Irish Geography 41 (1), 71–87.

CEC (Commission of the European Communities) (2006). The Leader approach: a basic guide. Brussels: CEC.

CEC (Commission of the European Communities) (2015). The Common Agricultural Policy: investing in rural Europe. ec.europa.eu/agriculture/rural-development-2014-2020/country-files/common/rdp-factsheet_en.pdf. Accessed 20 December 2015.

Dax, T., Strahl, W., Kirwan, J. & Maye, D. (2016). The Leader programme 2007–2013: enabling or disabling social innovation and neo-endogenous development? Insights from Austria and Ireland. European Urban and Regional Studies 23 (1), 56–68.

DAFF (Department of Agriculture, Food and Forestry) (2007). Ireland, CAP rural development programme 2007–2013. Dublin: Department of Agriculture.

DCRGA (Department of Community, Rural and Gaeltacht Affairs) (2003). Review of community and local development structures: discussion document. Dublin: DCRGA.

DCRGA (Department of Community, Rural and Gaeltacht Affairs) (2007). Guidelines on the governance of integrated local development companies and urban based partnerships. Dublin: DCRGA.

DECLG (Department of Environment, Community and Local Government (2012a). Final report of the local government/local development alignment steering group. Dublin: DECLG.

DECLG (Department of Environment, Community and Local Government (2012b). Putting people first: action programme for effective local government. Dublin: DECLG.

DECLG (Department of Environment, Community and Rural Government) (2014). Guidelines for the establishment and operation of Local Community Development Committees. Dublin: DECLG.

European Court of Auditors (2010). Implementation of the Leader approach for rural development, Special Report No. 5. Luxembourg: European Court of Auditors.

ENRD (European Network for Rural Development) (2015). Community-led local development. enrd.ec.europa.eu/en/themes/clld. Accessed 1 May 2015.

Faludi, A. (2001). European spatial planning: a contested field. Built Environment 27 (4), 245–252.

Fitzpatrick Associates (2005). Mid-term evaluation of LEADER+, Phase 2 Report. Dublin: Fitzpatrick Associates.

Geddes, M. (2006). Partnership and the limits to local governance in England: institutionalist analysis and neoliberalism. International Journal of Urban and Regional Research 30 (1), 76–97.

Geddes, M. & Benington, M. (Eds.) (2001). Local partnerships and social exclusion in the European Union. London: Routledge.

Goodwin, M. (1998). The governance of rural areas: some emerging research issues and agendas. Journal of Rural Studies 14 (1), 5–12.

Government of Ireland (1995). Operational programme for the implementation of the EU Leader 2 initiative in Ireland 1994–1999. Dublin: Government Publications.

Government of Ireland (1998). Task force on the integration of local government and local development systems, Report. Dublin: Government Publications.

Government of Ireland (1999). White paper on rural development. Dublin: Government Publications.

Government of Ireland (2002). National spatial strategy 2002–2020: people, places and potential. Dublin: Government Publications.

Government of Ireland (2014). Local Government Reform Act, No. 1 of 2014. www.irishstatutebook.ie/pdf/2014/en.act.2014.0001.pdf. Accessed 30 September 2014.

Healey, P. (2003). Institutionalist theory, social exclusion and governance. In Madanipour, A., Cars, G. & Allen, J. (Eds.), Social exclusion in European cities (pp. 53–73). London: Jessica Kingsley.

Holland, K. (2014). Thousands protest against plan to change Leader schemes. The Irish Times, 9th July 2014. www.irishtimes.com/news/social-affairs/thousands-protest-against-against-plan-to-change-leader-schemes-1.1860899. Accessed 23 April 2015.

ILDN (Irish Local Development Network) News (2014a). Local development companies express concerns over local government reforms, 7 March 2014. www.ildn.ie/news/2014/03. Accessed 23 April 2015.

ILDN (Irish Local Development Network) News (2014b). Galway companies express concerns over local government reform, 27 March 2014. www.ildn.ie/news/2014/03. Accessed 23 April 2015.

ILDN (Irish Local Development Network) News (2014c). What do local government reforms mean for rural communities? 11 March 2014. www.ildn.ie/news/2014/03. Accessed 23 April 2015.

Jessop, B. (1995). The regulation approach, governance and post-Fordism: alternative perspectives on economic and political change? Economy and Society 24 (3), 307–333.

Kearney & Associates (1997). Leader 1 initiative in Ireland: Report on final outcome of Leader 1. Dublin: Kearney and Associates.

Kearney & Associates (2000). Operational Programme for Leader 2 Community Initiative: Ex post evaluation, interim report. Dublin: Kearney and Associates.

Keating, M. (1998). The new regionalism in Western Europe: territorial restructuring and political change. Cheltenham, UK: Edward Elgar.

McDonagh, J. (2001). Renegotiating rural development in Ireland. Aldershot: Ashgate.

Moseley, M.J. (2003). Local partnerships for rural development: the European experience. Wallingford: CABI.

Moseley, M.J., Cherrett, T. & Cawley, M. (2001). Local partnerships for rural development: Irish experience in context. Irish Geography 34 (2), 176–193.

NESC (National Economic and Social Council) (1994). New approaches to rural development. Dublin: NESC.

O'Keeffe, B. (2009). Regional and local development in Ireland: the potential of Leader partnerships to provide municipal government. Lex Localis-Journal of Local Self-Government 7 (3), 257–269.

O'Keeffe, B. (2014). Supporting enterprise development in rural areas. In O'Donoghue, C., Conneely, R., Frost, D., Heanue, K., Leonard, B. & Meredith, D. (Eds.), Rural economic development in Ireland (pp. 429–453). Athenry, Galway: Teagasc Rural Economy Research Centre.

O'Malley, E. (1992). The pilot programme for integrated rural development 1988–1990, Broadsheet Series Paper No. 27. Dublin: Economic and Social Research Institute.

OECD (Organisation for Economic Co-Operation and Development) (1990). Partnerships for rural development. Paris: OECD.

OECD (Organisation for Economic Co-Operation and Development) (2006). The new rural paradigm: policies and governance. Paris: OECD.

Peck, J. & Tickell, A. (1994). Too many partnerships. The future for regeneration partnerships. Local Economy 9 (3), 251–265.

Ray, C. (2000). The EU Leader programme: rural development laboratory. Sociologia Ruralis 40 (2), 163–171.

Sabel, C. (1996). Ireland: local partnerships and social innovation. Paris: OECD.

Storey, D. (1999). Issues of integration, participation and empowerment in rural development: the case of LEADER in the Republic of Ireland. Journal of Rural Studies 15 (3), 307–313.

Swyngedouw, E. (2005). Governance, innovation and the citizen: the Janus face of governance beyond the state. Urban Studies 42 (11), 1991–2006.

Turock, I. (2001). Innovation in local governance: the Irish partnership model. In OECD (Organisation for Economic Co-operation and Development), Local Partnership for Better Governance (pp. 135–173). Paris: OECD.

Walsh, J., Craig, W. & McCafferty, D. (1998). Local partnerships for social inclusion? Dublin: Oak Tree Press.

Regional governance in the context of LEADER local action groups in the Warmińsko-Mazurskie Voivodship (Poland)

Marta Jacuniak-Suda

Abstract

The chapter examines the regional governance processes and structures which have arisen in the rural areas of the Warmińsko-Mazurskie Voivodship since the accession of Poland to the EU in 2004. The focus is on selected LEADER local action groups (LAGs) which were introduced in Poland with the pilot programme Leader+. The chapter will trace the setting-up processes, forms of organisation and composition, leadership and the difficulties of partnership-building. The results of expert interviews and a systematic analysis of documents show that there was already a process of partnership building for regional development projects in the rural areas of the Warmińsko-Mazurskie Voivodship before Polish accession to the EU. Specifically, the PHARE programme had given an important impetus towards the emergence of rural governance structures. These were initiated by local foundations or municipal associations, which then called into life the LEADER action groups. It is striking that the LAGs are strongly associated with the public sector, as their members are directly or indirectly connected with public administration. With respect to the composition of the LAGs, it is clear that the social sector is dominant, as evidenced by the extent of mobilisation and engagement of rural interest groups. However, its emancipation and the creation of sustainable partnerships involving equal representation of the various sectors is still a work in progress.

Introduction

In the last few decades, rural areas in Poland have undergone considerable changes. As a consequence of the decentralisation process, EU accession and the resulting implementation of European cohesion policies, rural areas have increasingly seen the introduction of informal and project-related forms of regulation. Since the middle of the 1990s, there has been a growing number of local and regional multi-sectoral partnerships complementing the existing economic and non-governmental organisations (agricultural producer groups, agri-tourism associations, country women's associations, volunteer fire services and cooperative banks, Kamiński and Budzisz-Szukała 2007, 17–18). These partnerships, composed from the representatives of the public sector, civil society and increasingly the private business sector, embody new forms of rural organisation which are linked to a general change in responsibilities, competences and levels of action, a phenomenon discussed under the term regional governance (Fürst 2001, 370; Fürst 2003, 46; Fürst and Knieling 2002; Benz and Dose 2004).

The various European support programmes provide the basis for the duty of potential project organisations to set up such multi-sectoral partnerships, which then conceive and implement the corresponding development programmes and projects. In this way, the recommendations of the White Paper on European Governance (European Commission 2001, 4) are implemented, addressing the issue of the opening of political decision-making and the need for more democracy and legitimacy. According to the White Paper, more citizens and organisations should be included in the shaping and implementation of European policy. The European Commission believes that policy should be based on "feedback, networks and participation at all levels" and should not be announced from the top down (European Commission 2001, 8).

These demands of the European Union for participative democracy and the application of the partnership principle are now expressed in all Polish planning documents at national, regional and local levels. In this way, there has been a rethinking of the role and significance of civil society and the private sector in the decision-making process with respect to participative democracy. This process is accompanied by the introduction of a perspective incrementalism in Polish planning practice. Parallel to the use of formal planning instruments, a trend towards project-based activity can indeed be seen in the area of land use planning in Poland. According to Knieling (2005, 813), this orientation towards project-based activity is expressed in a spatial, technical and temporal concentration of activities and resources by public stakeholders who wish to implement the objectives of land use planning. As in Germany in the 1990s, there is currently in Poland a paradigm shift from a hierarchical to a more cooperative system and a move from territorial to sub-regionally concentrated control, as well as a move from long-term to temporary short-term governance. In this way, various reform initiatives are being applied which have the purpose of activating endogenous potentials (Knieling 2005, 814).

New reform and governance initiatives in the context of regional rural development have been discussed since the middle of the 1980s, in connection with the introduction of concepts and programmes of the endogenous regional development in western and central Europe. The purpose was to identify and develop endogenous potentials in the regions so that through their strengths they could become more assertive in international competition (Mose 1993, 29; Mose 2010, 156). The implementation of the new reform and governance initiatives was accompanied in the 1990s by the introduction of the LEADER community initiative (1991) and later by "The Cork Declaration — A living countryside" (European Commission 1996) which established the concept of integrated rural development.

On the basis of regional-political action initiatives, integrated rural development can be characterised by the following features (Mose 2010, 156; see also Shucksmith 1999; Grabski-Kieron and Krajewski 2007):

- sectoral and cross-departmental procedures;
- territorial orientation;
- transfer of formerly centrally located political decision-making structures to the regional and local levels;
- setting up and use of networks involving state, private and civil society stakeholders;
- Participation of the affected population;
- Implementation of suitable forms of regional mobilisation and advice as well as regional management.

The transformation of Polish planning practice towards participative democracy, project orientation and an integrated rural development can be seen in Polish rural areas through the example of the LEADER action groups which seek to put into practice the European model of good governance in rural areas. Under the Regulation EC 11698/2005 of the European Council on the support for development of rural areas through the European agricultural funds (EAFRD EC 2005), the LEADER initiative is characterised by a number of features, shown in Figure 1.

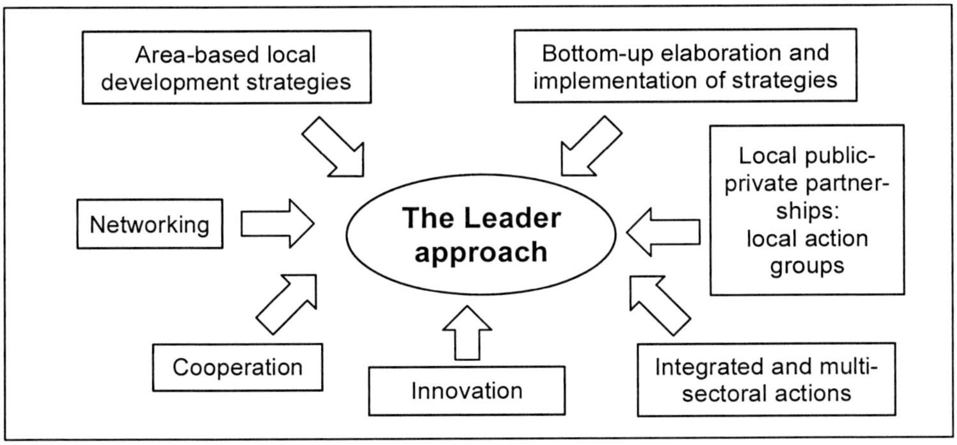

Figure 1: The LEADER principles

Source: European Commission 2006, 8

The characteristics of the LEADER initiative are put on an equal footing with those of regional governance by Fürst (2007, 8). According to his understanding, regional governance is connected with new types of regional management and cooperation which are characterised by, among other things, the following features:

- the desire of regional stakeholders to work together on problems and objectives,
- networks in which various stakeholders cooperate on economic and political issues,
- horizontal forms of interaction based on reasoning and negotiation,
- negotiated rules based on social norms, contracts and agreements,
- embedment in the existing institutional framework conditions.

Therefore, it is expected that the LEADER action groups will become a stage for local and regional decision-making processes, for the mobilisation of rural population groups, multi-sectoral cooperation and for the working out of 'rules of the game', with the intention of developing innovative initiatives on the development of rural areas. Through the inclusion of private and civil-society stakeholders at regional levels in areas which were formerly the responsibility of state authorities (Fürst 2001 and 2007; Benz and Dose 2004), the LEADER action groups appear to be a suitable practical example of regional governance. Thus, in the context of this chapter a narrow definition of governance will be chosen instead of the broad definition which implies "the totality of political-administrative management activities" (Fürst 2007, 9).

The analysis of the LEADER local action groups permits on the one hand insights into the development of the governance processes and on the other, into governance structures in the rural areas of Poland. Here, according to Zürn (2008, 557), governance processes comprise "the processes of political negotiating, coalition-building, threats and persuasion, which are associated with the production of governance". With governance structures, "stable ideal[s], institution[s] and framework conditions are addressed and created within the governance subject matter". Zürn includes here constellations of stakeholders which have a particular significance, and less so an individual stakeholder ("the intelligent political centre", that is, an institution, club etc.). Governance refers to the coordination and self-regulation of different stakeholders and organisations which are not related in a hierarchical structure to each other. Depending on the role of the governments in the governance processes, Zürn (2008, 558) differentiates between "governance by, with and without government".

While in both Germany and Great Britain several research studies have been carried out on the implementation of the LEADER initiative and its links to regional governance aspects (e.g., Shortall and Shucksmith 1998; Ray 2000; Moseley 2003a, 2003b; Lee et al. 2006; Shucksmith 2000; 2010; Shucksmith et al. 2005; Dargan and Shucksmith 2008; Bruckmeier 2000; Hahne 2004; Böcher 2003, 2008; Böcher et al. 2008), this subject has hardly been studied in Poland. Apart from some contributions on the general implementation of the LEADER community initiatives in Poland (Głuszyński 2005; Hałasiewicz 2008; Budzich-Szukała 2008) and some individual studies on initial experiences with the LEADER+ pilot programme as well as other local development initiatives (Błąd and Kamiński 2005; Gramzow 2006a, 2006b, 2006c; Kovách and Gorlach 2006; Brodziński 2006; Gorlach and Adamski 2007; Śpiewak 2009), there has been little research work on the LEADER action groups in the funding period 2007–2013. This also applies to regional comparative studies involving the various groups (Halamska et al. 2010; Czapiewska 2011; Kołodziejczak 2011). Above all however, there is limited application of the analytical approaches of regional governance to empirical research.

This chapter attempts to close these gaps by examining the regional governance processes and structures with the example of the selected LEADER action groups (referred to as LAGs) in Warmińsko-Mazurskie. The study will focus on the following research questions:

- Under which framework conditions did the LAGs in Poland come into existence and which organisations were involved?
- What are the key features of the LAGs in Warmińsko-Mazurskie with respect to the organisational structure, composition and leadership?
- What are the challenges that the selected LAGs in Warmińsko-Mazurskie are confronted with in the context of partnership principle?

Finally, the aim of this chapter includes the identification of particular values of regional governance which occur in the selected LAGs areas of the Warmińsko-Mazurskie Voivodship.

The research has emerged within the framework of the research project "Regional Governance in rural peripheries of Europe — regional stakeholder constellations and processes as key factors in regional development. A comparative analysis based on selected case studies in peripheral rural regions of the

EU", at the University of Oldenburg (Germany), supported by the German Research Foundation. The research objectives focus on the examination of concepts and instruments for the development of peripheral rural areas in the EU and to what extent these are influenced by the formation of new forms of control of so-called regional governance. There is a particular focus on the regional-political approach of an integrated rural development.

Description of the research area

There have been profound changes in several areas of the societal, economic and political life since the first democratic elections in 1990, resulting in an increase of regional disparities (Churski 2008, 582). In particular, there has been a growth in the urban-rural differences, e.g., in the year 2000, GDP per head in the poorest Polish rural regions was five times lower than in the economically stronger regions (Baum and Weingarten 2005, 211–212). Next to regions with strong urban centres and rapid economic growth (such as Poznań, Wrocław and Warszawa), there are rural regions which are economically and structurally weak and with respect to their locations are considered peripheral. An example of this is the Voivodship Warmińsko-Mazurskie which demonstrates the characteristic development problems of the peripheries of east Poland, located far from centres and near the border to the Kaliningrad district of Russia (Droth et al. 2000; Haase et al. 2004; Rosner 2010, 13; MIR 2014[1]). The Voivodship Warmińsko-Mazurskie with the main town of Olsztyn is situated in the north-east of Poland (Figure 2). With the exception of three regional centres — Olsztyn, Elbląg und Ełk — there is a low level of urbanisation and the region is largely rural. The average population density in Warmińsko-Mazurskie in 2014 amounted to 60 inhabitants per km² and is twice as low as the national average of 123 inhabitants per km². In rural areas, this figure is down to 25 (Table 1). The forecast is that up to the year 2035, the population numbers will continue to sink. Both regional centres and rural areas will be equally affected (USTAT 2010, 2015).

---

1 Voivodship Warmińsko-Mazurskie is part of the territory of the operational programme "East Poland" (2007–2013, 2014–2020) of the Polish Ministry for Infrastructure and Development, designed for the most thinly populated and structurally weak voivodships of Poland to promote innovative economic development and the modernising of infrastructure.

Figure 2: Location of the Voivodship Warmińsko-Mazurskie
Source: Author's compilation

The current problems in Warmińsko-Mazurskie are caused also by an exodus of younger and more active population groups and a lack of qualified employment provision in Warmińsko-Mazurskie. Since 2002, the number of unemployed has in fact reduced continuously, particularly since 2005, and reduced from 25.6% in 2002 to 16.8% in 2008 (EU 2008), probably because of emigration following EU accession and economic growth in Poland. The number of registered unemployed in the Voivodship was in 2014, however, at 18.9% higher than the national average (11.5%). The number in employment is highest in the processing industries (52.3%) (dairy industry, timber industry, furniture industry), followed by retail (14.9%) and construction (8.9%) (USTAT 2015).

Table 1: Key data on the Voivodship Warmińsko-Mazurskie

Area	24 173 km^2
Population (2014)	
Average population density	60 inhabitants/km^2
Population density in rural areas	25 inhabitants/km^2
Most important employment (2015)	
Processing industries /dairy, forestry, furniture	52.3%
Retail	14.9%
Construction	8.9%
Unemployment (2014)	
Average unemployment rate	18.9%
Unemployment rate in Olsztyn (regional capital)	6.9%
Unemployment rate in rural areas (e.g., Piski District)	29.8%

Source: USTAT 2015a

The Warmińsko-Mazurskie Voivodship is characterised by a large proportion of rented land previously belonging to state-owned farms, with a population consisting of a large proportion of former workers, largely unemployed and with lower education levels than the average of the population in rural areas. The unemployment and low education levels have reduced family incomes and resulted in a decrease in demand that again has a negative impact on the establishment and profitability of local firms (Piasecki and Rogut 2004, 271–272). As Grabski-Kieron and Krajewski (2007, 17–18) report, small subsistence farms operated privately represent an important collecting pond for unemployed workers but serve mainly for self-sufficiency. All these factors contribute to a low engagement of the population in community activities, such as the development of local strategies in the framework of LEADER+ (Brodziński 2006, 110), resulting in a very slow development of cooperation structures. Therefore, training to increase levels of rural leadership and educational programmes for enhancing professional skills have been identified by involved LEADER+ members in Warmińsko-Mazurskie as the most important activities (Brodziński 2006, 109).

As far as regional governance structures at the level of the Voivodship Warmińsko-Mazurskie are concerned, there are several active organisations working for the state, such as the Warmińsko-Mazurskie Chamber of Agriculture, the Regional Agency for Business and Innovation, the Regional Agency for Restructuring and Modernisation of Agriculture and the Warmińsko-Mazurskie Enterprise Zone. These organisations, however, have not focused specifically on an integrated regional development of rural areas and do not offer their own development concepts or support programmes. Therefore, in the rural areas of the voivodship there are hardly any regional partnerships and organisations, such as spatially delimited regional initiatives, regional management, round tables, regional marketing organisations or regional development societies (Danielzyk and Knieling 2011, 473) offering organisation forms and platforms for cross-sectoral cooperation comparable to LEADER local action groups. The Marshal Offices, equipped since the area reforms of 1999 with regional-political competences, are the most important institutions entrusted with the task of developing rural areas.

In fact, under the communist government in Poland, there were no support instruments for rural regional development (Kozak 2001; Bukowska and Łukaszewicz 2002, 67; Grosse 2004, 236; Sagan and Lee 2005, 173; Pieper 2006, 357), although during the 1980s and 1990s, non-governmental organisations such as foundations and associations had formed which sought to encourage integrated rural regional development (Pieper 2006, 357, 361). The catalyst for this is seen in the lack of an existing regional level up to 1999, which offered scope for experimental forms. Today, their tasks include regional business support, action against unemployment, social integration, further education and support measures as well as support for tourism. They have often gathered experience with the processing of EU subsidy money within the framework of the implementation of the EU pre-accession programme.

An example of such a foundation in the research region Warmińsko-Mazurskie is the development foundation "NIDA" (*Nidzicka Fundacja Rozwoju*) based in the administrative capital Nidzica. The foundation was set up in 1993 as a public-private partnership with the aim of comprehensively revitalising the region Nidzica through the creation of jobs, support for entrepreneurs and the social economy, support for modern production methods, support for culture and further education and the protection of monuments and the environment. One of the other aims of the foundation is in strengthening partnership creation between various public and non-public organisations in the region. Even among the founding members of the foundation, there were representatives from business groups and the public and third sectors, among others. Within the context of the introduction of the LEADER initiative in Poland, foundations and municipal associations, but less often municipalities[2], have worked together in taking on leadership roles in making applications and in the search for cooperation partners. Furthermore, they have organised events and public participation workshops and they were involved in the development of strategies in the early phase of LEADER+ (Jacuniak-Suda 2015, 49).

Methods of data collection and evaluation

Data collection and evaluation took place within the framework of the DFG research project "Regional Governance in rural peripheries of Europe — Regional stakeholder constellations and processes as key factors for regional development" (2008–2011) at the University of Oldenburg (Germany) (Jacuniak-Suda 2015). Following a document analysis, an evaluation of the relevant plans, development strategies and operational programmes in the context of regional policy development of rural areas at national and regional level (Warmińsko-Mazurskie) took place. Subsequently, LEADER local development strategies, statutes and internet pages of the LEADER action groups in Warmińsko-Mazurskie were analysed.

The document analysis was followed up by visits to Warsaw and Olsztyn — the regional capital of Warmińsko-Mazurskie — in order to conduct problem-centred expert interviews. In the first empirical phase, interviews were carried out with the representatives of relevant institutions associated with regional policies and development at national and regional level (departments in national ministries and in the voivodship administration responsible for regional policy and the development of rural areas; scientific institutes). These interviews made it possible to record the changes in framework conditions in Poland and in Warmińsko-Mazurskie with respect to the organisation of regional policy and development planning in rural areas.

---

2 The Polish administrative units and their chief officers are termed: district administration (*powiat*) — district councillor (*starosta*); municipality (*gmina*) — municipality director (*wójt*); village (*wieś*) — village director (*sołtys*).

In the second empirical phase, there were interviews with chairpersons and vice-chairs of the selected LEADER action groups in Warmińsko-Mazurskie. For a closer examination, eight from fourteen LAGs which existed in 2009 were selected (Figure 3): LAG "LGD 9", LAG "Łączy Nas Kanał Elbląski", LAG "Lider w EGO", LAG "Brama Mazurskiej Krainy", LAG "Partnerstwo dla Warmii", LAG "Kraina Drwęcy i Pasłęki", LAG "Warmiński Zakątek", LAG "Barcja". The interviews took place in Elbląg, Pieniężno, Łukta, Dobre Miasto, Węgorzewo, Stare Juchy, Kętrzyn and Nidzica. The journeys by local public transport and the subsequent site visits made it possible for the author to gain an impression of the rural settlement structure, supply of services as well as infra-structure in the LAG areas. In addition, the journeys have provided an insight into the selected local development projects of the LAGs.

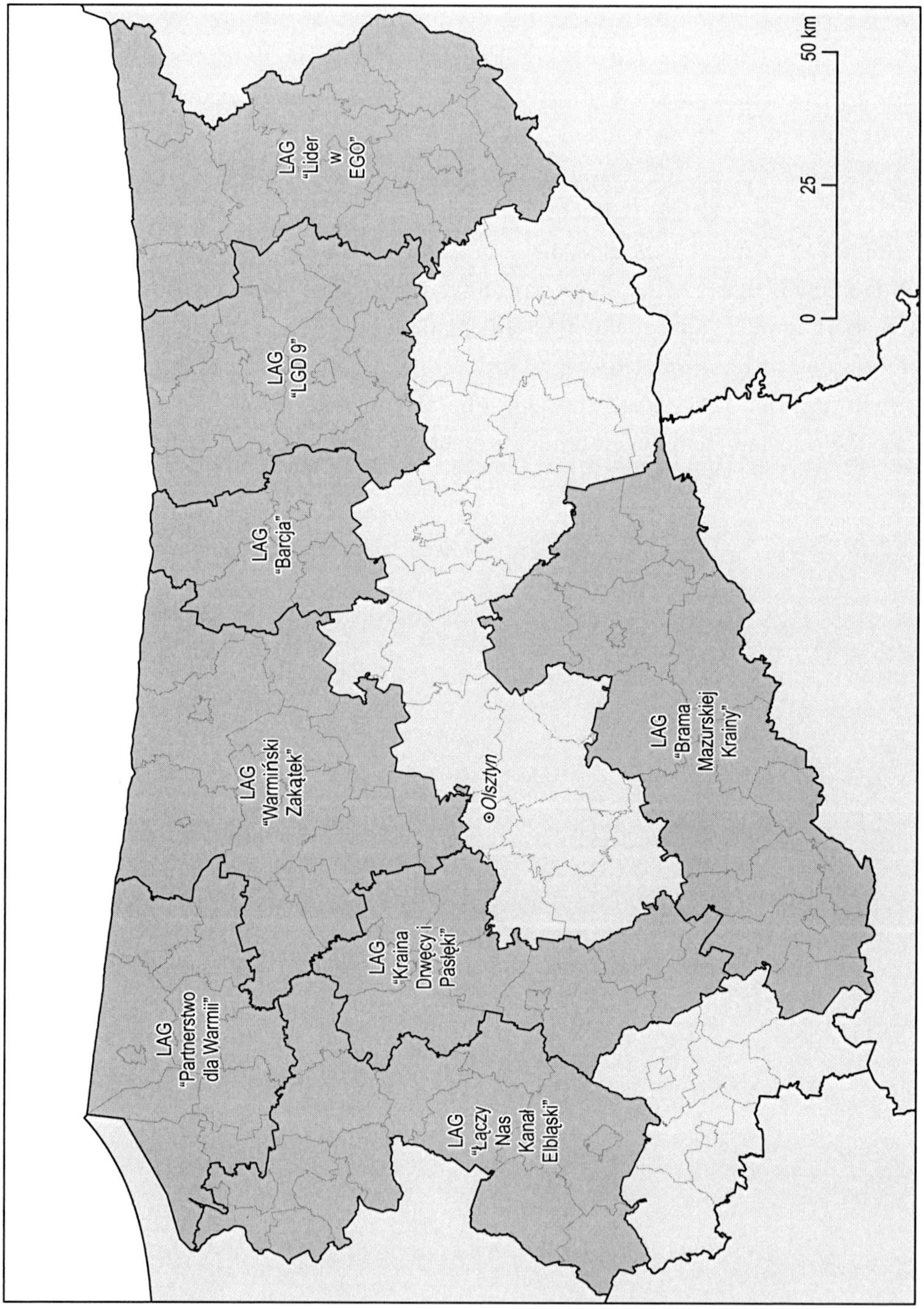

Figure 3: The location of the examined LAGs in the Voivodship Warmińsko-Mazurskie

Source: Urząd Marszałkowski Województwa Warmińsko-Mazurskiego 2015a, adapted.

Table 2 lists the various criteria used in the selection of the LAGs, which depict the geographically and economically heterogeneous picture of the Voivodship Warmińsko-Mazurskie. The selected LAGs show varying characteristics: while some were already set up within the framework of the pilot programme LEADER+, others only formed in the funding period 2007–2013. Some groups are strongly institutionalised and financially well-established while others are affected by limited financial resources and staff availability. In addition, the groups focus on different subjects in line with area-related potentials and regional problem priorities.

Table 2: Criteria set for selection of LAGs

Distance to the regional capital Olsztyn
Population density
Unemployment
GDP
Set-up in the phase of LEADER+
Set-up in the EU funding phase 2007–2013
Level of institutionalisation
Number of income streams
Thematic orientation
Distance to the regional capital Olsztyn
Population density
Unemployment

Source: Author's compilation

The applied case study approach and multi-case design in accordance with Yin (2003, 46) allowed for the observation of the phenomenon of regional governance in the field. The study was oriented to the principle of replication, that is, the attempt to identify similar or different characteristics of the research object, here the characteristics of regional governance, by way of the selection of the cases. As a result, each case served as an experiment that helped to confirm or reject the observations made in other cases. Finally, it was possible to identify similarities, differences and specificities between the cases. In total, 20 on average 1.5 hour long, guideline-supported expert interviews were conducted in April 2009.

The interviews were digitally registered and transcribed. With respect to the data evaluation, the processing of the guideline interviews and problem-centred expert interviews played a central role. These were evaluated based on the work of Mayring (1999, 73, 91) with the help of a qualitative analysis of contents. For the formation of a category system and for the identification of text materials, the software program MAX QDA was suitable and available (Jacuniak-Suda 2015, 30). To complement the methods, the system of participatory observation was used. For this purpose, the author took part in the members meetings of the LAG "Brama Mazurskiej Krainy" and LAG "LGD 9". Participation in the conference "National network of rural areas in the Voivodship Warmińsko-Mazurskie"[3] on 22.04.2009 made it possible to make contacts and meet important regional experts on rural development.

Results

Introduction of the LEADER community initiatives in Poland

As a new member of the EU having joined in 2004, Poland could not participate in the erstwhile operating community initiative LEADER+ 2000–2006. However, there was the possibility to apply measures which were comparable to the LEADER community initiative and which were financed from the European Agricultural Guarantee Fund (for the period 2000–2006). Poland made use of this possibility and integrated measures similar to LEADER+ in its operational sector programme "Restructuring and modernisation of the food sector and the development of rural areas"[4]. Within the framework of the "pilot programme LEADER+", as it was known, two types of measures similar to LEADER+ were in the main promoted. Phase 1 was the acquisition of competences to be made available to rural communities, the setting up of LAG groups and the preparation of local strategies for the development of rural areas. Phase 2 was the implementation of integrated development strategies with the character of pilot schemes, prepared by local action groups (LAGs). With the exception of the formation of networks, the LEADER+ programme in Poland comprised all the principles of the LEADER approaches (Figure 1) (European Commission 2010; MRiRW 2007, 113; Hałasiewicz 2008, 1).

---

3 Krajowa Sieć Obszarów Wiejskich w Województwie Warmińsko — Mazurskim.

4 Sektoralny Program Operacyjny "Restrukturyzacja i modernizacja sektora żywnościowego i rozwój obszarów wiejskich".

For the implementation of the measures similar to LEADER+ in Poland, support funding amounting to 18.8 million euros was made available. 80% of the funding was provided by the EU and the rest came from state finances (FAPA 2006, 10). The implementation of the pilot programme LEADER+ was intended to prepare the way for the actual LEADER initiatives in Poland after 2007.

The following institutions and organisations were involved in the programming and implementation of LEADER+ (MRiRW 2004, 109–110; 118–121, Table 3):

- Ministry for Agriculture and Village Development[5] as the administrating institution (acceptance of the newly formed LAGs and their projects),
- Foundation of the Support Programme for Agriculture[6] as the implementing institution (evaluation of applications, selection of LAGs, project selection, preparation and completion of contracts, processing of financial applications, assessments, preparation of citizen participation),
- Agency for Restructuring and Modernisation of Agriculture[7] as the institution responsible for the payment of support funds.

In addition, a steering committee from the Ministry for Agriculture and Community Development was appointed which was to support the implementation of the pilot programme LEADER+. It has the special task of participating in the process of project evaluation and to make recommendations on project implementation (MRiRW 2004, 122). The steering committee comprises the following institutions and organisations: state administration at the national level (Ministry for Agriculture and Village Development, Ministry for the Economy and Employment), state administration at the regional level (Marshal Offices), representatives from business and NGOs, representatives from trade unions, representatives from public agricultural advisory services (Chambers of Agriculture) and educational establishments.

---

5 Ministerstwo Rolnictwa i Rozwoju Wsi (instytucja zarządzająca).
6 Fundacja Programów Pomocy dla Rolnictwa (instytucja wdrażająca).
7 Agencja Restrukturyzacji i Modernizacji Rolnictwa (instytucja płatnicza).

Since the funding period 2007–2013, the LEADER approach has been integrated as a fourth pillar into the Programme for Development of Rural Areas[8]. In contrast to the previous funding period, the implementation was moved from the national to the regional level. In this way, the Marshal Offices have taken on the tasks of selecting LAGs, arranging the completion of contracts and assessing the eligibility of local development strategies. In fact, the Marshal Offices process and assess the funding applications of the LAGs although for the payment of funding, the Agency for Restructuring and Modernisation of Agriculture continues to be responsible. Other important changes include the transfer of decision-making competences for project selection and the granting of project funds to the level of the LAGs (MRiRW 2007, 304; MRiRW 2014, 235; Urząd Marszałkowski Województwa Warmińsko-Mazurskiego 2015b). Table 3 shows the responsible institutions which were or still are involved in the administration, implementation and financing of the hitherto LEADER programmes in Poland.

Table 3: The institutions involved in the LEADER programmes

	Programme	*Administering institution*	*Implementing institution*	*Financing agency (state institution)*
Pilot programme LEADER+ (2004–2006)	Operational Sector programme: Restructuring and Modernisation of the food sector and development of rural areas (2004–2006)	Ministry for Agriculture and Community Development	Foundation of the support programmes for agriculture (FAPA)	Agency for Restructuring and Modernisation of Agriculture, Ministry of Finance
LEADER (2007–2013, 2014–2020)	Programme for Development of Rural Areas (PROW) (2007–2013, 2014–2020)	Ministry for Agriculture and Community Development	Marshal Offices (public administration at regional level), LAGs	Agency for Restructuring and Modernisation of Agriculture, Ministry of Finance

Source: Author's compilation

8 Program Rozwoju Obszarów Wiejskich na lata 2007–2013 (PROW).

Initiation of the LAGs

In Poland, the LAGs are allowed to function in the following legal forms: as special purpose municipal associations (*związek gmin*), umbrella organisations (*związek stowarzyszeń*) or as foundations (*fundacje*). In addition, they may act as not-for-profit bodies (*organizacja pożytku publicznego*) and therefore be economically active, on condition that profits are used to fulfil the aims laid down in the rules of procedure. Consequently, the following laws and documents provide the legal basis for the establishment and functioning of LAGs in Poland:

- the Law on Foundations[9] of 06.04.1984,
- the Law on Associations[10] of 07.04.1989,
- the Law on Not-for-Profit Activity and Voluntary Service[11] of 24.04.2003,
- Council Regulation (EC) No. 1698/2005 on the funding support for lopment of rural areas by the European Agricultural Fund of 20.09.2005,
- the Law on the funding of development of rural areas from the European Agricultural Fund[12] of 07.03.2007,
- Programme for the Development of Rural Areas for 2007–2013[13] and 2014–2020,
- rules of procedure of the LAGs.

Most of the LAGs in Warmińsko-Mazurskie were set up in 2006 within the framework of the pilot programme LEADER+, with the exception of a group (LAG "Barcja") which was founded in 2008, that is, in the new funding period 2007–2013. The initiation of the researched LAGs was carried out through the municipalities, municipal associations and foundations, the two latter set up in the middle of the 1990s and later (Table 4). The foundations, in particular, follow an integrated development approach and focus on regional business devel-

[9] Ustawa z dnia 6 kwietnia 1984r. o fundacjach (Dz.U. z 1991 r. Nr 46, poz. 203 z późn. zm).

[10] Ustawa z dnia 7 kwietnia 1989 r. — Prawo o stowarzyszeniach (Dz. U. nr 79 z 2001 r., poz.855 z późn. zmianami).

[11] Ustawa z dnia 24 kwietnia 2003 r. o działalności pożytku publicznego i o wolontariacie (Dz.U. z 2003 r. Nr 96, poz. 873.)

[12] Ustawa z dn. 7 marca 2007 r o wspieraniu rozwoju obszarów wiejskich z udziałem środków Europejskiego Funduszu Rolnego na rzecz Rozwoju Obszarów Wiejskich (Dz. U. z 2007 r. nr 64, poz.427).

[13] Program Rozwoju Obszarów Wiejskich na lata 2007–2013.

opment, loans and credit transactions for small and medium-sized enterprises, action to combat unemployment, educational and cultural projects, social integration, student grants programmes and support for tourism. Through the wide spectrum of activities, the foundations can be considered to be 'local development centres'. Because of the frequent implementation of national and European projects such as PHARE, SAPARD and EQUAL, these organisations have gathered appropriate experience with EU procedures, project management and the set up of cooperation initiatives. In addition, they have professional staff with university degrees and on a well-established contact network which extends outside of the region, including (national level) politicians and scientists. The role of the foundations in the setting up of the LAGs includes the protracted persuasion process in the municipalities in establishing the LAGs, making applications, searching for suitable cooperation partners and also the coordination and organisation of the overall process. Additionally, the foundations were responsible for start-up financing, credit-worthiness and all administrative issues. On top of this, the foundations have carried out citizen participation actions, in which they organised a number of workshops and meetings giving the various interested parties from the public, social and private sectors the chance to make representations and exchange ideas.

In addition, individual municipalities and municipal associations were involved in the initiation of the LAGs. The founding of the municipal associations goes back to the start of the 1990s when the necessity of managing the inter-municipal tasks such as support for tourism, business assistance environmental protection, credit assistance for SMEs and the marketing of the region became an increasingly important issue in the awareness of local politicians. In some cases, the leaders of the foundations and municipal associations became the chairpersons or members of the LAGs. Often, the projects of the LAGs represent a continuation of the thematic orientation of founding organisations.

The interviewed LAG chairpersons gained motivation for the founding of the LAGs from the difficult economic situation within the LAG areas, high unemployment and a general lack of opportunity and the passivity of the local population, as well as a long-standing neglect of tourist sites.

Table 4: Founding organisations of the LAGs (with founding year)

Designation of the LAG	*Founding organisation*
LAG "Barcja" (2008)	Municipal association "Barcja" (1991)
LAG "Brama Mazurskiej Krainy" (2006)	NIDA foundation for development of Nidzica (1994)
LAG "Kraina Drwęcy i Pasłęki" (2006)	Foundation for development of the region Łukta (1995)
LAG "LGD 9" (2004)*	Association "Wiedza, Rozwój, Praca" (2004) Regional Development Association "Puszcza Borecka" (2003) and other local associations
LAG "Lider w EGO" (2006)	Municipality of Ełk, later municipal association of Ełk, Gołdap and Olecko (1997)
LAG "Łączy Nas Kanał Elbląski" (2006)	Municipal association "Związek Gmin Kanału Ostródzko-Elbląskiego" (1990)
LAG "Partnerstwo dla Warmii" (2006)	Foundation "Fundacja Wieś w XXI wiek" (1999)
LAG "Warmiński Zakątek" (2006)	Municipality Dywity and municipal office of Dobre Miasto

Source: Author's compilation based on LEADER-local development strategies 2007–2013 and LAG websites

Organisation structure

The organisation structure and competences of the LAGs are laid down in their rules of procedure, as well as in other documents listed above. Each group consists of the following bodies: a business office, managing board, council, plenary meeting and an audit commission. Although the details of the terms and conditions vary from group to group, there are many points in common. The office assumes administrative tasks and is staffed with well-qualified personnel. It is noticeable that staff who are responsible for project management and implementation have university qualifications in law and business studies as well as years of professional experience in the implementation of EU programmes. The processing and updating of local development strategies is the responsibility of the managing boards of the LAGs and are usually commissioned to planning offices or scientific institutes.

The managing board is a central controlling committee which leads the work of the LAG, recruits new members, leads meetings and prepares resolutions and applications for funding. It represents the local action groups in dealings with the public. In some cases, the board is responsible for the preparation, updating and implementation of local development strategies and in addition, it functions as a financial director. The actual decision-making committee is the council. The selection of projects which are to be carried out within the LEADER framework is decided in the council meeting. In addition, the council decides on the overall strategy of the LAG, reports on the local development strategy and is responsible for changes and evaluation. Also, the council members must sign a statement of impartiality and confidentiality before each meeting. This statement also obliges the council members to keep confidential all information and file contents that have been evaluated during the process of project selection and evaluation. Council members may not take part in the council meeting if they are related to the applicant or are members of the organisation making an application and where a suspicion of partiality may exist. The audit commission oversees the continuing activities of the LAG and those of the board. In addition, it oversees the financial plan. The plenary meeting of members advises on the overall strategy of the LAG, expresses opinions on the planned projects and on the local development strategy and gives them an official approval. The general meeting is empowered with decision-making where there is a majority of members present and entitled to vote. There is no requirement that at least 50% of the present members must come from the circle of business and social partners or from the other representatives of civil society.

The composition of the LAGs

According to Article 62b of the EAFR regulation (EC No 1698/2005), the local action groups are composed of partners from different socio-economic backgrounds in the respective areas. The regulation states: "At the level of decision-making, the business and social partners as well as other representatives of civil society such as farmers, country women and young people and their associations must comprise at least 50% of the local partnership" (Council Regulation (EC) No 1698/2005, 26). The study of the selected LAGs has shown that this requirement has been met, as Figure 4 illustrates. It is striking that the third sector is the most strongly represented while the number of business partners is small.

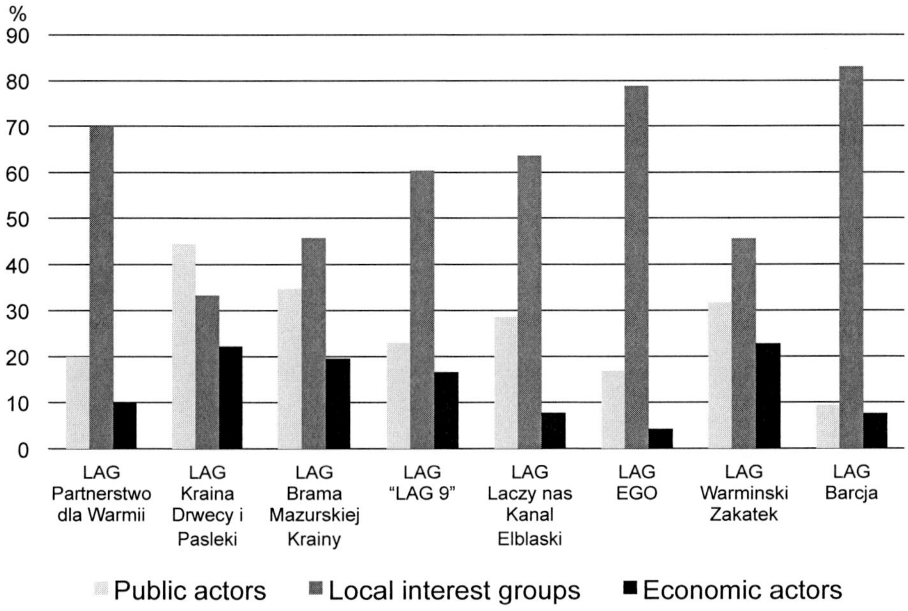

Figure 4: Composition of the LAGs according to sector

Source: Author's compilation based on LEADER local development strategies 2007–2013

There is greater differentiation to be seen when one considers the composition of the council which makes decisions on project selection and funding allocation (Figure 5). Apart from in two cases in which the council was clearly dominated by NGOs (LAG "EGO", LAG "Łączy nas Kanał Elbląski"), the public sector is well-represented. It was clear from the talks with experts and from general observations that the mayors often form an unofficial regulation body. They were described as the most active and reliable partners who regularly took part in the assemblies. This can have effects on the decision-making process as in the council meetings the social and business partners are not always present. Additionally, the interviews have shown that the representatives of public administration in the LAGs become chairpersons or members of associations belonging to the LAG. In this way, many LAG members are directly or indirectly connected with the public sector. Anyhow, in most cases, the social and business partners are represented in the LAGs at more than 50% and therefore fulfil the LEADER requirement on the composition. Figure 5 shows that except for LAG "Brama Mazurskiej Krainy", the private sector is poorly represented.

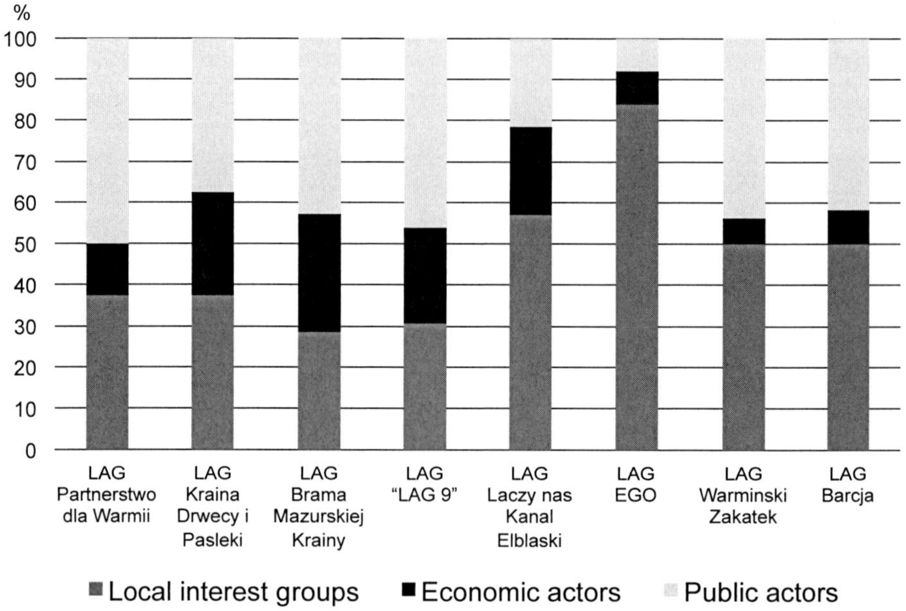

Figure 5: Composition of the council according to sector

Source: Author's compilation based on LEADER-local development strategies 2007–2013

With respect to the representatives of civil society, the village citizen associations dominate in all the studied LAGs. Other groupings such as employer associations, trade unions, farmer associations and youth groups are under-represented or missing completely. Specifically, the number of young people engaged in the LAGs is also very low. The chairpersons notice that young people were initially active but have lost interest over the course of time. Still, the integration of the different interest group representatives in the LAG bodies was accompanied by a large number of training events focusing on, among other things, information about the LEADER approach, the develop-ment of rural areas, effective use of funding, support for tourism, partnership building and project management.

Leadership

According to the interviews with chairpersons of the LAGs and with the staff in the Warmińsko-Mazurskie Marshal Office responsible for LEADER implementation, the municipality directors (*wójt*) and other representatives of the public sector were considered as the real drivers (*liderzy*) of the groups. They participate regularly in the assemblies, are disciplined, ambitious and take part actively in the work of the groups. In addition, they provide financial resources and knowledge and municipal personnel, if necessary. Often, they are strong leaders with high levels of persuasiveness and charisma and were described as visionary. At the same time, it is striking that 50% of the chairpersons in the LAGs are originally from the town council or municipal administration (Table 5). As a result of their long period of employment in the administrative or policy bodies, they have a good knowledge of the region, a network of contacts in all sectors and therefore corresponding sources of information. Many of them have founded their own not-for-profit associations or are members of an association. Representatives of the private sector or NGOs are likewise politically active as members of the council.

Table 5: Characteristics of the LAG chairpersons

Name of the LAG	*Chairperson*	*Activity in local administration or council*
LAG "Barcja"	W. Bogdanowicz	District councillor in Kętrzyn, previously also community councillor in Barciany
LAG "Brama Mazurskiej Krainy"	K. Margol	Previously city mayor of Nidzica 1990–1994
LAG "Kraina Drwęcy i Pasłęki"	M. Ofierska	Divisional head of tourism marketing, local development and structural funds, Dobre Miasto town council
LAG "Łączy Nas Kanał Elbląski"	S. Pańczuk	Inspector, department of tourism, culture, education and sport marketing in Elbląg district council
LAG "LGD 9"	B. Dawcewicz	Commissioner of Węgorzewo district council for tourism marketing and international cooperation
LAG "Lider w EGO"	J. Franczuk	District secretary in Stare Juchy
LAG "Partnerstwo dla Warmii"	J. T. Skiba	None
LAG "Warmiński Zakątek"	W. Łaszkowska	District councillor in Ostróda (Commission for Business and Finance)

Source: Author's compilation (as of 2010)

The close connection to public administration is seen as a positive issue by the interview partners as it enables the LAGs to have financial stability in the transitional phase from the funding period or long period of project cost repayment, or from the start of the implementation of the local development strategy. This is because the municipalities can take over some of the running costs of the LAG offices or increase the contributions of their members in order to cover these costs.

Challenges of the multi-sectoral partnership creation

Since cross-sectoral partnership building is regarded as a central feature of the governance phenomenon this chapter deals specifically with challenges facing the LAGs in the context of cooperation between the various societal groups as pointed out by the interviewees. The expert interviews indicated that the process of multi-sectoral partnership creation runs into many difficulties, which are described briefly below.

First of all, the mobilising of the village residents to take part in LAG activities represents a serious challenge. In villages where up to 90% of the residents were employed in former state-owned production cooperatives, the passive attitudes to life and employment, as well as social exclusion, is so high that this situation cannot be remedied in the lifetime of the project. Therefore, regular and repeated activities within the framework of social work are of major importance for the LAGs as part of a process of capacity and engagement building.

In addition to this, the interviews indicated that the representatives of the three sectors were not sufficiently integrated. Also following the participative observation at the plenary meetings, it was noticeable in the seating arrangements that the public stakeholders, associations and business representatives all formed separate groups. Furthermore, there was a lack of projects initiated between, for example, enterprises and municipalities or enterprises and associations. The best at integrating projects are the municipality directors who know each other well and who are active together in other bodies. As a result, multi-sectoral partnership creation develops only very slowly. An LAG summarises this deficit in their local development strategy as follows: "Despite the ever-increasing advertising for local partnership creation and for activities in this direction, the three-sectoral cooperation is one of the weakest aspects of local development" (LAG Warmiński Zakątek 2009, 3). The majority of interviewees do not believe in partnerships which have been created from above. Only where partner relationships were formed before LEADER+ and where other types of contact are cultivated will the groups be able to have stability. In the regions themselves, there must be a supportive climate for the creation of partnerships. The interviewees desired more trust, understanding of their objectives and tasks and a depoliticising of the issues. There is a conviction that the LEADER concept and LEADER funding alone cannot promote partnership creation.

A further challenge is the indirect competition between members of the LAGs. The information about the planned projects is rather discretely dealt with in the municipalities in order that the neighbouring municipality does not copy the idea. There is a lack of 'thinking beyond the district borders' and an effective experience exchange with respect to best and worst project cases. Some LAG members feel that they have not been treated fairly, for example because part of their project was not implemented on the area of their district. This is attested by a lack of rules on dealing with conflict or negotiating alternative solutions and compensating measures.

Additionally, there were several voices expressing concerns that the LEADER pillar in the funding period 2007–2013 was extremely bureaucratic and complicated compared to the pilot programme LEADER+. Small, inexperienced rural interest groups are overwhelmed by tasks such as making applications or completing the documentation. Without the support of LAGs offices, municipalities or other organisations with respect to advice, application procedures and credit granting, only very few local associations would be able to independently process and implement a project. It is claimed that rather than promoting the mobilisation of rural interest groups, the system is promoting their disinclination towards participation in partnership projects. As a consequence, third sector organisations leave the group or withdraw applications as the chance of approval seems low compared with the amount of effort required.

Discussion

Institutional processes and conditions for the emergence of LAGs in Poland

As a consequence of political negotiating processes at European and national level, a paradigm shift can be seen in Poland from a hierarchical approach towards regional governance, based on the example of the introduction of the LEADER approach (Fürst 2007, 8). Its implementation in Poland since the funding period 2007–2013 has no longer been centrally administered, but regionally through the Marshal Offices of the Voivodships. Additionally, the LAGs have gained decision-making freedom with respect to the selection and implementation of projects and the amount of funding. Here, a clear degree of decentralisation of state power can be seen, though this is not accompanied by a decentralisation of financing (Grosse 2004, 295). Pieper (2006, 362–63), however, concedes that the voivodships, strengthened through the implementation of European structural funds, prevent other organisations such as regional development agencies, chambers of industry and other socio-economic stakeholders from participating in the planning processes and their implementation.

According to the scientific literature, LAGs can be understood as an expression of the regional governance processes in rural areas, based on the creation of cooperation forms between local and regional stakeholders, networked forms of interaction on a voluntary basis and on the motivation to tackle regional problems and issues (Fürst 2007, 9; Böcher et al. 2008, 15). The results of selected LAGs in the Warmińsko-Mazurskie Voivodship demonstrate that the involved actors pursue various aims, however the aim to address local challenges such as unemployment, passivity of some rural population groups and a neglect of tourist potentials is predominant. The LAGs were set up by organisations such as foundations and municipal associations concerned with regional development and business support in rural areas, which had been operating in the region since the 1990s. Without the financial help and organisational support of these organisations, both the founding and functioning of the LAGs would not have been possible. These organisations can be regarded as an expression of predecessors of governance structures existing in Poland before EU accession. Gorzelak (2000, 101) notes with the example of the implementation of the programme for local initiatives in Poland (*Program Inicjatyw Lokalnych, Phare 1993–1996*) that local establishments (e.g., job centres, foundations, municipal associations) have driven on regional development with concrete projects, training, resources, loans and in this way have supported the activities of local leaders and elites. In particular, the role of municipalities and municipal associations needs to be highlighted, as they contributed largely to the setting up of the LAGs in Warmińsko-Mazurskie, being no exception in Poland. For example, in the LAG in the municipality Parzęczew in the Voivodship Lódzkie (central Poland), which was studied by Kovách and Gorlach (2006, 58), the initiative to set it up came likewise from the local public administration. Hence the validity of the statement from Gorzelak (2000, 100): "Local development is less and less dependent on decisions of the national state administration [...]" remains relevant.

Key governance features of the LAGs

The coalition building processes, negotiations and convincing processes as a part of governance (Zürn 2008, 9) are significantly influenced by the public sector representatives, mainly the political leaders of rural municipalities. The results of the analysis along with other research in the Warmińsko-Mazurskie Voivodship (Jacuniak-Suda 2015; Mose et al. 2014) point to the individual person-centred (*personenzentriert*) nature of governance structures rather than a collective-like actor constellation. Thus, the notion of "governance by government" (Zürn 2008, 558) can be used to describe the governance of the selected LAGs in the Warmińsko-Mazurskie Voivodship. Identifying the specific bargaining rules or social norms (Fürst 2007, 8) which underlie the decision making processes and the composition of relevant bodies in the LAGs presents a difficulty, as „Local rural development is a long and often nebulous process" (Moseley 2003, 14). Their identification would therefore require the undertaking of a long-term study at the micro-level of individual LAG networks. The top-down given rules on the LEADER implementation inherent in European and national regulations regulate largely the work of the LAGs, but at the same time make the production of problem-tailored own rules and solutions difficult.

The organisation structure of the LAGs is set out in the programme for the development of rural areas for 2007–2013, in the Laws on Associations and Foundations (1984, 1989) and in the rules of procedure of the individual LAGs, so there is no room for loose governance structures (Benz 2001, 55). Whereas the members assembly is named as the highest organ in the LAG terms and conditions, in practice it plays a less important role which is as a rule limited to voting on organisational procedures and general decisions which have previously been decided by the board or the council. The council is the most important organ as it makes decisions on the selection and financing of projects and in this way influences regional development. Although the business and social partners represented in the council officially exceeded the 50% level, in reality, the public sector is most involved in decision-making. This assumes greater importance at the council meeting, at which mayors regularly participate in line with their duties, while business and social partners are not always present as they must attend to professional duties. Although there is a formal requirement for a 50% presence of business and social partners in council assemblies (Urząd Marszałkowski Województwa Warmińsko-Mazurskiego 2015b), it is not always practical to assure this requirement. Therefore the decisions are made with a majority vote of those present.

As concerns the composition of the LAGs in Warmińsko-Mazurskie, the social sector is numerically predominant. This is evidenced by a successful mobilisation and engagement of the local associations (tourism groups, rural women clubs, sport clubs). With the under-representation of enterprises, farmers and young people, the demand for a multi-sectoral partnership is however only partially met. This situation is also confirmed by Furmankiewicz (2008, 232) in the LAG (LEADER+) studied by him in Poland. The representatives of the public and social sector were found to be the best represented, while the engagement of enterprises was very low. The study conducted by Kołodziejczak (2011, 211) of the LAGs in southern Poland (Lubelskie and Podlaskie Voivodships) also showed the dominance of the social sector in the composition and the lowest representation was that of the business partners. It remains to be seen whether the low involvement of the business community in LAGs is a result of an under-representation of enterprises in Warmińsko-Mazurskie (USTAT 2015, 16) and/or is because of a mutual or one-sided disinclination to cooperate with enterprises, which in communist times had acquired a negative image (Osowska 2014, 39–40).

According to Fürst (2007, 9), governance is characterised by "a priori undefined guidance" as the control is based on network-similar interactions. However, guidance is required in regions where there is a lack of social capital. This is relevant in Warmińsko-Mazurskie where there is hardly any social capital, reflected in a low level of interest and engagement in local issues as well as in weak links between social groups with citizens appearing to be more prepared to hand over responsibility to other people (Dziemianowicz and Schlachta 2012, 46). This allows for the interpretation that in the LAGs studied in Warmińsko-Mazurskie, strong leaders have been identified among the LAG chairpersons. In addition, a charismatic, individualistic style of governance in one of the researched groups was identified in the course of the subsequent qualitative and quantitative network analysis (Mose et al. 2014, 13; Jacuniak-Suda 2015, 56–57). Similarly, Gorzelak (2000, 101) points to strong local leaders which form the head of the stable and dynamic local partnerships in Poland, who have developed a long-term vision for the region implemented by way of the local development strategy. Additionally, they possess the ability of being able to win over the local public administration through their personality and as a result of their own professional experience. Furthermore, it was noticeable within the framework of the analysis of the LAGs in Warmińsko-Mazurskie that 50% of the LAGs chairpersons were recruited originally from town councils or municipalities. The directors which originated from the foundations or from business are or were also politically active in the community or district councils. These (politically) direct or indirect connections to public administration are advantageous in terms of the network of contacts and also because of the financial benefits of starter funding from the municipalities or districts.

Challenges

It can be established that governance structures in Poland are strongly anchored in public administration. This is underlined also by Hałasiewicz (2008, 3) who points to the dominating position of the district councils and municipalities in the LAGs, which engage them because of their competences and organisational resources. He recommends that in the course of the strengthening of competences of the business and social stakeholders of the LAG, they should draw back. Also, Fürst (2007, 10) notes the tendency of political leaders and civil servants to 'occupy' governance structures for better control.

This is related to another challenge for the LAGs — the still inadequate level of multi-sectoral partnership creation. The integration and active cooperation between the various sectors are not strong enough as the level of influence and power relationships are unequally distributed. Gramzow (2006b, 18), who has investigated several local initiatives in the rural area of the Voivodship Podkarpacke (south Poland), also points to weak cooperation or even a lack of cooperation, particularly between the public authorities and the associations, which can be attributed to the frequent rotation of officials. Additionally, the authorities fear losing control and influence in the face of the increasing number of associations and their high level of acceptance in the region. Augustyn (2009, 11) suggests that the old control paradigms which are an inheritance of the communist system still dominate the decision-making processes in the LAGs. In the selected LAGs in the Warmińsko-Mazurskie Voivodship, however, this was not clear as the questions on specific political views or political party membership were not answered.

Civil society is accorded a special role in the processes of regional development, "which often serves as an early warning system, steering the political discourse in a particular direction" (European Commission 2001, 19). Thereby, the low degree of emancipation of NGOs was apparent in the study of the LAGs. Although they were numerically dominant in the LAGs, they do not have sufficient assertiveness, as was observed in the members assemblies and also confirmed in the interviews by the LAG chairpersons. On the one hand this can be ascribed to the low levels of social capital in Polish society in general, that is low levels of trust, unwillingness to be active and excessive demands toward the state, as Błąd and Kamiński (2005) report and also as the conducted interviews show. On the other hand, a rising number of rural associations is observed across Poland. This paradox of the low social capital on the one hand and the "boom" of rural associations on other hand (Bukraba-Rylska 2007, 12) is related to the complex history of regions which were part of Germany before 1945, such as Warmińsko-Mazurskie. Here, two types of behaviour can be identified among rural society groups: the isolated and passive behaviour due to the former state owned farms and a pro-active behaviour of groups open for new initiatives, being actively involved in local activities (Bednarek-Szczepańska 2011, 224; based on Karwacki and Antonowicz 2003 and Wendt 2007). Still, as the interviews indicate, the bureaucracy regarding the process of filling in applications presents for many rural associations an obstacle and creates discouragement. In this context, Hałasiewicz (2008, 2) notes that excessive formalisation in the implementation of the LEADER initiative makes the achievement of its goals impossible.

The lack of assertiveness of civil society groups can be explained, by among other things, the regulations on criteria for the selection of LAGs in Poland in 2008, created by the Ministry for Agriculture and Village Development[14]. According to the regulations, in the process of selection, those groups gained more points in which half of the members of the council "could show documented knowledge on the development of rural areas and the LEADER initiative", as well as in "project implementation and processing in the area of the development of rural areas which were financed from EU funds" (Dziennik Ustaw 2008, 5671). The boards of the LAGs were therefore influenced to take in representatives of public administrations who had more competences in these areas than non-governmental organisations. In this way, the conditions for the dominance of the local public administrations in the LAGs were created.

[14] Regulation of the Ministry for Agriculture and Community Development (No. 659) of 23.05.2008 concerning the detailed criteria and the voting procedure of local action groups for the implementation of local development strategies in the framework of the programme for the development of rural areas in the years 2007–2013.

Conclusion

The introduction of the LEADER initiative as a pilot programme LEADER+ and its further implementation through the programme for the development of rural areas 2007–2013 in Poland led to the consolidation of existing, albeit very loose, governance structures in the rural areas of Warmińsko-Mazurskie. It was primarily about the cooperation between local foundations, municipalities and municipal associations which were mainly initiated in the course of the EU accession programmes of the 1990s. These processes of partnership building were strengthened and institutionalised by LEADER LAGs, as representatives of public administrations, NGOs and private enterprises had the opportunity through their participation in the LAG bodies to meet one another in person and exchange experiences. However, it was of greater significance that the various social and private enterprise stakeholders have been invited to participate in the decision-making processes and in the shaping of rural regional development, which was limited before 1990. For the first time, the targeted mobilisation of the previously passive societal groups could take place. From the example of the LAGs studied in the Warmińsko-Mazurskie Voivodship, it can be established that the implementation of the LEADER programme has led to a strengthening of endogenous potentials in the respective regions (Gorzelak 2000, 101).

It is desirable that LAGs select as a central theme the real problems of the rural population groups and avoid falling into 'project automatism'. LEADER creates a good 'bottom-up' complement to state spatial planning which is worth retaining. Regional governance is, of course, "no nostrum for a better future" (Fürst 2007, 10) but when a balanced multi-sectoral local elite is created, it can put considerable pressure on regional decision-making, to the benefit of an effective rural development in line with the needs of local people. The question remains, however, as to what extent the LEADER- legal regulations, for example, bureaucratic procurement rules, make it possible to achieve these aims. Another contested issue is the multi-sectoral local partnership, an underlying principle of LEADER approach and regional governance. Due to a lack of trust, interest divisions and low emancipation of local interest groups, the multi-sectoral partnership lacks stability and the shaping of decision making processes on a level playing field.

With respect to the transferability of the lessons learnt from regional governance in the Warmińsko-Mazurskie Voivodship to other peripheral rural areas in Poland (Mose 2010, 157–58), it can be concluded that this is limited as in every rural district, different stakeholder constellations, cooperation cultures and development potentials predominate. Baumgarten and Lahusen (2006, 186) also point to the problem of a limited ability to generalise from the results, "as the network structures are only a snapshot". As the rural areas of Poland have historically different societal and economic types of development (having been parts of Prussia, Austria and Russia) and had also undergone an exchange of population after the Second World War, it is hard to establish generally valid development tendencies or explicit data interpretation (Bednarek-Szczepańska 2011, 224). In principle however, a growing importance in network-oriented cooperation can be observed, in which representatives of state institutions as well as non-state interest groups are increasingly active. It is also apparent, however, that the (local) public sector actors are playing a key role in the development of peripheral rural areas.

Acknowledgement

I am very grateful to the interviewees for their time and important information and also for their cooperation and enthusiasm in supporting this research. In addition, I would like to thank the editors of this book for their invaluable comments on the earlier version of this chapter.

References

Augustyn, A. (2009). The role of rural civil society in the governance reforms. Example of the LEADER approach in Poland. Paper presented at the Changing Europe Summer School 2009 "Civil Society in Central and Eastern Europe before and after the End of Socialism". Kyiv (Ukraine), 27–31 July 2009.

Baum, S. & Weingarten, P. (2005). Interregionale Disparitäten und Entwicklung ländlicher Räume als regionalpolitische Herausforderung für die neuen EU-Mitgliedstaaten. Agrarwirtschaft 54 (4), 210–224.

Baumgarten, B. & Lahusen, C. (2006). Politiknetzwerke — Vorteile und Grundzüge einer qualitativen Analysestrategie. In Hollstein, B. & Straus, F. (Eds.), Qualitative Netzwerkanalyse. Konzepte, Methoden, Anwendungen (pp. 177–197). Wiesbaden: VS Verlag für Sozialwissenschaften.

Bednarek-Szczepańska, M. (2011). Organizacje pozarządowe na obszarach pozametropolitalnych na przykładzie wybranych regionów. In Wesołowska, M. (Ed.), Wiejskie obszary peryferyjne — uwarunkowania i czynniki aktywizacji (pp. 217–231). Studia Obszarów Wiejskich, 26. Warszawa: Instytut Geografii i Przestrzennego Zagospodarowania PAN.

Benz, A. (2001). Vom Stadt-Umland-Verband zu „regional governance" in Stadtregionen. Deutsche Zeitschrift für Kommunalwissenschaften 40 (2), 55–71.

Benz, A. & Dose, N. (Eds.) (2004). Governance — Regieren in komplexen Regelsystemen. Eine Einführung. Wiesbaden: Springer.

Błąd, M. & Kamiński, H. (2005). Social capital enhancement in the Polish countryside: experience from the implementation of LEADER-type programmes. In Zawalińska, K. (Ed.), Rural Development in the Enlarged European Union (pp. 235–248). Warsaw: Polish Academy of Sciences.

Böcher, M. (2003). Die politische Steuerung nachhaltiger Regionalentwicklung. Das Beispiel der EU Gemeinschaftsinitiative LEADER+. In Grande, E. & Prätorius, R. (Eds.), Politische Steuerung und neue Staatlichkeit (pp. 235–258). Baden-Baden: Nomos.

Böcher, M. (2008). Regional Governance and rural development in Germany — the implementation of LEADER+. Sociologia Ruralis, 48 (4), 372–388.

Böcher, M., Krott, M. & Tränkner, S. (Eds.) (2008). Regional Governance und integrierte ländliche Entwicklung. Ergebnisse der Begleitforschung zum Modell- und Demonstrationsvorhaben „Regionen Aktiv". Wiesbaden: VS Verlag für Sozialwissenschaften.

Brodziński, Z. (2006). Uwarunkowania zintegrowanego rozwoju obszarów wiejskich w opiniach przedstawicieli lokalnych grup działania. Zeszyty Naukowe Akademii Rolniczej we Wrocławiu 540, 105–111.

Bruckmeier, K. (2000). LEADER in Germany and the discourse of autonomous regional development. Sociologia Ruralis 40 (2), 219–227.

Budzich-Szukała, U. (2008). Program LEADER w Polsce — sposób na aktywizację wsi. In Wilkin, J. & Nurzyńska, I. (Eds.), Polska Wieś 2008. Raport o stanie wsi (pp. 117–130). Warszawa: Fundacja na rzecz Polskiego Rolnictwa.

Bukowska, G. & Łukaszewicz, A. (2002). Obszary wiejskie i rolnictwo polskie w programach przedakcesyjnych Unii Europejskiej. Ekonomia 6, 55–67.

Bukraba-Rylska, I. (2007). O wsi inaczej. Uwagi polemiczne na temat jakości kapitału społecznego i kulturowego polskiej wsi. Nowe Życie Gospodarcze 21 (Wyzwania przed obszarami wiejskimi i rolnictwem w perspektywie 2014–2020), 12–14.

Churski, P. (2008). Structural funds of the European Union in Poland — Experience of the first period of membership. European Planning Studies 16 (4), 579–607.

Council Regulation (EC) No 1698/2005 of 20 September 2005 on support for rural development by the European Agricultural Fund for Rural Development (EAFRD). OJ L 277, 21.10.2005, 1–40.

Czapiewska, G. (2011). Wpływ środków Unii Europejskiej na pobudzenie inicjatyw społeczności wiejskich obszarów peryferyjnych województwa zachodniopomorskiego. In Wesołowska, M. (Ed.), Wiejskie obszary peryferyjne — uwarunkowania i czynniki aktywizacji (pp. 233–248). Studia Obszarów Wiejskich 26. Warszawa: Instytut Geografii i Przestrzennego Zagospodarowania PAN.

Danielzyk, R. & Knieling, J. (2011). Informelle Planungsansätze. In Akademie für Raumforschung und Landesplanung — ARL (Ed.), Grundriss der Raumordnung und Raumentwicklung (pp. 473–498). Hannover: ARL.

Dargan, L. & Shucksmith, M. (2008). LEADER and innovation. Sociologia Ruralis 48 (3), 274–291.

Droth, A., Grimm, F.-D. & Haase, A. (2000). Polen aktuell. Daten, Fakten, Literatur zur Geographie Europas 6. Leipzig: Institut für Länderkunde.

Dziemianowicz, W. & Schlachta, J. (2012). Konkurencyjność Warmii i Mazur – diagnoza problemowa. Zlecenie Zarządu Województwa Warmińsko-Mazurskiego, Olsztyn 2012. www.strategia2025.warmia.mazury.pl/artykul/25/49/zaktualizowane-opracowanie-pn-konkurencyjnosc-warmii-i-mazur-diagnoza-problemowa.html. Accessed 22 May 2016.

Dziennik, U. (2008). Rozporządzenie Ministra Rolnictwa i Rozwoju Wsi (Nr. 659) z dnia 23 maja 2008 r. w sprawie szczegółowych kryteriów i sposobu wyboru lokalnej grupy działania do realizacji lokalnej strategii rozwoju w ramach Programu Rozwoju Obszarów Wiejskich na lata 2007–2013. Dziennik Ustaw 103 (659). www.dziennikustaw.gov.pl/du/2008/s/103/659. Accessed 29 June 2016.

European Commission (1996). The Cork Declaration — A Living Countryside. European Conference on Rural Development, November 1996, Cork, Ireland. ec.europa.eu/agriculture/rur/cork_en.htm. Accessed 5 May 2016.

European Commission (2001). European Governance. A White Paper, OJ C 287, 12.10.2001, S. 1–29.

European Commission (2006). The LEADER approach. A basic guide. Luxembourg, Office for Official Publications of the European Communities. ec.europa.eu/agriculture/publi/fact/leader/2006_en.pdf. Accessed 21 May 2016.

European Commission (2010). LEADER+ Member States files. ec.europa.eu/agriculture/rur/leaderplus/memberstates/index_en.htm. Accessed 21 May 2016.

FAPA, Fundacja Programów Pomocy dla Rolnictwa (2006). Polityka Rozwoju obszarów wiejskich. Bieżące informacje i analizy funkcjonowania Wspólnej Polityki Rolnej UE. Opracowanie kwartalne. Warszawa, sierpień 2006 r.

Furmankiewicz, M. (2008). Enhancing endogenous development in rural areas: the implementation of the LEADER Pilot Programme in Poland. Proceedings of the 4th International Conference of ASECU. 22–24 May, 2008, Bucharest. www.asecu.gr/files/RomaniaProceedings/24.pdf. Accessed 22 May2016.

Fürst, D. (2001). Regional Governance — ein neues Paradigma der Regionalwissenschaften? Raumforschung und Raumordnung 59 (5–6), 370–380.

Fürst, D. (2003). Regional Governance. In Benz, A. & Dose, N. (Eds.): Governance — Regieren in komplexen Regelsystemen. Eine Einführung (pp. 45–64). Wiesbaden: Springer.

Fürst, D. (2007). Regionale Governance. Implikationen für LEADER. LEADER+ Magazine 8, 8–11.

Fürst, D. & Knieling, J. (Eds.) (2002). Regional Governance. New Modes of Self-Government in the European Community. Studies in spatial development 2. Hannover: Akademie für Raumordnung und Landesplanung.

Głuszyński, J. (2005). Inicjatywa LEADER w Polsce. Warszawa: Instytut Badania Opinii i Rynku Pentor.

Gorlach, K. & Adamski, T. (2007). Neo-endogenous development and the revalidation of local knowledge. Polish Sociological Review 4 (160), 481–498.

Gorzelak, G. (2000). Zewnętrzna interwencja jako czynnik rozwoju lokalnego (na przykładzie Programu Inicjatyw Lokalnych). Studia Regionalne i Lokalne 3 (3), 99–119.

Grabski-Kieron, U. & Krajewski, C. (2007). Ländliche Raumentwicklung in der erweiterten EU. Geographische Rundschau 59 (3), 12–20.

Gramzow, A. (2006a). Local partnership as an incubator for rural development: The case of Debrzno, north-western Poland. IAMO Discussion Paper No. 90. Halle (Saale).

Gramzow, A. (2006b). Doswiadczenia oddolnych inicjatyw rozwoju regionalnego oraz perspektywy dla programu LEADER+ w regionie Doliny Strugu w Polsce. IAMO Discussion Paper 100. Halle (Saale).

Gramzow, A. (2006c). Endogenous initiatives as a chance to improve rural livelihood? Results of a case study in Baltow, South-Eastern Poland. IAMO Discussion Paper 95. Halle (Saale).

Grosse, T. G. (2004). Polityka Regionalna Unii Europejskiej. Przykład Grecji, Włoch, Irlandii i Polski. Warszawa: Instytut Spraw Publicznych.

Haase, A., Wust, A., Knappe, E. & Grimm, F. D. (2004). Wandel in osteuropäischen Grenzregionen. Auswirkungen der zunehmenden Durchlässigkeit der polnischen Ostgrenze auf Grenzregionen und Grenzbeziehungen. Beiträge zur Regionalen Geographie 59. Leipzig: Institut für Länderkunde.

Hahne, U. (2004). Ländliche Regionalentwicklung mit LEADER+. Raumplanung 116, 199–204.

Halamska, M., Śpiewak, R. & Michalska, S. (2010). LEADER w Polsce. Drogi implementacji programu. Wieś i Rolnictwo 149 (4), 104–119.

Hałasiewicz, A. (2008). LEADER w Polsce. Wykorzystana szansa i nowe możliwości. Warszawa. www.fundacjaprym.pl/dokumenty/func-startdown/1/. Accessed 22 May 2016.

Jacuniak-Suda, M. (2015). Regional Governance im Kontext der Regionalentwicklung in peripheren ländlichen Räumen am Beispiel von regionalen Netzwerken in Ermland-Masuren (Polen) und Western Isles (Schottland). Kumulative Doktorarbeit, Universität Oldenburg.

Kamiński, R. & Budzisz-Szukała, U. (2007). Samoorganizacja i samorządność wsi. Nowe Życie Gospodarcze 21 (Wyzwania przed obszarami wiejskimi i rolnictwem w perspektywie 2014–2020), 17–19.

Karwacki, A. & Antonowicz, D. (2003). Underclass i kultura biedy w środowiskach popegeerowskich w kontekście sporów teoretycznych i interpretacyjnych. Studia Socjologiczne 3 (170), 69–111.

Knieling, J. (2005). Projektorientierung in der Raumordnung. In Akademie für Raumordnung — ARL (Ed.), Handwörterbuch der Raumordnung (pp. 813–818). Hannover: ARL.

Kołodziejczak, A. (2011). Lokalne Grupy Działania jako czynnik rozwoju wiejskich obszarów peryferyjnych w województwach lubelskim i podlaskim. In Wesołowska, M. (Ed.), Wiejskie obszary peryferyjne – uwarunkowania i czynniki aktywizacji (pp. 203–215). Studia Obszarów Wiejskich, 26. Warszawa: Instytut Geografii i Przestrzennego Zagospodarowania PAN.

Kovách, I. & Gorlach, K. (2006). Innovation and sustainability in Central Europe. Corason Research Project. Budapest: The Political Science Institute of the Hungarian Academy of Science.

Kozak, M. (2001). Wsparcie Unii Europejskiej na rzecz rozwoju regionalnego w Polsce w latach 1989–1999. Studia Regionalne i Lokalne 1, 145–158.

LAG Warmiński Zakątek (2009). Lokalna strategia rozwoju 2009–2015. www.warminskizakatek.com.pl/wz/index.php?site=strategia/strategia. Accessed 22 May 2015.

Lee, J., Arnason, A., Nightingale, A. & Shucksmith, M. (2006). Networking: social capital and identities in European rural development. Sociologia Ruralis 45 (4), 269–283.

Mayring, P. (1999). Einführung in die qualitative Sozialforschung. 4th edition. Weinheim: Beltz.

MIR, Ministerstwo Infrastuktury i Rozwoju (2014). O programie. www.polskawschodnia.gov.pl. Accessed 22 May 2016.

Mose, I. (1993). Eigenständige Regionalentwicklung — neue Chancen für die ländliche Peripherie? Vechtaer Studien zur Angewandten Geographie und Regionalwissenschaft 8. Vechta.

Mose, I. (2010). Integrierte ländliche Entwicklung — Vergleichende Analyse unterschiedlicher konzeptioneller Ansätze der Entwicklung ländlicher Peripherien in Europa. In Belina, B. & Miggelbrink, J. (Eds.), Hier so, dort anders. Raumbezogene Vergleiche in der Wissenschaft und anderswo (pp. 153–171). Münster: Westfälisches Dampfboot.

Mose, I., Jacuniak-Suda, M. & Fiedler, G. (2014). Regional Governance-Stile in Europa. Eine vergleichende Analyse von Steuerungsstilen ausgewählter LEADER-Netzwerke in Extremadura (Spanien), Warmińsko-Mazurskie (Polen) und Western Isles (Schottland). Raumforschung und Raumordnung 72 (1), 3–20.

Moseley, M. J. (2003a). Rural development. Principles and Practice. London: Sage.

Moseley, M. J. (2003b). Local partnerships for rural development: the European experience. Wallingford: CABI Publishing.

MRiRW, Ministerstwo Rolnictwa i Rozwoju Wsi (2004). Sektoralny Program Operacyjny „Restrukturyzacja i Modernizacja Sektora Żywnościowego i Rozwój Obszarów Wiejskich na lata 2004–2006. Warszawa.

MRiRW, Ministerstwo Rolnictwa i Rozwoju Wsi (2007). Program Rozwoju Obszarów Wiejskich na lata 2007–2013. Warszawa.

MRiRW, Ministerstwo Rolnictwa i Rozwoju Wsi (2014). Program Rozwoju Obszarów Wiejskich na lata 2007–2013. Warszawa, 12 grudnia 2014. www.minrol.gov.pl/Wsparcie-rolnictwa/PROW-2014-2020. Accessed 22 May 2016.

Osowska, R. (2014). The meaning and development of entrepreneurial culture and its relationship to start-up behaviour: the case of Poland. Doctoral thesis. University of the West of Scotland.

Piasecki, B. & Rogut, A. (2004). Poland: The Żary and Białystok Regions. In Labrianidis, L. (Ed.), The future of Europe's Rural Peripheries (pp. 271–297). Aldershot, Burlington: Ashgate.

Pieper, K. (2006). Regionale Auswirkungen der EU-Entwicklungspolitik. Flexible Kooperationsformen in ländlichen Gebieten am Beispiel Ungarn und Polen. In Kleinfeld, R., Plamper, H. & Huber, A. (Eds.), Regional Governance. Volume 1 (pp. 349–364). Göttingen: V&R unipress.

Ray, C. (2000). The EU LEADER programme: rural development laboratory. Sociologia Ruralis 40 (2), 163–171.

Ray, C. (2006). Neo-endogenous rural development in the EU. In Cloke, P.; Marsden, T. & Mooney, P. (Eds.), Handbook of Rural Studies (pp. 278–291). London: Thousand Oaks.

Rosner, A. (2010). Zróżnicowanie przestrzenne rozwoju społeczno-gospodarczego polskiej wsi i tendencje jego zmian. In Kancelaria Senatu (Ed.), Opinie i Ekspertyzy OE-148 (pp. 1–16). Warszawa.

Sagan, I. & Lee, R. (2005). Spatialities of Regional Transformation in Central Europe and Administrative Spaces of the EU. In Sagan, I. & Halkier, H. (Eds.), Regionalism Contested: institution, society and governance (pp. 163–176). Aldershot, Burlington: Ashgate.

Shortall, S. & Shucksmith, M. (1998). Integrated rural development: issues arising from the Scottish experience. European Planning Studies 6 (1), 73–88.

Shucksmith, M. (2000). Endogenous development, social capital and social inclusion: perspectives from LEADER in the UK. Sociologia Ruralis 40 (2), 208–218.

Shucksmith, M. (2010). Disintegrated Rural Development? Neo-endogenous Rural Development, Planning and Place-Shaping in Diffused Power Contexts. Sociologia Ruralis 50 (1), 1–14.

Shucksmith, M., Thomson, K. J. & Roberts, D. (2005). The CAP and the regions: the territorial impact of the common agricultural policy. Wallingford: CABI Publishing.

Śpiewak, R. (2009). Kierunki rozwoju wsi prezentowane przez wybrane lokalne grupy działania, realizujące program LEADER+ w Polsce. Wieś i Rolnictwo 4 (145), 212–225.

Urząd Marszałkowski Województwa Warmińsko-Mazurskiego (2015a). Mapa LGD Warmii i Mazur. www.sporol.warmia.mazury.pl/files/463/55ffef16c9874/2015%20LGD.jpg. Accessed 19 May 2016.

Urząd Marszałkowski Województwa Warmińsko-Mazurskiego (2015b). Wybór LSR. www.sporol.warmia.mazury.pl/463-Aktualnosci_PROW_2014_2020-5257.htm. Accessed 29 April 2016.

Urząd Statystyczny w Olsztynie — USTAT (2010). Przemiany demograficzne w Województwie Warmińsko-Mazurskim w latach 2002–2009. olsztyn.stat.gov.pl. Accessed 22 May 2016.

Urząd Statystyczny w Olsztynie — USTAT (2015). Raport o sytuacji społeczno-gospodarczej Województwa Warmińsko-Mazurskiego w 2014 r. olsztyn.stat.gov.pl/download/gfx/olsztyn/pl/defaultaktualnosci/847/2/4/1/raport_2014_0.pdf. Accessed 22 May 2016.

Wendt, J. (2007). Wymiar przestrzenny struktur i aktywności społeczeństwa obywatelskiego w Polsce. Prace Geograficzne 208. Warszawa: IGiPZ PAN.

Yin, R. K. (2003). Applications of case study research. 2nd edition. Applied Social Research Methods Series 34. Thousand Oaks: Sage.

Zürn, M. (2008). Governance in einer sich wandelnden Welt — eine Zwischenbilanz. In Schuppert, G. F. & Zürn, M. (Eds.), Governance in einer sich wandelnden Welt (pp. 553–580). Politische Vierteljahresschrift, Special Issue 41. Wiesbaden: VS Verlag für Sozialwissenschaften.

Nature conservation-oriented governance processes in dynamic border areas in Germany and the Netherlands: reflections on regional development in formerly peripheral regions

Ulrike Grabski-Kieron

Abstract

Many border areas at both the internal and external borders of the European Union belong, or used to belong, to the rural peripheries of Europe. In the wake of European integration they have undergone changes in their dynamics which, also with regard to an activating EU structural policy, are still being felt today and are continuing. This change in structure and function is being accompanied in many regions by increased pressure for land use in open spaces. Numerous conflicts arise due to the fact that many of these regions are also seen as cultural landscapes which are worthy of conservation — valuable, even — and which assume a high degree of importance for the preservation of biodiversity. Integrative strategies for the future are necessary with a view to any sustainable development of these regions. Nature conservation, with its task of establishing a Europe-wide biotope network (NATURA 2000), must position itself within such strategies. This article looks more closely at problem situations and approaches for action in current nature conservation-oriented governance processes in the German-Dutch border area, being the results of a research project. As part of a realisation of cross-border biotope network projects at the German-Dutch border, this project was designed to determine backgrounds and factors on both sides of the German-Dutch border which have an effect or influence on whether — and, if so, how — the project objectives can be reached. Beyond this it asked for future strategies and methods to utilise the potential of joint nature conservation work in a better way. Some of the results of the project are outlined.

Introduction

The general issue of “European peripheries between stagnation and a new beginning” is particularly applicable to many inner-European border areas. The post-war order of Europe had made clear the borders existing between states. This meant that for a long time many rural border areas remained as peripheral areas in the shadow of prospering centres of development and regions marked by economic strength. It was only the formation of the European Union that provided a decisive impetus for opening the borders and, as a result, for cross-border ties and increasing cooperation. The fundamental aims of a common EU policy are also supported by the INTERREG funding programmes introduced in the 1990s. Together with other EU structural funds, these programmes, with their various areas of focus, have been providing important incentives for cross-border development which, going beyond the development of business, infrastructure and labour markets, have increasingly turned to, and provided support for, other areas of activity.

As a result, many peripheral border areas have undergone a new dynamic. In many places they have put an end to any peripherization and have focused increasingly on the potential of these regions for a coordinated — indeed, a jointly undertaken — development in the spirit of European cohesion. Thus "overcoming borders" has triggered profound transformation processes in these border areas. Lastly, European planning culture, characterized by governance and having undergone a change since the 1990s, has also influenced these spatial transformation processes: Despite all the differences in national planning systems, this planning culture contains the justification for social and democratic conventions and shared values, e.g., participation in planning processes which open up joint areas of activity across borders.

Transformation processes in the German-Dutch border area

Regardless of the different natural conditions which originally existed in different regions and of the developments which individual sections have undergone, the German-Dutch border area represents a rural area which has progressed, and is still progressing, from being an out-and-out periphery to a dynamic — in some cases highly dynamic — area. In the last decades of the 20th century, all the attributes of a periphery still applied to large parts of the area on both sides of the border. The characteristics of a periphery were given by its remoteness and isolation — underlined by the demarcation provided by the border — and by the specific natural conditions which originally existed in the moors and wetlands, which for a long time were not amenable to use as farmland and which, acting as barriers, also impeded cross-border interactivities. At the same time, however, some of these sub-regions — the Emsland, for example — also drew on their endogenous potentials and developed their own endogenous development dynamic. Today, supported by European policy on promoting border regions and rural areas, they are at last in the process of having their regional identities rediscovered beyond their own borders. They are developing new approaches to this, even going so far as to form genuine areas of cooperation in which common historical roots are perceived more strongly and are seen as having the potential for further development in the future.

Seen from the point of view of research into border areas, "classic" development paths can be recognized in these transformation processes in border regions of the past decades. Accordingly, the German-Dutch border region too has transformed itself from being an "alienated", non-interacting area (Martinez 1994), passing through various phases of development to become a cooperative type of area — possibly, indeed, already an integrative one today. For the field of governance research, they present themselves today, in the sense defined by Blotevogel (1996), as real regions (in contrast to virtually perceived ones) with growing cross-border links as well as political and administrative areas: each with their own systems of institutions and planning, but also as areas with shared perceptions and a shared sense of their own identity — leading to areas of cooperation and to shared activities.

The current transformation processes are decisively influenced by

- a change in interrelationships and in relations with metropolitan areas and urban agglomerations on both national and European scales,
- a concomitant embedding in trans-European development axes,
- a change in the evaluation of locations and the utilisation of the potential of abandoned areas and, last but not least,
- increasing pressure for land use as a result of the demand for open space in the fields of tourism and recreation in the countryside.

At the same time, a rich nature potential has been preserved in these areas which is currently assuming major significance for the implementation of the European biodiversity strategy (European Commission 2011) and for building up the European "NATURA 2000" biotope network in accordance with the EU's Flora and Fauna Habitats Directive (Council Directive 92/43/EEC) and the Directive on the conversation of wild birds (Directive 2009/147/EC). The tasks in hand are aimed at safeguarding and improving these regionally important biotope and biotope corridor functions, even when the pressure to utilise land is increasing. Accordingly, cross-border nature conservation does not only mean biotope management on a broad scale. Rather, it is to comprehend as a governance task, because what needs to be done, more than ever before, is to put into practice area's nature conservation concerns as an integrative part of rural development. Concurrently acceptance for this has to be promoted. At the same time emphasis has to be put on strategies to create or even to extend viable examples of cross-border cooperation with local and regional stakeholders on both sides of the border.

Depending on the view being taken, interdisciplinary border area research looks comprehensively at transformation processes in their various facets, either as spatial phenomena in cultural landscapes, or as an issue relating to changed regional identities as well as to changed social perceptions and regional constructions (Jacobs and van Assche 2014; Johnson et al. 2011; Kühne and Meyer 2015;

Schönwald and Kühne 2012; Wille et al. 2014; Wille 2015). Approaches taken in research centring on governance and border areas currently work together to understand, and further develop, changed patterns of steering and cooperation with their emphasis on hybrid steering structures (Fürst 2010), with their public-private development partnerships and partnerships between stakeholders, as well as in their orientation towards planning strategies and planning projects.

In the context of cross-border nature conservation, there must of necessity be a convergence of the views taken in rural spatial research, governance research and applied landscape ecology. Basing their findings on case studies in German border areas, Hager (2003), Harteisen et al. (2010), Liebenhardt et al. (2010) and Knippschild (2009) have already pointed out not only problems which exist in cross-border nature conservation work, but also possible fields of activity in cross-border governance processes. In doing so, they have made it clear how important it is that any models for action work together on the various planning and decision levels. Their work emphasized the role of cross-border cooperation in different planning systems and even more associated with different understandings of planning. Their findings were important for nature conservation-oriented governance research. However, they were not quite sufficient to make a decisive difference to policy consultation. As a result, for governance research in this thematic context it remained relevant how nature conservation concepts for building up a European biotope network in rural border areas with landscape potential worthy of protection can be improved and implemented in the best possible way as part of any cross-border development.

Taking a broader view of regional development, these tasks fit directly into the field of activity involving integrated rural development: its strategic and rational objectives aim at concentrating and bringing together a broad cross-section of sectoral needs in such a way that the greatest possible potential for synergies regarding rural spatial development can be utilised and implemented (Grabski-Kieron 2012). Stakeholders involved in nature conservation can be partners in public-private development partnerships at regional and local levels. The leveraging of the potentials that landscapes have in connection with rural tourism and with land management of man-made landscapes on one side, and with regional added-value or with creating a diversification of income in farming on the other are proven integrated fields of activity in rural development in Germany and Europe (Laquement et al. 2012). In this sense many other examples could be quoted. In practice, many regional development concepts show that conservation issues are not neglected but, rather, can gain in sustainability through being embedded in such holistic planning. Researchers working in both nature conservation and spatial development agree that large-scale protected areas and, in particular, nature parks prove themselves to be important institutional structures having a responsibility for projects: they offer opportunities for pushing forward and implementing sustainable regional development in the sense of integrated development (Liesen 2015; Hammer et al. 2016). However, the institutional framework alone is not what is decisive. Findings from governance research show beyond dispute that

- building up a network of stakeholders involving representatives from nature conservation,
- drawing up development concepts with a strategic orientation and a view to implementation,
- the interplay of informal and formal/legal planning instruments and models for action,
- the concentration of different instruments of promotion and incentive, and finally also
- a stronger acceptance and a positive perception of any processes initiated

are decisive for any successful regional development, including any with a nature-conservation orientation (Grabski-Kieron and Kötter 2013). This is also basically true for development processes in rural border regions. However, here they take place under the special conditions dictated by the border location. These conditions include, to name just a few, language barriers to a common understanding, working in different planning systems with different planning instruments, different perceptions of problems, and working practices (Heintel and Weixlbaumer 2016).

Research accompanying the "Green Border Network" project

It was against this background that the development and trial project (D+T Project 2012–2015) of the Federal Agency for Nature Conservation entitled "Green Border Network — Nature without Borders is embedded between North Rhine-Westphalia and the Netherlands" was undertaken. The project (Grabski-Kieron et al. 2016) aimed at making it possible, taking the example of the German-Dutch border, to realise a model of cross-border biotope network projects. The aim was thus primarily one of nature conservation, which was also reflected in the character of the case studies: the overall project area was divided into four independent sub-project areas — "Heathland network in the border triangle in the District of Steinfurt"; "Zwillbrocker Venn heathland areas" in the District of Borken; "Bog-heathland complex at the southern edge of the Reichswald forest" in the District of Kleve; and "Meinweg oak and birch network" in the District of Heinsberg (Figure 1). These case study areas are evaluated as components of the European biotope network Natura 2000 in the cross-border region, but different requests exist at improving their particular ecological functions within this biotope network (Figure 2).

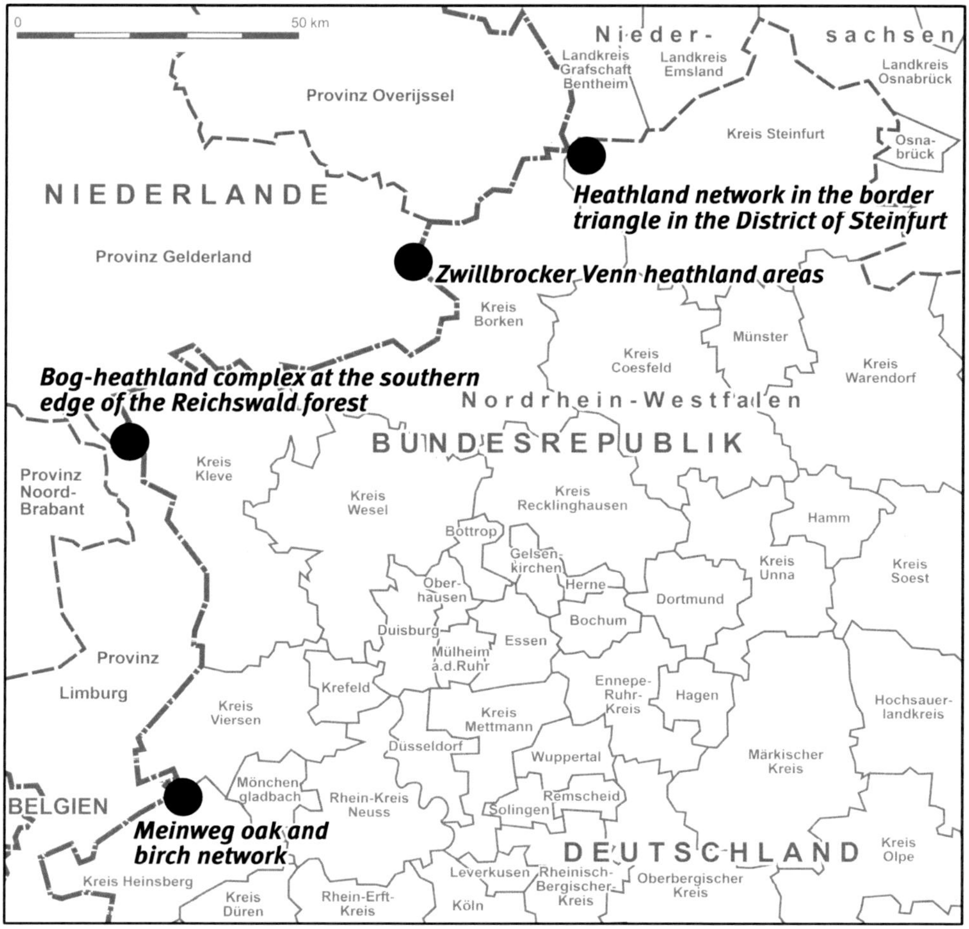

Figure 1: Overview of the geographical locations of the four sub-projects of the D+T Project "Green Border Network"

Source: Cartography by C. Schroer

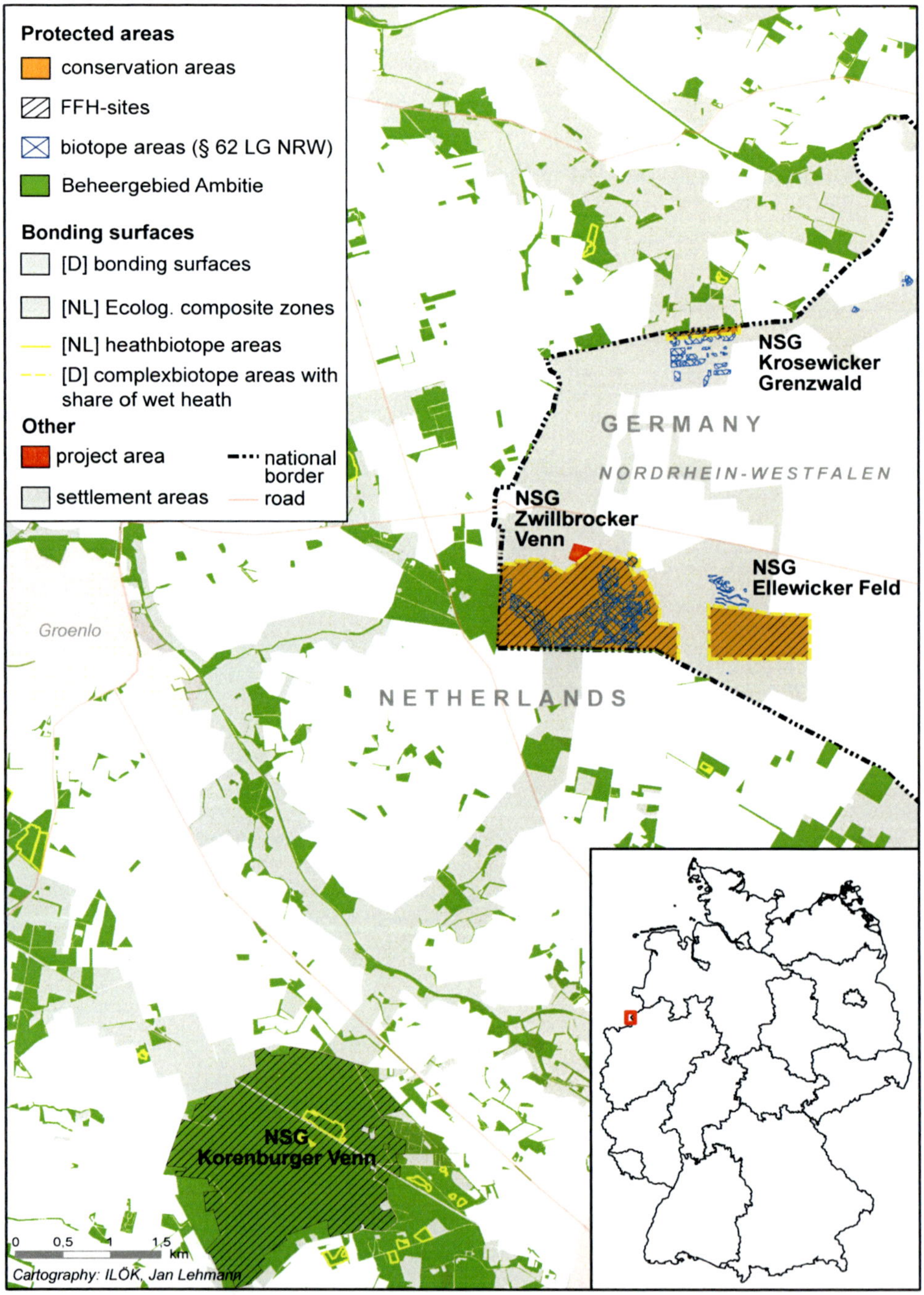

Figure 2: European Biotope Network Natura 2000 — Border Region Germany-Netherlands — Biotope bounding surfaces and nature protection area "Zwillbrocker Venn"

Source: Cartography by J. Lehmann

Comprehensive scientific research accompanied the D+T Project, focused on the regional project contexts. It was divided into the two branches of governance research and applied landscape ecology. While an evaluation of the biotope management measures specific to certain areas was at the forefront of the accompanying ecological research, the governance research aimed at systematically noting the particular planning challenges in German-Dutch cooperation to build up a cross-border biotope network against a background of different national planning cultures and the use of different incentive and steering instruments — but also taking into account different perceptions of problems and strategies on the part of the stakeholders involved on both sides of the border. On the basis of an analysis of the conditions and factors influencing not only behaviour in decision-making and cooperation but also the use of incentive and planning instruments, the idea was to produce suggestions for measures which could both shape such planning and implementation processes and also be transposed to other border regions.

This meant that an analysis not only of stakeholders and networks but also of instruments was in the focus of the governance research carried out. The sets of questions outlined were operationalised in the three areas of analysis and evaluation (Figure 3): in the area of "Use of formal/legal and informal planning and incentive instruments", a study was carried out — having recourse to the different planning systems in Germany and the Netherlands — of the extent to which these instruments contribute to reaching common objectives.

The area of evaluation entitled "Stakeholder constellations and networks" gathered information on the project-relevant stakeholders and groups of stakeholders, the motives for their actions, acceptance and project perception. One central question focused on how the planning and implementation processes are shaped in their local and regional — i.e., in particular in their cross-border — forms of cooperation, and on which methods of cooperation — using which organizational structures — were conducive or inhibiting for the development of the project. The different mentalities and planning cultures, as well as the linguistic differences, existing on both sides of the border were taken into account.

In the third area of evaluation "Perception and acceptance on the part of the public and those involved" it was examined whether — and, if so, how — the projects were perceived in each of their local environments and how the measures carried out were received. If there were found to be any success factors, deficits or conflicts impacting upon the realisation of the cross-border projects, they were to be recognized and taken into account when extrapolating any measures to be recommended.

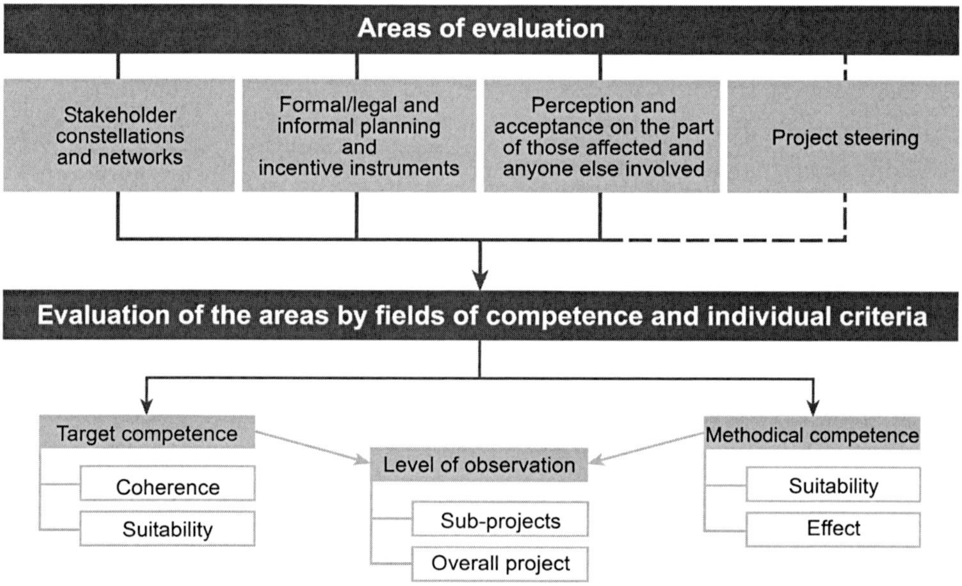

Figure 3: Areas of evaluation in governance research
Source: Grabski-Kieron et al. 2016

For the purpose of further operationalisation, the above-mentioned areas of analysis and evaluation were further sub-divided into two fields of competence — target competence and methodical competence — whose contents were drawn up with the aid of individual criteria (see, for example, Table 1). In doing so, a coherent framework for analysis and evaluation was developed which formed the basis of the empirical work carried out.

Table 1: Example of the content drawn up for the evaluation framework

Criterion	*Indicator*
Coherence of objectives	Coherence of objectives on German side
	Coherence of objectives on Dutch side
	Cross-border coherency of objectives
Suitability of objectives	Relation between the stakeholders' objectives and the objectives of the project
	Degree to which stakeholder constellations fit
Suitability of methods	Framework conditions supporting the establishment of a cross-border network
	Framework conditions inhibiting for the establishment of a cross-border network
Effectiveness of methods	Effectiveness of network

Source: Grabski-Kieron et al. 2016

In the following part of this article, some results of the research project will be outlined, and the contributions made by the nature conservation-oriented governance processes to rural regional development in the German-Dutch border area will be shown.

Use and significance of formal/legal and informal planning and incentive instruments

Area-related issues of nature conservation are included as obligatory planning requirements both in the multi-level system of governance in German planning and in that employed in the Netherlands (Table 2 and 3). The interdisciplinary and coordinating role of spatial planning is undisputed. Both planning systems integrate both formal/legal and informal instruments to allow the generation of impulses for spatial development.

Table 2: System of spatial planning in the Netherlands showing relevance for environmental protection and nature conservation

Level	Relevant planning instruments	Explanation
Rijk	Wet ruimtelijke ordening (Wro) 2006 Nota Ruimte-Ruimte voor ontwikkeling (2006)	Spatial planning legislation and national regulatory framework, general guidelines for spatial development
	Structuurvisie Infrastructuur en Ruimte (SVIR, 2012) Meerjarenprogramma Infrastructuur, Ruimte en Transport (MIRT)	Concept for the future of infrastructure and spatial planning; multiannual programme for infrastructure, spatial planning and transport as a basis for further project-based development in the Provinces
	Groene Ruimte 1995 Natuur-Netwerk Nederland — Ecologische Hoofdstructuur (EHS) NL	Model and guiding principles for national system for protected areas and for a supraregional biotope network with the implementation of Natura-2000 (EHS); stipulations for specification and implementation at Province level
Provincie	Provinciale structuurvisie (PRSV),	Models for future development in the Provinces with specification of the models set out by the "Rijk"
	Ecologische Hoofdstructuur (EHS)	Specification of the spatial objectives and contents of the EHS NL and planning for implementation to develop the system of protected areas and biotope networks at Province level (responsibility since 2014)
	Provincial Omgevingsplan	Objectives and principles of spatial development in the Provinces relating to planning regulations, also for regional open spaces/environmental protection and nature conservation (Natuur, Landschap, Landbouw)
	Provincial Streekplan	Regional plan based on planning regulations to specify the Provincial Omgevingsplan
Gemeente	Gemeentelijke Structuurvisie	Municipal concept for spatial models and development, also as a masterplan
	Bestemmingsplan	Municipal zoning plan relating to planning regulations to define allocation of sub-regional functions within the municipal area

Source: Grabski-Kieron et al. 2016

Table 3: System of spatial planning in Germany

Statutory			Informal/Non-statutory	
Spatial Planning	**Nature conser-vation/ landscape man-agement**		**Spatial Planning**	**Nature conser-vation/ landscape man-agement**
Structural Policy/ Structural Funds	Habitats-/ Birds Directives, Fund-ing program (LIFE)	**EUROPEAN UNION**	Territorial Agenda 2020	Biodiversity Strategy
Federal Spatial Planning Act	Nature conserva-tion act/ National Biodiversity Strat-egy	**FEDERAL**	Guiding principles for spatial devel-opment	Biodiversity strategy Guiding principles for nat. resources protec-tion and man-agement
States Planning Acts	States Nature conservation acts and programs	**STATES**	Objectives and programs	Stakeholder conferences
Regional spatial structure plans	Regional Land-scape plans	**REGIONS**	City Networks Regional Confer-ences Regional Devel-opment Strategies Initiatives, pro-jects	Stakeholder conferences Public-private partnerships Habitat manage-ment Initiatives, projects, habitat management
Federal Building Code/Urban-Land-Use Plans	Landscape Plans/ gen. Urban green space plans	**MUNICIPALITIES/ LOCAL GOV-ERNMENTS**	Initiatives, projects	Initiatives, pro-jects, habitat management

Source: Author's compilation

Behind this display of commonalities, however, there are some clear differences which impede any coherent cross-border development as far as safeguarding areas for nature conservation and any follow-up measures are concerned. Clearly differing perspectives arise, for example, regarding the basic agreement of goals for drawing up options for implementation in regional spatial planning (Germany, Table 3) and in the Provinciaal Omgevingsplan (Netherlands, Table 2). Regional planning in the Netherlands , in contrast to that in Germany, is clearly linked to project implementation and is directly reflected not only in the project- and implementation-related plan entitled Ecologische Hoofstructuur (EHS), but also in the tied allocation of funds to project implementation. The statutory requirements for German regional planning (Table 3) do not provide for such an operational orientation. Analyses carried out during research showed that these differences of principle in planning legislation lead, among other things, to the fact that cross-border work processes at regional level are restricted to a simple exchange of information and that no further coordination of measures to be implemented occurs. It comes as no surprise, therefore, that many of the german stakeholders involved in nature conservation look upon the formal/legal instruments as being of rather secondary importance and that the role played by these instruments in coordinating and preparing the safeguarding of areas is often not apparent to any such stakeholders involved.

In view of the weaknesses that present themselves in agreeing on the content of planning processes, as well as in the way these proceed, the stakeholders involved in nature conservation value the informal instruments and approaches to cross-border nature-conservation work as being much more constructive and practicable. These tools enable other groups of stakeholders, such as those involved in agriculture, to be brought in at an early stage and strategies to be drawn up for the minimization of any conflict. Joint goals can continue to be pursued with reference to specific projects and with possibilities of adaptation. This is not a completely novel insight.

The study also showed clearly the ways in which the two planning instruments — used in landscape design in Germany and in the Ecologische Hoofstructuur in the Netherlands (Table 2), which has a similar orientation — develop different types of spatial effectiveness. The landscape plan (Table 3) — legitimized as a statute in the State of North Rhine-Westphalia — specifies the contents of the regional plan and creates the conditions for a binding implementation of measures relating to nature conservation and landscape conservation over a wide area. This status of being a statute is seen on the one hand as being effecttive, but on the other as problematic because the planning process is burdened by a high degree of conflict as well as by significant time requirements. Overall, its effectiveness lags behind its possibilities. By contrast, the Dutch Ecologische Hoofstructuur directly incorporates the stipulations laid down in the Province's planning, taking into account synergies, e.g., through being linked to recreation in the countryside, to the cultivation of cultural landscapes, to agriculture and forestry or to the extraction of raw materials. Overall, it finds stronger cross-border points of contact in landscape development. Changing financial frameworks being set in Dutch policy-making are the only thing currently restricting the effectiveness of the Ecologische Hoofstructuur.

The EU's nature conservation funding programme "LIFE+" (Regulation (EC) 614/2007, Table 3) plays a decisive role on both sides of the border in safeguarding areas and implementing projects. If — particularly in the border regions — link-ups with other EU funding programmes (INTERREG, EFRE, environmental measures as part of the EU agricultural fund) present themselves, they do not, however, happen because of the regional scenarios and funding conditions currently defined. Added to this are uncertainties, or even a lack of knowledge, regarding agreed, common funding possibilities. As a result, opportunities for promoting actual synergy effects for regional development that include nature conservation are utilised only to a limited extent.

Stakeholders and networks

A large number of stakeholders were included in the study — either as groups, as individuals or as public or private institutions. Depending on the extent of their functional connection with the field of activity of biotope protection on the one hand, and with spatial development on the other, they were each allocated to an inner and outer circle of stakeholders (Table 4). This produced different types of questionnaire for the empirical study.

Table 4: Stakeholders and institutions included

Level	External stakeholders	Internal stakeholders
superordinated	Bund für Umwelt u. Naturschutz Deutschland e.V. (BUND) — Landesverband NRW Deutsch-niederländische Raumordnungskommission (Staatskanzlei Düsseldorf) Ministerium für Klimaschutz, Umwelt, Landwirtschaft, Natur- und Verbraucherschutz (MKULNV) NRW Landesamt für Natur, Umwelt und Verbraucherschutz (LANUV) NRW Landesforstverwaltung NRW Natuurmonumenten (total: 6)	Biologische Station Kreis Steinfurt e.V. Biologische Station Zwillbrock NABU Naturschutzstation Niederrhein NABU Naturschutzstation Haus Wildenrath e.V. (total: 4)
local	Zweckverband Naturpark Schwalm-Nette Bezirksregierung Köln Kreis Heinsberg Stadt Wassenberg Stadt Wegberg Euregio Rhein-Maas-Nord Provinz Limburg Natuurmonumenten (Limburg/N.-Brabant) Gemeente Roerdalen Bezirksregierung Düsseldorf Kreis Kleve Landesbetr. Wald u. Holz/RFA Niederrhein Stadt Goch Gemeinde Kranenburg Euregio Rhein-Waal Gemeente Gennep Regionaal Bureau voor Toerisme Arnhem Nijmegen Bezirksregierung Münster Stadt Vreden EUREGIO Landwirtschaftskammer NRW-Kreisstelle Borken Regio Achterhoek Gemeente Berkelland Gemeente Winterswijk Gemeente Oost-Gelre Münsterland e.V. Kreis Steinfurt Stadt Ochtrup Gemeinde Wettringen Stadt Gronau Provinz Overijssel Nationaal Landschap Twente Gemeente Losser (total: 34)	Biologische Station Kreis Steinfurt e.V. Bertha Jordaan van Heek-Stiftung Biologische Station Zwillbrock Stichting Staring Advies NABU Naturschutzstation Niederrhein Teunesen Zand en Grint NABU Naturschutzstation Haus Wildenrath e.V. Landesbetrieb Wald und Holz — RFA Jülicher Börde Bundesforstbetrieb Rhein-Weser Stadt Wassenberg Natuur en Landschap Winterswijk Zweckverband Deutsch-Niederländischer Naturpark Maas-Schwalm-Nette Landesbetrieb Wald und Holz – RFA Niederrhein u. RFA Ruhreifel-Jülicher Börde FBB Schwalm-Nette und FBB Wassenberg Stadt Wassenberg Bundesforstbetrieb Rhein-Weser Staatsbosbeheer (Beheer Meinweg) Bundesanstalt für Immobilienaufgaben (total: 18)

Source: Grabski-Kieron et al. 2016

However, it was the institutional and structural/situational work context which proved to be decisive. In other words, bottlenecks regarding personnel, equipment and finance in public authorities at all levels on both sides of the border prevent any strengthening of more thoroughgoing cooperation. Added to this are differences in people's perceptions of their own tasks. Cross-border institutions which have been in existence for a long time, such as the German-Dutch Spatial Development Commission or the EUREGIOs, confined themselves — as far as the case-study regions were concerned — to their tasks as stipulated in the statute or made inadequate use of potential for further cooperation. Nature conservation issues were not, by and large, taken any further.

The municipalities on both sides of the border are also mandated to include nature conservation issues in their planning. They can, in equal measure, contribute to safeguarding areas as well as assuming a mediating role to reduce conflicts. In the case studies undertaken, however, the municipalities showed themselves to be passive. They did not play any role in helping to promote cross-border nature conservation concepts. This is in complete contrast to many activities which municipalities in border areas are currently rolling out as co-ordinated developments of location or infrastructure.

Over and above institutional contacts, the study confirmed that decisive factors for success include personal contacts, continuity on a personal level, and a similarly personal commitment within the networks. This is true not only for specific project work, but also for the integration of nature conservation issues in other regional activities. The project work itself is the starting point for everyone involved in building up networks. This is augmented by having clear objectives, overcoming problems of understanding, and ensuring transparent contents. Regular meetings are indispensable.

However, one thing that became evident when the case studies were compared with one another was that networking must not be allowed to "get stuck" in informal forms of cooperation. The existence of cross-border organizations, or building them up, has a positive effect because such organizations can be a focus for activities and can provide contact persons. Also, they can initiate — better than others can — connections to other partners working in regional development. This was clearly shown in the "Meinweg oak and birch network" case study (Figure 1). The cross-border "Maas-Schwalm-Nette" nature reserve and the biological station responsible on the German side adopted an important position as far as being multipliers was concerned, as well as acting as mediators and supporting the implementation of measures in the project area and incorporating them in cross-sector issues of regional landscape development.

Perception and acceptance

As has already become apparent, there were differing perceptions and assessments of all the development and trial projects (with their related sub-projects) on the part of representatives from the areas of politics and administration on the one side, and of stakeholders involved in nature conservation on the other. Although cross-border nature conservation issues relating to a particular area are everywhere caught between different requirements for land use, this problem was in fact "evaluated" differently. Many people representing higher planning levels, as well as practical nature conservation in Germany, emphasize the constant latent conflict triangle of "nature conservation, municipal interests and agriculture". This problem was also seen on the Dutch side, but many of those questioned saw the cross-border nature conservation projects more as activities presenting options for leveraging natural potentials in a non-harmful way, e.g., for local recreational purposes. In doing so, they interpreted the projects on the basis of an indirect economic use — much more strongly than was the case on the German side.

The orientation for implementation peculiar to the projects in the case studies was seen as being decisive on both sides of the border for promoting acceptance on the part of third parties. At the same time, it was also emphasized how important it is that this project orientation be integrated into regional concepts of landscape development. Almost all the case studies showed deficits here. Also taking these points into consideration, the cross-border "Maas-Schwalm-Nette" nature reserve was assigned a special role: as a "working platform", so to speak, nature conservation issues can basically be well positioned here and blended in with other regional issues from the point of view of sustainable development. As far as cross-border public relations work, environmental education and marketing measures are concerned, the nature reserve was the only institution in the study to have built up communication channels in several languages. But this was the exception. Overall, the PR work in the case studies was largely confined to their national contexts with an emphasis on the local and regional levels. Any communication requirements regarding cross-border nature conservation work necessarily took second place to this.

Summary and conclusions

Many border regions, for which the german-dutch border area serves as an example, show remarkable regional changes in terms of socio-economic performance and infrastructural improvement. Their current spatial functional patterns are to be measured in cross-border spatial coherence largely concomitant of leaving behind peripheralisation. Due to their pathways of development these regions have nevertheless preserved a rich natural heritage until today. Its endangering as well as the increasing requests of habitat network and of the preservation of biodiversity are among others to be comprehended as core components of an integrated rural development. As such cross-border nature conservation and habitat management are at the same time also embedded in issues of resurgent common regional identity and cross-border regional governance.

The study confirmed many findings resulting from governance research into regional development. Focused on the requirements of nature conservation for individual areas in rural development, it showed deficits both in cross-border and in integrated regional development. It pointed out problems in cross-border nature conservation work within the multi-level system inherent in spatial planning. This is all the more serious because in future nature conservation issues will have to be integrated into cross-border planning measures more closely than in the past when requirements for preserving biodiversity and for climate protection are taken into account.

The study showed to the stakeholders involved in administrative and organized nature conservation on the one side, and in spatial planning on the other, the strengths and weaknesses of their positions in a binational working environment. Factors influencing any cross-border cooperation geared towards nature conservation were identified. Whenever, on the German side, the project concentrated on regions in the state of North Rhine-Westphalia, where there were specific conditions in the administrative structure and planning system, it was possible to note indications of a need to act in various cases — which could be transferred to nature conservation-oriented processes in other border regions. In this sense, cues for action were grouped into the following seven fields of activity and specified through numerous recommendations (Grabski-Kieron et al. 2016):

- Field of activity 1: Use and extend existing communication and co-operation structures.
- Field of activity 2: Focus resources through cooperation with third parties.

- Field of activity 3: Utilise the potential to be found in the formal/legal planning instruments existing in Germany.
- Field of activity 4: Make stronger use of the potential to be found in informal incentive and control instruments.
- Field of activity 5: Create joint systems and forms of cross-border planning.
- Field of activity 6: Mitigate the conflict triangle of “nature conservation — agriculture — municipalities“.
- Field of activity 7: Build up supra-regional acceptance for nature conservation issues.

References

Blotevogel, H. (1996). Auf dem Weg zu einer Theorie der Regionalität. In Brunn, G. (Eds.), Region und Regionsbildung in Europa. Konzeptionen der Forschung und empirischen Befunde (pp. 44–68). Baden-Baden: Nomos.

Council Directive 92/43/EEC of 21 May 1992 on the conservation of natural habitats and of wild fauna and flora. OJ L 206, 22.7.1992, 7–50.

Directive 2009/147/EC of the European Parliament and of the Council of 30 November 2009 on the conservation of wild birds. OJ L 20, 26.1.2010, 7–25.

European Commission (2011). The EU biodiversity strategy to 2020. Brussels: Publications Office of the European Union.

Fürst, D. (2010). Raumplanung, Herausforderungen des deutschen Institutionensystems. Planungswissenschaftliche Studien zur Raumordnung und Raumentwicklung 1 (1). Detmold: Rohn.

Grabski-Kieron, U. (2012). Le développement rural intégré en Allemange: une demarche de planification et d'action prise entre ambitions et réalités. In Lacquement, G., Born, K. M. & von Hirschhausen, B. (Eds.), Réinventer les campagnes en Allemange (pp. 185–198). Lyon: École normale supérieure de Lyon.

Grabski-Kieron, U. & Kötter, T. (2013). Regionalisierte Entwicklungsansätze und zentrale Handlungsfelder. In Kummer, K. & Frankenberger, J. (Eds.), Das deutsche Vermessungs- und Geoinformationswesen. Themenschwerpunkt 2013: Landesentwicklung für ländliche Räume, Analysen und Antworten zu Demographiewandel, Planungszielen und Strukturveränderung, Regionalentwicklung, Entwicklung von Dörfern und Städten, Bodenordnung und Landmanagement, Arbeitsprozess Flurbereinigung (pp. 81–132). Berlin: Wichmann.

Grabski-Kieron, U., Buttschardt, T. K., Foerster, J., Böhmichen, J. & Lehmann, J. R. K. (2016). Grenzüberschreitender Biotopverbund — Handlungsansätze und Herausforderungen für Planung und Naturschutzpraxis, Ergebnisse aus dem Erprobungs- und Entwicklungsvorhaben (E+E): „Netzwerk Grüne Grenze — Natur ohne Grenzen zwischen Nordrhein-Westfalen und den Niederlanden". In Bundesamt für Naturschutz (Ed.), Naturschutz und biologische Vielfalt, Vol. 146. Bonn.

Hager, G. (2003). Auf dem Weg zu einer grenzüberschreitenden Region — das Beispiel Pamina. In Pröhl, M. & Adamaschek, B. (Eds.), Regionen erfolgreich steuern, Regional Governance — von der kommunalen zur regionalen Strategie (pp. 139–154). Gütersloh: Verlag Bertelsmann-Stiftung.

Hammer, T., Mose, I., Siegrist, D. & Weixlbaumer, N. (Eds.) (2016). Parks of the Future. Protected Areas in Europe Challenging Regional and Global Change. München: oekom.

Harteisen, U., Neumeyer, S., Schagbauer, S., Bizer, K., Hensel, S., Krüger, L. (2010): Grünes Band — Modellregion für Nachhaltigkeit. Göttingen: Universitätsverlag Göttingen.

Heintel, M. & Weixlbaumer, N. (2016). Protected areas at the interface of political systems. Regional governance arrangements using the example of the Austro-Hungarian Nature Park Írottkő-Geschriebenstein. In Hammer, T., Mose, I., Siegrist, D. & Weixlbaumer, N. (Eds.), Parks of the Future. Protected Areas in Europe Challenging Regional and Global Change (pp. 177–190). München: oekom.

Jacobs, J. & van Assche, K. (2014). Understanding Empirical Boundaries: A Systems-Theoretical Avenue in Border Studies. Geopolitics 19 (1), 182–205.

Jessel, B. & Tobias, K. (2002). Ökologisch orientierte Planung. Stuttgart: Eugen Ulmer.

Johnson, C., Jones, R., Paasi, A., Amoore, L., Mountz, A., Salter, M. & Rumford, C. (2011). Interventions on rethinking 'the border' in border studies. Political Geography 30 (2), 61–69.

Knippschild, R. (2009). Außer Spesen nichts gewesen? Nutzen und Erfolgsfaktoren grenzüberschreitender Kooperation in der Stadt- und Regionalentwicklung im deutsch-polnisch-tschechischen Dreilandereck. Raumforschung und Raumordnung 67 (3), 228–238.

Kühne, O. & Meyer, W. (2015). Gerechte Grenzen? Zur territorialenm Steuerung von Nachhaltigkeit. In Kühne, O. & Weber, F. (Eds.), Bausteine der Regionalentwicklung (pp. 25–40). Wiesbaden: Springer.

Lacquement, G., Born, K. M. & von Hirschhausen, B. (Eds.). Réinventer les campagnes en Allemange. Lyon: École normale supérieure de Lyon.

Leibenath, M., Blum, A. & Stutzriemer, S. (2010). Transboundary cooperation in establishing ecological networks. The case of Germany's external borders. Landscape and Urban Planning 94 (1), 84–93.

Liesen, J. (2015). Naturparke stärken nachhaltige Entwicklung in ländlichen Räumen. In BBN (Bundesverband Beruflicher Naturschutz e.V.) (Eds.), Verantwortung für die Zukunft — Naturschutz im Spannungsfeld gesellschaftlicher Interessen. Jahrbuch für Naturschutz und Landschaftspflege 60 (1), 116–123.

Martinez, O. (1994). The Dynamics of Border Interaction. New approaches to border analysis. In Scheffield, C. H. (Eds.), World Boundaries (pp. 1–15). Global Boundaries 1 (1). London, New York: Routledge.

Regulation (EC) No 614/2007 of the European Parliament and of the Council of 23 May 2007 concerning the financial instrument for the environment (LIFE+). OJ L 149, 9.7.2007, 1–17.

Schönwald, A & Kühne, O. (2012). Landschaften und Identitäten in Zeiten offener Grenzen und fortgeschrittener Hybridisierung — das Beispiel der Großregion. Europa Regional 20 (1), 3–13.

Wille, C. (2015). Spaces of the Border — a Practice-theoretical Cultural Studies Perspective in Border Studies. Europa Regional 20 (1–2), 60–70.

Wille, C., Reckinger, R., Kmec, S. & Hesse, M. (Eds.) (2014). Räume und Identitäten in Grenzregionen. Politik – Medien – Subjekte. Bielefeld: transcript.

Securitisation from below? Civil engagement within processes of bordering in internal EU borderlands

Kristine Beurskens, Judith Miggelbrink and Christoph Creutziger

Abstract

In spatial research, the regions along the internal EU borders have generally been associated with rising cross-border co-operation within various social, economic and cultural fields. Among local inhabitants and politicians, however, there appears to be a growing perception of some problematic developments in these formerly well secured state border regions.

Since the establishment of a common Schengen space, most of the visible control functions have been moved to the external borders of the EU. Along some internal EU borders, occurrences of theft and crime are increasingly perceived as a failure of the state in providing security for their citizens. In consequence, community and private actors have begun to launch neighbourhood initiatives, apply technical devices of securitisation and generally establish their own solutions of controlling and bordering.

By looking at the rise of civil engagement in establishing security in borderlands and the discourses and practices involved, we aim to discuss some inconvenient questions around the functionalities of the internal EU borders and address reconfigurations of the relation of state, society and territory within these processes. The empirical studies for this article were carried out in German borderlands before the drastic rise in numbers of refugees in 2015. The current events demonstrate even more strikingly the urgency of border politics and issues of humanity involved all across Europe.

Introduction

For more than six decades now, there has been a tale of border regions within the European Union depicting them as core sites and agents of European integration. These tales have been coming along with various expectations of peaceful neighbourhood, cultural contacts and exchanges and administrative co-operation. Moreover, cross-border relations have been expected to enhance capacities especially of rural and small- to medium-sized town regions to overcome problems and deficiencies resulting from their often peripheral location within the domestic territory. Especially the institutionally organized Euroregions (e.g., 'Euregios') have become the epitome of establishing and consolidating cross-border co-operation of administrations, cultural institutions, associations and various other organizations. The social relations and interactions at and across the inner borders, that have risen in spite of numerous experiences of hostilities, war, occupation and eviction, have been an essential step towards the realisation of the 'utopian project Europe', in the depth of which a whole new understanding of democracy and even the idea of a transnational civil status could grow (Lissandrello 2006).

Benefits, especially economic advantages, have also been expected from other spatial strategies of the EU, namely the formation of the internal market from the late 1970s onwards (van Apeldoorn 2000). This strategy was aimed at creating a unified economic space to ensure global competitiveness. As a precondition, border functions had to be shifted to the outer border of the emerging common, 'macro-territorial' (Jureit and Tietze 2015) European space, thus taking over sovereignty rights from countries participating in the Schengen regime. Many regions along the now internal borders can be considered peripheral in a number of aspects; infrastructures are not as developed as in other parts of the country, employment chances are dim, the population is small in numbers and shrinking. Their position along the margins of the political East and West and various economic spheres has been shifted through ongoing transformations of the last century — and just a decade ago again through the integration of Poland and the Czech Republic into the European Union. For many, the final opening of these newly declared internal EU and Schengen borders in 2007 was therefore associated with expectations of intensified cross-border exchanges and a general growth in importance of these regions.

In some respect, both strategies have been approaching their limits. Euroregions have increasingly been being disenchanted due to the limited democratic potential of trans-border co-operation (Kramsch 2001 and 2011). Scholars have also framed Euroregions as elitist projects that lack a close connectedness to the people living in cross-border regions (Trillo-Santamaria 2014). Critiques of the macro-territorialisation of the EU have been arising not only because of its failures in the most recent crisis of migration and asylum. There is also a growing scepticism as regards the potential positive effects freedom of movement has exerted on border regions (Lederer 2009; Zielcke 2015).

Nevertheless, the transformations of internal EU borders remain being associated with the positive expectations of co-operation and exchange in political arenas based on their 'open' character enabling movements without (formal) hindrances. But even this openness itself has not remained uncontroversial. Since then, numerous press articles, media reports and web forum contributions covered stories about crime and insufficient security in border regions, openly connoting the opening of borders in a negative way. Regarding the eastern borders of Germany, for example, these discourses are typically related to an assumed "invasion of organized crime" and so-called "eastern European gangs of thieves" into the German territory. In addition, it is assumed that criminal activities in borderlands are directly related to the fact that potential suspects can move freely from one national juridical space to the other, thus easily withdrawing themselves from being tracked down and prosecuted. This assumption is sustained regardless of the actual juridical conditions and practical handling of the so-called hot pursuit by law enforcement officers on foreign territory, under which prosecution might be impeded in some cases, but not impossible.

Whether the barrier-free internal EU/Schengen borders in concurrence with an increasingly expanded cross-border infrastructure actually do favour crime, remains a controversial issue. What is undisputed though is a certain awareness among the German border population of seeing themselves as victims of so called border crime. Though not empirically proven so far, there is some evidence that this (self-) positioning might also result from people's feeling of being left behind at the outer margins of the state, peripheralised not only in economic or social terms, but left alone with the security problems of a border area, with too little support by police and other state agencies.

> *"We have to help ourselves, if the state doesn't come up against theft of such magnitudes." (resident, in "topagrar online" 09.12.2012)*
>
> *"In the whole area, there are Poles, Russians and Lithuanians cruising through the villages, nosing around. If nothing changes, we will start vigilantism." (resident, in the newspaper "Die Welt", 25.07.2012)*

Starting from the current urgencies that are expressed in such perceptions of and discourses on border crime, we will discuss the current relations of actors in questions regarding the production of borders and security in the regions along the German state borders to its eastern neighbours. At first, we will place the notions of "open borders" into a context of complementarities of external and internal EU/Schengen border regulations. We will then take a closer look at the term *Grenzkriminalität* (border crime) that within a very short time has risen to be the central significant for a negative perception of the barrier-free borders in everyday and mass communication. By analysing perceptions, emotions and their related discourses, we attempt to add a new perspective to current debates on relations of state and non-state actors and the effects of their intertwined practices on the development of rural-peripheral regions.

Schengen towards the outside: transformation of European state borders

The Schengen border regime, originating from agreements between the Benelux states, France and Germany, has been developed continuously since the 1980s. By now it has become an integrated part of the "polycentric system of the European administrative organization" (Fischer-Lescano and Tohidipur 2007) and belongs to the so-called "acquis" that all joining states have been and are obliged to accept and adopt (Table 1). We consider three aspects of this border regime as especially relevant; three traits that we want to discuss under the terms of re-scaling, securitisation and displacement.

Table 1: Evolution of European integration and border policies

European Integration	*Border policies*
1980s Preparation of a common market	1985 Schengen Treaty (Schengen I) between France, Belgium, the Netherlands, Luxemburg and Germany regarding future abolishment of border controls between the member states
1992 European Monetary Union (Maastricht Treaty)	1995 Schengen Convention (Schengen II) Abolishment of border controls within the union and relocation of border functions to the borders shared with third states
1999 Treaty on European Union (Amsterdam Treaty)	1999 Proclamation of the "common space of freedom, security and justice"
2003 Introduction of the European Neighbourhood Policy (ENP)	2002 Ongoing communications about an integrated border management
2004 Accession of additional 10 states to the European Union (Estonia, Latvia, Lithuania, Poland, Czech Republic, Slovakia, Slovenia, Hungary, Cyprus, Malta)	2005 Prüm Convention (Schengen III) Agreement on intensified cross-border cooperation particularly in combating terrorism, cross-border crime and illegal migration; Foundation of the EU border management agency FRONTEX
2007 Bulgaria and Romania become new member states of the European Union	2007 Full integration of the accession states of 2004 into the Schengen border regime, abolishment of border controls also at these internal EU/Schengen borders
2013 Croatia becomes a new member state of the European Union	2013 Introduction of new border politics by the European Commission: the "smart-border package"

Re-scaling

The redefinition of meaning of the territorial borders of EU member states has certainly been one of the most striking changes of the geopolitical order of Europe over the last decades. Gradually, an external border regime was developed, based on common standards of surveillance and control. This was the precondition for the emergence of an internal transnational space, also called the "area of freedom, security and justice" (Treaty of Amsterdam, Part 1, Article 1), which distinguishes itself by the absence of internal state border controls. Mostly based on economic motivations (van Apeldoorn 2000), the aim was to create a space that was competitive on a global economic level (Heeg and Ossenbrügge 2002). In consequence of the political transformations of the 1990s and the at least formal removal of the political-ideological delineations between "East" and "West", more and more European states were integrated into this complementary system of opened inner and homogenized outer borders (Miggelbrink 2013).

This process implies a re-scaling of sovereignty: the standards of surveillance and control — a fundamental part of nation state sovereignty — are no longer only handled by the states alone, but are argued and defined on a transnational level, and only implemented by nation state powers.

Securitisation

The technical-territorial changes of European borders are embedded in contested, ever anew negotiated orders of the relations of sovereignty, territoriality and security (Belina 2014). Within this context, "(in-)security" has grown to be a discursively established grid of interpretation, by which phenomena, persons, actions and decisions are perceived and treated mainly/primarily according to their possibly implied threat to security. Derived from the work of Foucault on the dispositives of security (Foucault 2004, 20–21), the term securitisation is used for this discursive filing and indexing/classification. At the same time, the range of what is considered to be an individual, collective or social risk has been considerably extended (Müller-Mahn et al. 2013). The practices in handling such risks, mainly regarding organized crime and terrorism, have direct consequences for the (re-)alignments of border regimes. Securitisation therefore can be understood as a dispositive aiming at the defence against criminal and terrorist threats coming from 'the outside' (Neal 2009; Andreas 2003) as well as aiming at ensuring welfare and prosperity and other interests in the name of an inner sphere (Huysmans 2000).

The production of security, in the sense of blocking violations of the territorially defined sovereignty, has long been the function of state borders, in their task of the territorial marking of state sovereignty. However, with the rise of security as a leading motive of contemporary society, and the related perception of all kinds of social subdomains, groups and individual orientations in the light of their relevance for security, the traditional security functions of state borders have been redefined. New requirements for state borders have emerged, and the people and goods crossing the border have increasingly been observed within a framework of politics of securitisation. Multiple and complex filters have been established that grant people differentiated rights, according to their citizenship and to current geopolitical assessments of their countries of origin (Grabbe 2000; Müller 2013). However, these selectively securitising regulations collide with the economic interest in ever faster and cheaper cross-border exchange of co-operations and institutions. Jones and Rosière (2012, 231) describe this contradiction of the simultaneous occurrence of hardening security efforts and essential global economic need for flows and movements as a paradox of modern geopolitical developments.

Current border regimes therefore can be perceived as strategic attempts to solve this contradiction and create "limited permeability" (Komlosy 2005; see also Bigo and Guild 2005). While in the Northern American regions, this has already for a longer time been pursued by smart border policies (Côté-Boucher 2008), in Europe the attempt was first made by a complementary principle of low-threshold internal and securitised external borders (Belina and Miggelbrink 2013), with the external border regime also increasingly orientated at smart-border principles (Table 1). All of such efforts in the end aim at a very precise filtering of cross-border flows, a filtering that abides to powerful interests within the society. Some of the effects of such securitisation processes are a growing concern for identity and identity markers (Muller 2004), the expansion of biopolitics (Sparke 2006), but also — as an outcome of the structuring capacity of territorial logics — a latent criminalisation of 'outsiders' (Belina and Miggelbrink 2013).

Displacement

Re-scaling and securitisation have been accompanied by displacement meaning that border control and surveillance are not (longer) primarily bound to the territorial location of the border but have been spread spatially. Databases, consulates, police control etc. are located at different places and related to a wide range of different technologies. Measures of control and surveillance have increasingly become "downstream" and "upstream" of the borders, a-spatially and ubiquitous. The Schengen border implemented through "scattered security checks" (Atger 2008) is "everywhere". Moreover, the displacement of the border has come along with an increasing number of non-state actors involved in the development of technical equipment and organisation but also more directly in tasks of securitisation along the border.

Schengen to the inside: De-securitisation of internal EU/Schengen borders?

With the relocation of security-related border functions to the external borders of the territory of Schengen member states, internal EU/Schengen borders have emerged as a new type of state borders. According to the Schengen acquis, the internal borders are defined as "the common land borders of the contracting parties, their airports for internal flights and their sea ports for regular ferry connections exclusively from or to other ports within the territories of the contracting parties and not calling at any ports outside those territories" (Schengen-Acquis, Title 1, Article 1). Due to their internal location between the Schengen member states, border controls along these state borders have been gradually removed since the mid 1990ies. These internal borders "may be crossed at any point without any checks on persons being carried out" (Schengen-Acquis, Title 2, Article 2, Paragraph 1). The related infrastructures at border crossing points have been abandoned, partly dismantled and in some cases converted to other functions. As long as they do not coincide with physical or material elements in space (e.g., rivers), these borders have vanished as landmarks almost entirely (Figure 1 and 2).

Figure 1: The new internal EU/Schengen border: abandoned border infrastructure at the border crossing point of Schwedt/Oder (GER, Brandenburg) — Krajnik-Dolny (PL, Województwo Zachodniopomorskie)

Source: C. Creutziger

Figure 2: Minimal landmarks of the border near Jonsdorf (GER, Saxony) — Mařenice (CZ, Liberecký kraj)

Source: C. Waack

This obvious physical disappearance of measures of surveillance and control constitutes one of the major consequences of the relocation of border security functions for the local perception of internal EU/Schengen borders in border regions. It comes along with the paradox effect that a growing securitisation of external borders can well be accompanied by a perception of de-securitisation within internal parts of the Schengen territory. In fact, however, what we are witnessing is first and foremost not more than a relocation of security measures to other places:

Firstly, along with the re-scaling of the EU border regimes, the modes of operation of national agencies of border police were changed towards cooperation of several national border police agencies at the external borders. Depending on the specific national law, they are empowered or required to carry out checks also at the insides of the Schengen territory (Laube 2010). Thus, checks of persons and goods are continued to be carried out at main roads, in international trains and on railway stations, albeit in a more mobile and less visible way. The German border police, the 'Bundespolizei', continues to carry out control functions. In contrast to their earlier modes though, these checks do not necessarily happen at constant locations at the German state borders, but are spread over the whole territory of Germany, over neighbouring countries as far as the external EU borders (Bundespolizeigesetz §2, §58, §68). Another re-scaling of these formerly national control functions can be seen in the shift of some surveillance tasks towards lower levels, the levels of singular federal states (Bundespolizeigesetz §2; Brandenburgisches Polizeigesetz §11/12), thus involving a wider variety of actors into the production of security in these regions.

Secondly, at some internal EU/Schengen borders, a new regionalised type of bilateral cross-border co-operation is emerging regarding security tasks. This co-operation had already been envisaged with the so-called 3rd pillar of the Maastricht treaty, the pillar on "Police and Judicial Co-operation in Criminal Matters" (Treaty on European Union, Title VI). In concrete terms, this means for instance, that along a number of sections of the internal EU/Schengen borders agreements have been signed between the respective police agencies to guarantee and regulate cross-border police co-operation in the case of crime incidents. Throughout the transformations of the internal EU/Schengen borders, these agreements have in many cases been extended and intensified, for instance through the signing of new contracts of police co-operation between Germany and Poland (2014) and Germany and the Czech Republic (2015).

Despite the effects of growing securitisation that are obvious in everyday life, the described functional re-organisations of the establishment of security inside the Schengen space often get associated with de-securitisation.

Discursive contextualisation of border crime

Analysing the discourses and the activities of people within the border regions it becomes obvious, that there is a feeling of insufficiency concerning the state's production of security. This impression of a lack is mainly articulated through growing discussions involving the notion of border crime (*Grenzkriminalität*). This term underwent a considerable transformation within the last 25 years. In the 1990s it was exclusively used regarding crimes directly related to the border like smuggling or illegalised immigration (Schultz 2004). Today the term serves as a phrase that summarises events, fears and threats that either take place near the border or are imagined to be related to the border. However, the term as such will not be found (defined) in any legal framework or police statistics.

The current use of the term in a problematic way attaches responsibility for any irregular activities to the border itself. The discursive conjunction of *Grenzkriminalität* as a category combines in an undifferentiated way different phenomena that are being criminalised. It also includes the idea that the other (side) of the border is the one to blame.

According to Hornung (2012, 102), *Grenzkriminalität* is a long established term in politics, media and science. Because of the lack of a clear definition of the term, Hornung typifies a range of definitions and associations that were collected as part of a student seminar in the Oberlausitz (a rural border region in eastern Germany). In this case defining bears the problem: There is no set definition of *Grenzkriminalität* within the nomenclature of police and border protection. Therefore, within the current discourses around the term, mainly the associations of those that are expected to know or/and use term are being inquired. As a problematic effect of this, a certain everyday phenomenology of crime related interpretations is reconstructed. Those interpretations include racist connotations as well as associated causality assumptions. Hornung explicitly demands a definition that is intersubjective and extends single situations to serve practical political application. Prosecution services and its higher ministries should mediate an integral image of *Grenzkriminalität* (Hornung 2012, 110).

While keeping the problematic use of the term in mind, we use the expression as a way of understanding the discursive construction of border in particular contexts. In our case we try to understand the term in the context of current everyday speech, (party) political and mass medial practices of constituting border. Those constituted borders are associated with insecurity, threat and state failure.

Grenzkriminalität appears as a discursive marker for the representation of unwanted incidents, actions and phenomena. In this form it generates political/normative positions and initiates political action. We therefore read the discourse as an important part of the present transformation of the internal EU/Schengen border and negotiation of its safety function among different actors in the regions along these borders. To approach the observed narrations we have analysed in what context and how often the term *Grenzkriminalität* was used to describe incidents along the border. The data for our lexicometrical analysis was generated from a media corpus of national daily and weekly newspapers from 1992 until 2015. Namely we included Süddeutsche Zeitung (daily), Die Welt (daily), taz — Die Tageszeitung (daily), BILD (daily, rainbow press), Die Zeit (weekly) and Der Spiegel (weekly) to cover a large variation of political backgrounds.

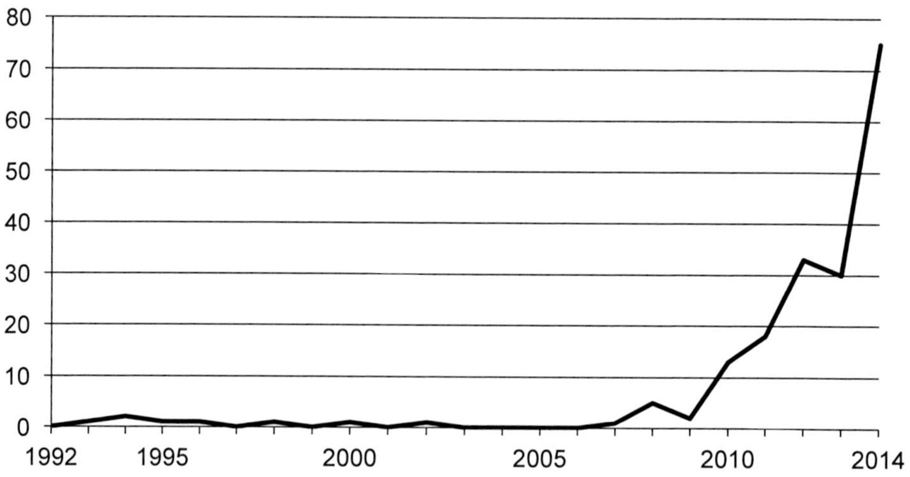

Figure 3: Absolute frequency of articles on border crime (*Grenzkriminalität*) in German newspapers and journals (1992–2014)

Source: Authors' analysis

The analysis shows a significantly increasing use of the term in 2012 largely due to the coverage of Süddeutsche Zeitung and Die Welt. Only BILD has long been using the term regularly. The term also keeps occurring in relation to names of parties and politicians. Therefore we assume that this takes place mainly in the setting of election campaigns for communal and regional elections.

One of the remarkable results is the fact that 97% of the articles in BILD using the term *Grenzkriminalität* are linked to the federal states Brandenburg or Sachsen which contain border regions to Poland and the Czech Republic. At the same time, the internal borders to older EU/Schengen member states such as Switzerland, France or the Netherlands only rarely get associated with this topic.

Most frequent words:

AfD (newly founded eurosceptic populist right wing party)

NPD (far-right political party)

Sachsen (German federal state)

CDU (conservative political party)

Brandenburg (German federal state)

Stanislaw Tillich (Ministers-President of Sachsen)

Prozent (percentage)

Polen (Poland)

Linke (democratic socialist political party)

Recht (right or law/justice)

Dietmar Woidke (Ministers-President of Brandenburg)

Figure 4: Most frequent words in press articles about *Grenzkriminalität* (border crime) from Süddeutsche Zeitung in the years 1992–2015. Larger letter types relate to higher frequency

Source: Authors' analysis

Populist parties such as NPD (National Democratic Party of Germany) and AfD (Alternative for Germany), a newly founded populist right wing and Euro-sceptic party) have strategically been using the topic *Grenzkriminalität* in their election campaigns in Sachsen and Brandenburg. Only later, the term became popular through all political parties and even by the German chancellor Merkel.

The medial presence of the term *Grenzkriminalität* neither reveals anything about the extent of the phenomenon (criminal acts at or near the border) nor about the relation between open borders and local crime incidents. Instead it serves as an explanation which ascribes the reasons of those local incidents to the open borders and calls to action can easily be connected with it.

Within the discourses about the border, statistics play an important role. As shown in Figure 5, 6 and 7 it is easy to find statistics that can be used to prove various opinions or personal viewpoints and can therefore be considered as a discursive effect. Even while thinking within the logic of statistics one must be aware of the various ways of data collection and the differences in meanings that the resulting data can support: As an example, the frequently used police statistics on automobile-related crime incidents in border regions often combine the registered cases of car thefts as well as reported car theft *attempts*. On the other hand, the police considers a case as solved when finding a suspect — no matter whether a court finds this person guilty or not. Another statistical weakness can be seen in the police presence: the more active police is in an area the more illegal acts and incidents they will be able to register — if police presence drops, so will the statistics on crime. The link between police statistics and the feeling of security can thus be a complicated one. Therefore, we are not looking for 'true' statistics but need to analyse in which way statistics influence the production of popular ideas.

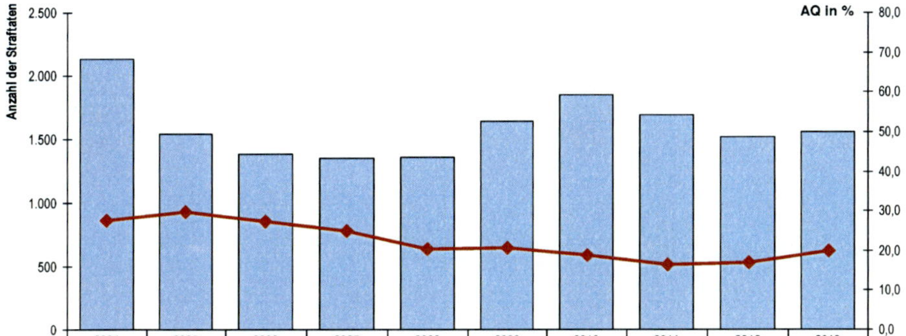

	2004	2005	2006	2007	2008	2009	2010	2011	2012	2013
Erfasste Fälle	2.132	1.541	1.383	1.356	1.361	1.645	1.855	1.696	1.517	1.562
Entwicklung	58	-591	-158	-27	5	284	210	-159	-179	45
HZ	218	155	138	134	133	160	179	163	145	150
AQ in %	27,7	29,9	27,5	24,9	20,5	20,8	19	16,6	17,1	19,9

Häufigkeitszahl: Straftaten je 100.000 Einwohner

Figure 5: Car theft in Brandenburg 2004–2013
Source: Land Brandenburg — Ministerium des Innern Brandenburg 2014

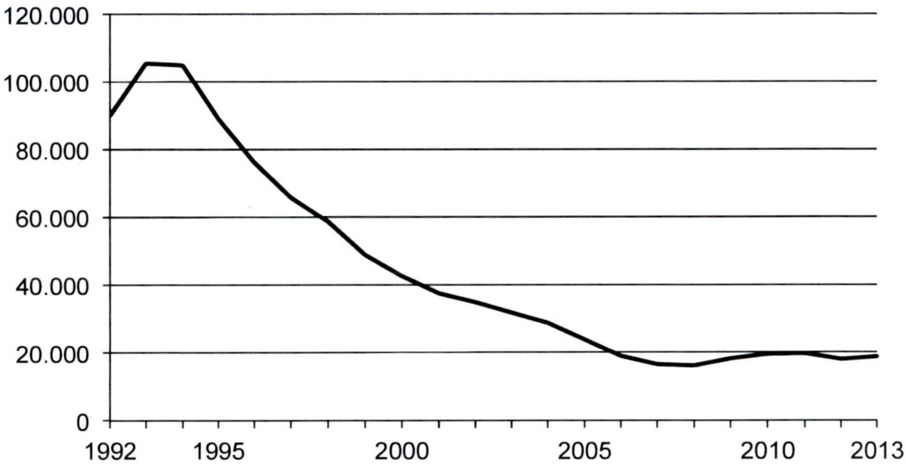

Figure 6: Total amount of stolen vehicles in Germany 1992–2013
Source: Based on data from German Insurance Association (GDV 2014)

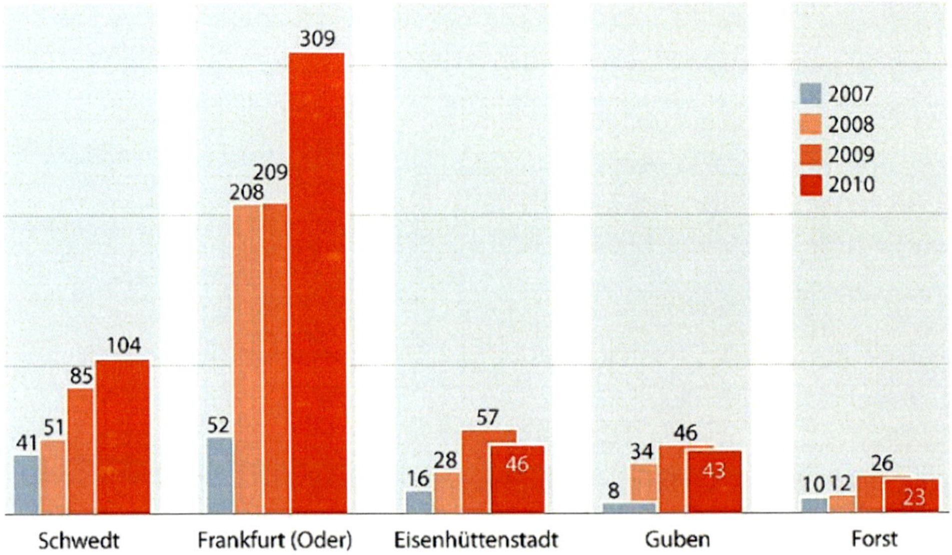

Figure 7: Car theft in particular areas of Brandenburg 2007–2010
Source: Berliner Morgenpost Online 2011

Effects of discourses

Finally, we want to draw attention to the effects on the political and social developments of these peripheral border regions that are associated with these discourses. Firstly, the responsible members of regional governments, e.g., the government of the federal state of Brandenburg, began to see a growing need for action in border regions in recent years. Special units of the police should be increased and roadside inspections for the border regions are planned. Such actions are taking place on several political levels. This re-scaling of security efforts is not an easy task, as different political frameworks cause major contradictions; competences and capacities seem unclear and need to be resolved with a variety of partners.

Secondly, the proclaimed growth in *Grenzkriminalität* has caused a constant rise in private engagement within the establishment of security. This involvement of non-state actors has taken on different forms of practice and organisation in these regions along the border. One mode of practical engagement can be observed in neighbourhood watch groups (*Bürgerwehren*) that control the public space and carry out nightly patrols especially in small towns and villages. These groups are mainly self-administered initiatives, an engagement that not rarely is connected with a disbelief in the abilities and competences of the police in the establishment of security. The groups include a range of actors that vary between rather unpolitical, concerned citizens and representatives of far right political ideas and xenophobic notions.

Another type of private engagement involves a close co-operation with the federal police. In Brandenburg, these public-private groups of securitisation are called security partnerships (*Sicherheitspartnerschaften*). The participants are regular inhabitants of the villages but the police provides basic equipment, a brief legal introduction and even finances part of their activities. The basic idea of these groups is to show presence in the settlements at night time and inform the police in case of irregular situations. Federal police in Brandenburg supports the general idea of neighbourhood watch groups because they notice the lack in their own ability to be present in all the peripheral rural areas. This example shows the difficulties in the establishment of security in these regions on this level and the interference between state and private responsibilities for safety issues.

Figure 8: Security partnership in borderlands: Federal police (right) and neighbourhood watch participant — both carry the label *police* (Polizei) — just that there is small additional *security partner of* (Sicherheitspartner der) on top

Source: Still from documentation about crime in Germany #Beckmann, 27.05.2015, ARD

In yet another form of non-state engagement in security establishment, technologies can be observed to play a major role, one that is yet to be analysed in their effects on security practices: On the one hand internet forums and blogs (re)produce much knowledge that keeps the topos of *Grenzkriminalität* alive and associates xenophobia and racism with it. On the other hand, materialised security techniques are increasingly used in local practices of securitisation. These include for instance the automatic vehicle location of objects with risk of theft such as agricultural machinery, in order to follow its whereabouts and to localise the crime as such. Another technology in widespread use is an artificial DNA called selectaDNA — a property marking system which can be added for example to tractors, cars, cables, bikes or smaller items. Every selectaDNA has its own code and is registered in a database. This enables the control of ownership of circulating items in these regions. Such alarm and security technology is offered by private suppliers, but often promoted in co-operation with local authorities and police. It heightens the risks of marketing these stolen products and thus aims at the prevention of thefts, often supported by announcements on public warning signs. Some villages and companies even post this information at the entrance of their estates or settlements (Figure 9).

Figure 9: Members of a neighbourhood watch group and police present a sign of DNA protection for their village in north-eastern Brandenburg

Source: spz / M. Schmidt 2013, Stadt Prenzlau

All those technologies and forms of action aim to prevent crime, localise a problem or even to hold somebody responsible for it (which can also be seen by the selection of languages on the sign in Figure 10). However, these practices do not reveal anything about the reasons of crimes and the origins of the criminals but merely reduce them to a local problem that calls for spatial answers such as practices that ultimately lead to a reactivation of territorialising border demarcations.

Figure 10: "We protect ourselves" — information in German, Polish and Russian at the entrance of a village near the German-Polish border

Source: Still from documentation "7 Tage... Bürgerwehr", 09.04.2016, rbb

Conclusion

A closer look at the regions along Germanys eastern state borders reveals contradictory aspects of their development: Burdened (and maybe even overburdened) with expectations, these mostly peripheral and rural regions are not only profiting from EU and cross-border integration but are also experiencing some problematic developments that have arisen in the context of border regime transformation. The re-scaling of processes of bordering and, thus, new balances of border securitisations in the EU including a relocation of border functions has produced a new positioning of these regions. They have reached a more central position in terms of their location within an emerging macro-territory but have not always gained the welfare level they might have expected. Against this background, everyday perceptions of the border region might be imbued by a growing scepticism towards the idea of an integrated Europe. In this content, perceptions of insecurity and de-securitisation can be observed. The further integration of the EU has therefore grown to be a controversial topic in these regions, a security topic. For the study of borders, border regions and their development, these phenomena call for our attention and analysis in several aspects:

State-non state involvement in establishment of security — a former state task
In the effect of border regime changes, discourses and practices have arisen in these border regions, which increasingly involve a wider variety of non-border police and non-state actors in the establishment of security. Apparent signs for such fundamental changes in processes of bordering and security are: an increasing involvement of local and regional politicians, neighbourhood watch groups and *Sicherheitspartner* (Security Partner of the Police) and security techniques as described above. These changes lead from an exclusively state-organised security system to a system of personal and private involvement which at times co-operates with the state, at times ignores it and at times is even working against it. Thus, the ensemble as such is becoming more complex and border protection will not primarily be a subject of governmental sovereignty anymore but of controversial negotiating between state and private players.

Changing perceptions, attitudes and discourses towards processes of bordering
The fact that the arising discourses on borders and crime not seldom are accompanied by movements of compartmentalisation and nationalisation shows, that the topic as such involves more and more emotional dimensions of "feeling border". The border is present and not present at the same time, you could almost unknowingly cross it or always focus on it. Local connotation and European perspectives on bordering and security are easy to clash here. Security as such in this area is to a large extent discussed in terms of perceptions and affects. Through the power of discourses, these emotional and affectual aspects give the background to actual practices of securitisation and bordering. Therefore, the integration of discourse analyses within media ranging from newspapers to internet blogs can be crucial for a deeper understanding of processes of bordering as well as general social developments in these regions.

Questioning EU integration and the current border regime of internal EU/Schengen borders from the perspective of the border regions themselves

Our first observations of discourses around the term *Grenzkriminalität* may be seen as a starting point for deeper analysis in order to bring more light into the currently pressing issues around social and political developments in regions that need to negotiate their in many aspects complex position at the fringes of states but at the inside of a so-called "space of freedom, security and justice", of a European integration utopia.

References

Andreas, P. (2003). Redrawing the Line. Borders and Security in the Twenty-first Century. International Security 28 (2), 78–111.

Atger, A. F. (2008). The Abolition of Internal Border Checks in an Enlarged Schengen Area: Freedom of movement or a scattered web of security checks? CEPS Challenge Paper 8, Brussels.

Belina, B. (2014). Sicherheit durch Grenzen überall? Geographische Rundschau 66 (9), 30–36.

Belina, B. & Miggelbrink, J. (2013). Risk as a technology of power. FRONTEX as an example of the de-politicisation of EU migration regimes. In Müller-Mahn, D. (Ed.), The Spatial Dimension of Risk. How Geography Shapes the Emergence of Riskscapes (pp. 124–136). London: Earthscan.

Berliner Morgenpost Online (2011). Wo Autodiebe besonders häufig zuschlagen. www.morgenpost.de/brandenburg/article104849150/Wo-Autodiebe-besonders-haeufig-zuschlagen.html. Accessed 4 May 2015.

Bigo, D. & Guild, E. (2005). Policing at a Distance: Schengen Visa Policies. In Bigo, D. & Guild, E. (Eds.), Controlling Frontiers. Free Movement Into and Within Europe (pp. 233–263). Aldershot: Ashgate.

Brandenburgisches Polizeigesetz — Gesetz über die Aufgaben, Befugnisse, Organisation und Zuständigkeit der Polizei im Land Brandenburg (BbgPolG).

Bundespolizeigesetz (BpolG) vom 19. Oktober 1994 (BGBl. I S. 2978, 2979).

Ceccato, V. & Haining, R. (2004). Crime in Border Regions: The Scandinavian Case of Öresund, 1998–2001. Annals of the Association of American Geographers 94 (4), 807–826.

Côté-Boucher, K. (2008). The Diffuse Border: Intelligence-Sharing, Control and Confinement along Canada's Smart Border. Surveillance & Society 5 (2), 142–165.

Council Decision 1999/435/EC of 20 May 1999 concerning the definition of the Schengen acquis for the purpose of determining, in conformity with the relevant provisions of the Treaty establishing the European Community and the Treaty on European Union, the legal basis for each of the provisions or decisions which constitute the acquis, OJ L 176, 10.07.1999, 1–16.

Council Decision 1999/436/EC of 20 May 1999 determining, in conformity with the relevant provisions of the Treaty establishing the European Community and the Treaty on European Union, the legal basis for each of the provisions or decisions which constitute the Schengen acquis, OJ L 176, 10.07.1999, 17–29.

Fischer-Lescano, A. & Tohidipur, T. (2007). Europäisches Grenzkontroll-regime. Rechtsrahmen der europäsichen Grenzschutzagentur FRONTEX. Zeitschrift für ausländisches öffentliches Recht und Völkerrecht 67, 1219–1276.

Foucault, M. (2004/1978). Geschichte der Gouvernementalität I. Sicherheit, Territorium, Bevölkerung.Vorlesung am Colleges de France 1977/1978. Frankfurt a.M.: Suhrkamp.

Gesamtverband der Deutschen Versicherungswirtschaft (2015). Autodiebstahl 2014 — Alle Zahlen auf einen Blick. www.gdv.de/2015/10/autodiebstahl-2014-alle-zahlen-auf-einen-blick. Accessed 4 May 2015.

Grabbe, H. (2000). The sharp edges of Europe: extending Schengen eastwards. International Affairs 76 (3), 497–514.

Groenendijk, K. (2003). New Borders Behind Old Ones: Post-Schengen Controls Behind the Internal Borders and Inside the Netherlands and Germany. In Groenendijk, K., Guild, E. & Minderhout, P. (Eds.), In Search of Europe's Borders (pp. 131–146). The Hague: Brill.

Heeg, S. & Ossenbrügge, J. (2002). State Formation and Territoriality in the European Union. Geopolitics 7 (3), 75–88.

Hornung, M. (2012). Auf den Spuren unerwünschter Freizügigkeit. Grenzkriminalität an der sächsisch-tschechischen Grenze. In Garsztecki., Waack, C. & Scherm, I. (Eds.), Regionale Identität und transnationale Räume in Ostmitteleuropa (pp. 97–114). Dresden: Thelem.

Huysmans, J. (2000). The European Union and the Securitziation of Migration. Journal of Common Market Studies 38 (5), 751–777.

Jones, R. & Rosière, S. (2012). Teichopolitics: Re-considering Globalisation Through the Role of Walls and Fences. Geopolitics 17 (1), 217–234.

Jureit, U. & Tietze, N. (2015). Postsouveräne Territorialität. Die Europäische Union und ihr Raum. Hamburg: Hamburger Edition — Verlag des Hamburger Instituts für Sozialforschung.

Komlosy, A. (2005). Begrenzte Durchlässigkeit. Personenfreizügigkeit an den Binnen- und Außengrenzen von Habsburgermonarchie und Europäischer Union. In Eurozine. www.eurozine.com/articles/2005-01-11-komlosy-de.html. Accessed 4 May 2015.

Kramsch, O. (2001). Towards Cosmopolitan Governance? Prospects and Possibilities for the Maas-Rhein Euregio. In Bucken-Knapp, G. & Schack, M. (Eds.), Borders Matter: Transboundary Regions in Contemporary Europe (pp. 173–192). Aabenraa: IFG Press.

Kramsch, O. (2011). Along the Borgesian Frontier: Excavating the Neighbourhood of 'Wider Europe'. Geopolitics 16 (1), 193–210.

Krätke, S. & Borst, R. (2004). EU-Osterweiterung als Chance. Perspektiven für Metropolräume und Grenzgebiete am Beispiel Berlin-Brandenburg. Münster: Lit-Verlag.

Land Brandenburg — Ministerium des Inneren (2014). Zur Kriminalitätslage in den Grenzgemeinden des Landes Brandenburg. Polizeiliche Kriminalstatistik 2013. www.mik.brandenburg.de/media_fast/4055/Praesentation_PK_Polizeiliche_Kriminalstatistik_2013.pdf. Accessed 4 May 2015.

Laube, L. (2010). Wohin mit der Grenze? Die räumliche Flexibilisierung von Grenzkontrolle in vergleichender Perspektive. TranState Working Papers 112. Bremen and Oldenburg: SFB 597 Staatlichkeit im Wandel.

Lissandrello, E. (2006). The Utopia of Cross-border Regions: Territorial Transformation and Cross-border Governance on Espace Mont-Blanc. Unpublished PhD thesis. Nijmegen: Department of Human Geography, Radboud University.

Miggelbrink, J. (2013). Vom „Eisernen Vorhang" zum Schengen-Abkommen — Entwicklung der Grenzen und Grenzregime in Europa und in der EU. In Gebhardt, H., Gläser, R. & Lentz, S. (Eds.), Europa — eine Geographie (pp. 208–211). Berlin, Heidelberg: Springer Spektrum.

Müller-Mahn, D., Everts, J., Doevenspeck, M. (2013). Making sense of the spatial dimensions of risk. In Müller-Mahn, D. (Ed.), The Spatial Dimension of Risk. How Geography Shapes the Emergence of Riskscapes. London: Earthscan.

Muller, B. J. (2004). (Dis)Qualified Bodies: Securitisation, Citizenship and 'Identity Management'. Citizenship Studies 8 (3), 279–294.

Müller, K. (2013). Yet another layer of peripheralization. Dealing with the consequences of the Schengen treaty at the edges of the EU territory. In Fischer-Tahir, A. & Naumann, M. (Eds.), Peripheralization: The Making of Spatial Dependencies and Social Injustice (pp. 187–206). Wiesbaden: Springer VS.

Neal, A. W. (2009). Securitisation and Risk at the EU Border. The Origins of FRONTEX. Journal of Common Market Studies 47 (2), 333–356.

rbb — Rundfunk Berlin-Brandenburg (2016). 7 Tage… Bürgerwehr. Documentation first broadcasted 14.02.2016/NDR. www.rbb-online.de/doku/0-99/7-tage-----buergerwehr.html. Accessed 15 April 2016.

Schultz, H. (2004). Zwillingsstädte an europäischen Grenzen. In Becker, J. & Komlosy, A. (Eds.), Grenzen weltweit. Zonen, Linien, Mauern im historischen Vergleich (pp. 61–183). Wien: Promedia.

Sparke, M. (2006). A neoliberal nexus: Economy, security and the biopolitics of citizenship on the border. Political Geography 25 (2), 151–180.

The Schengen acquis as referred to in Article 1(2) of Council Decision 1999/435/EC of 20 May 1999, OJ L 239, 22.09.2000, 1–473.

Treaty of Amsterdam amending the Treaty on European Union, the Treaties establishing the European Communities and certain related acts, OJ C 340, 10.11.1997, 1–144.

Treaty on European Union, signed at Maastricht on 7 February 1992, OJ C 191, 29.7.1992, 1–111.

Trillo-Santamaria, J. M. (2014). Cross-Border Regions: The Gap Between the Elite's Projects and People's Awareness. Reflections from the Galicia-North Portugal Euroregion. Journal of Borderlands Studies 29 (2). 257–273.

Van Apeldoorn, B. (2000). Transnationale Klassen und europäisches Regieren: Der European Round Table of Industrialists. In Bieling, H.-J. & Steinhilber, J. (Eds.), Die Konfiguration Europas (pp. 189–221). Münster: Westfälisches Dampfboot.

Zielcke, A. (2015). Grenzen — Unter uns Fremden. Süddeutsche Zeitung no. 250, 30 October 2015.

3 Exogenous and endougenous impulses to rural revitalisation

The village as a hotel. Tourism-oriented revitalisation of rural settlements: a good practice concept for European peripheries?

Anja Reichert-Schick

Abstract

In view of sustained processes of regression in the peripheral rural regions of Europe there is an ongoing search for proven approaches capable of stabilising and revitalising rural settlements. Particularly pressing is the question of how to deal with long-term vacant buildings in villages and to determine the basis on which these villages can be further developed. This trend has reached different stages of advancement in various European countries.

The paper examines the example of hotel villages that have seen a large proportion of their existing building stock transformed to cater for tourism, and thus analyses the way in which shrinking or indeed completely abandoned villages can be revitalised by tourist development. The focus here is on the Italian concept of the Albergo diffuso, which emphasizes the exploitation of endogenous potentials, an approach that seems especially promising. The analysis is based on three case studies that, on the one hand, present various revitalisation philosophies and, on the other hand, demonstrate the opportunities and risks connected to the "village as a hotel".

Introduction

For numerous areas close to the settlement frontiers of European and North American industrial nations, the twentieth century represents a period of sustained population loss. This is connected to regression processes, which can be accompanied by dereliction, a loss of built fabric, the degradation of the cultural landscape and, in the final stages, processes of abandonment. Settlement regression is a dynamic process concept that refers to the emptying and decline of villages, hamlets and farms in a specific area. It can proceed until settlements are abandoned, but it can also be halted and, indeed, reversed. The following processses of degradation are associated with the regression (Reichert-Schick 2013):

- Physiognomic degradation, which is seen in the abandonment and decay of property;
- Functional degradation, which refers to the increasing depletion of settlement functions culminating in the survival of only the residential function;
- Demographic degradation, which usually involves not only a decline in population figures but also the deformation of population structure,
- Economic degradation, which denotes the weakening of economic performance.

However, in many places the areas affected by these regressive processes simultaneously experience an increase in demands for leisure and recreational opportunities, as peripheral regions often possess a natural potential for recreation, offering possibilities for rambling and winter and water sports in attractive landscapes. In the course of the twentieth century, the socioeconomic transformation of society thus led to a revaluing of the periphery with reference to recreation and tourism. Lichtenberger describes this process as a succession from the agricultural to the recreational society, and Ehlers refers to it a transformation of the cultural landscape (Ehlers 1984b, 58–70, Ehlers 1984a, 172–177, Lichtenberger 1979).

The revaluation of the qualities of the periphery usually stems from the centres and is seldom the result of endogenous initiatives. The benefits also mostly accrue externally, as it is primarily the urban population who profit from the assumption of a new role in the field of recreation and tourism. Areas with suitable landscapes have thus been discovered to cater "for the growing recreational and leisure requirements of urban-industrial society that accompany rising living standards"[1] (Nitz 1982, 166).

Numerous areas that were regarded as being almost completely abandoned have experienced a revival due to recreational and tourist functions. This can be seen in parts of the Alps in particular, which since the mid-twentieth century have gained great significance for winter sports and have thus been provided with a completely new economic base. The various forms of tourism development include commercial tourist centres dominated by large-scale enterprises; developments with small-scale businesses like village guesthouses, small hotels and inns; other guest accommodation including holiday houses and second-homes; and holiday complexes and retirement communities.

The large-scale transformation of peripheral regions for leisure and tourism purposes occurred primarily in the 1970s and 1980s, although it continues to a lesser extent today. However, Land believes that overall a certain level of saturation in terms of tourism revitalisation strategies has been reached, and that this path cannot be a successful model for all. Indeed, he suggests that there is a danger that further expansion, accelerated by public funding, could lead to earlier, previously successful projects being withdrawn from the market (Land 2005, 8).

---

1 Original quote in German: „für das mit dem wachsenden Lebensstandard ansteigende Erholungs- und Freizeitbedürfnis der urban-industriellen Gesellschaft".

Nonetheless, even today there are villages and regions that seem predestined to find a sustainable function in the field of leisure and tourism. Here, the new "desire for the more or less ideal world of authentic living in harmony with nature [...], which can be best imagined 'in the countryside'"[2] (BBSR 2010, 1) appears advantageous. The wishes awakened by this desire encourage not only retired persons, drop-outs and the wealthy fleeing from the city to purchase a second-home in a rural area, but may also further promote rural tourism (Hahne 2011, 17).

This paper uses the example of hotel villages that have been transformed by tourism to analyse the way in which rural settlements in peripheral locations can be revitalised by tourist use. Particular attention is paid to the Italian concept of the *Albergo diffuso*, which is described and analysed using selected examples. The investigation is methodologically based on an analysis of secondary data and documents. The analysis of the German case study Rattey was further supported by numerous site visits and an expert interview with the owner of the country-house hotel.

Hotel villages and the Italian *Albergo Diffuso* concept

A comparatively recent tourism approach that addresses the desire of people for an authentic way of life in the countryside involves initiatives that utilise large areas or indeed the entirety of rural settlements for tourism, transforming them into hotel villages. Villages and the centres of smaller towns can be revitalised by the implementation of this new visitor-oriented function and thus saved from dereliction.

The pioneer of this idea is the Italian *Albergo diffuso*, a term which is best translated as "scattered hotel". It does not refer to a conventional hotel concentrated in a single built complex, but rather to an entire village in which individual houses come together to function as a hotel. In contrast to the usual vertical arrangement of hotels, an *Albergo diffuso* is based on a horizontal concept.

This kind of accommodation developed in the 1980s. The philosophy is "to bring the benefits, but not the blights of tourism to small villages" (Nixey 2011). The idea was developed by the current president of the national *Associazione Alberghi diffusi* (ADI), Giancarlo Dall'Ara, when the Carnic mountain villages in Friuli Venezia Giulia fell into decline due to high levels of out-

---

2 „Sehnsucht nach der mehr oder weniger heilen Welt ursprünglichen Lebens im Einklang mit der Natur [...], das man sich am ehesten ‚auf dem Lande' vorstellen kann".

migration. The concept was first put into practice in Comeglians (Friuli Venezia Giulia) and in Bosa (Sardinia). The *Albergo diffuso* concept is intended to provide exclusive and simultaneously distinctive and authentic guest accommodation, bring visitors into contact with the inhabitants still resident in the village, and to offer the village new development prospects (Dall'Ara 2007; Baltes 2011). It is supposed to be explicitly suitable for settlements off the beaten tourist track and therefore seems predestined for peripheral regions. Despite its dispersed character, the *Albergo diffuso* is meant to provide all the services of a traditional hotel and so to cater for everything that the guests require, such as reception, various services and gastronomy. The rooms are usually in authentically restored properties in the historical centre and are supposed to be situated within about 200 metres of the central hotel building. The guests should feel like locals, and any remaining inhabitants should also have the feeling that the accommodation is a real part of the village (Schmitt 2015). The inhabitants should participate directly in the tourism activities, produce goods required for the business and acquaint guests with their way of life. Furthermore, an *Albergo diffuso* is intended to pick up on the cultural and historical characteristics of the region, and to maintain high-quality standards similar to those of hotels with at least three stars. In terms of organization, the *Albergo diffuso* is a coherent unit with a central administration (ADI 2015).

In Italy in the last few decades a number of investors have bought parts of settlements or even entire villages and transformed their building stock into hotels. There are now 81 *Alberghi diffusi* throughout the whole country, and in 2006 they came together to form a national association, *Associazione Alberghi diffusi* (ADI 2015; Dall'Ara 2010). As can be seen in Figure 1, the *Alberghi diffusi* are now located in all the regions of Italy. A spatial focal point is found in the areas between Tuscany in the north and Molise in the south, regions that are part of the natural landscapes of the central Apennines and the Tyrrhenian foreland. Here there are 32 *Alberghi diffusi*, about 40% of all those that belong to the association.

Figure 1: Distribution of the Alberghi diffusi in Italy by region

Source: Author's representation of data from Associazione Nazionale Alberghi Diffusi 2015

Apart from the Italian *Alberghi diffusi* there are also numerous initiatives in other European countries that do not entirely correspond to this ideal type, differing for instance in terms of administration, the efforts made to preserve authenticity, and the extent and goals of revitalisation.

Among the other initiatives that are related to the *Albergo diffuso* concept is the French Crillon-le-Brave in the département of Vaucluse. Here a Canadian investor bought much of the vacant property in 1988 and established a small hotel in country-house style. The number of rooms available has grown from an initial 11 to 28, and the country inn has evolved into a five-star hotel that encompasses seven historical buildings in the centre of the village (Figure 2) (Polz 2014; Crillon le Brave 2015). The journal Wirtschaftskurier judges the hotel as "a Best of Provence", as it offers everything that guests expect from a holiday in Provence (Anonymous 2014).

Another example is found in Kapsaliana, a village in the municipality of Amnatos on the Greek island of Crete. In the mid-twentieth century the village had 50 inhabitants, having originally developed around an oil press that belonged to the nearby Arkadi Monastery. The oil press was abandoned in 1955 and all but two residents left the village. A local architect bought the almost derelict settlement and transformed it into a hotel. Figure 3 shows the built fabric after the restoration, which took over 20 years. In 2008 the Kapsaliana Village Hotel finally opened with 21 residential units, 17 of them in the old stone houses (Schüle 2014).

Figure 2: View of Crillon-le-Brave with the hotel complex in the centre of the village
Source: Wilson 2015

Figure 3: Alley in Kapsaliana after the restoration
Source: Kapsaliana Village Hotel 2015

The initiatives that aim to transform derelict or partly abandoned settlements into hotel complexes are numerous and widespread, particularly in Italy but also in other European countries. However, it is not always possible to put the basic concept successfully into practice or to initiate a process that can be made economically sustainable. An example of an unsuccessful attempt is the hamlet Courbefy, part of the Bussière-Galant municipality in the French region of Limousine, which consists of 21 buildings and was completely abandoned in the 1970s. In 2003 a married couple bought the settlement with the intention of setting up a restaurant and a hotel. After only four years, however, the owners were insolvent and had to relinquish the project (Bouchez 2012). By 2008 the buildings were once again abandoned. Four years later the hamlet was put up for auction. A South-Korean photographer bought Courfeby for 520 000 euros, intending to realise a concept that embraced environment, art and culture (Mallevoüe 2012). To date the plans have been neither substantiated nor implemented.

Presentation and explanation of selected case studies

This section presents a description and comparative analysis of three different case studies in order to, on the one hand, illustrate various revitalisation philosophies and, on the other hand, demonstrate the opportunities and risks connected to the "village as a hotel".

The first example, Santo Stefano di Sessanio, lies in the Italy region of Abruzzo and is an example of a classical *Albergo diffuso*. The second case study is the "Toscana Resort Castelfalfi", which has been run by a subsidiary of TUI since 2007 and has been the object of considerable financial expenditure. The third example is the village of Rattey in West Pomerania (Vorpommern), Germany, which is developing around the country-house hotel "Schloss Rattey" to increasingly resemble an *Albergo diffuso*.

Santo Stefano di Sessanio — an Albergo diffuso

Santo Stefano di Sessanio lies roughly 30 kilometres east of Aquila. Like many villages in Abruzzo this settlement was affected by the rural exodus of the twentieth century. Population figures reached a maximum in 1901 with 1489 inhabitants. In 1951 there were 791 people still resident in the village, but by 2015 the population had sunk to 117 (Istituto Nazionale di Statistica 2015). This considerable decline in population meant that many of the houses in the centre of the village were vacant.

In the year 2000 Daniele Kihlgren from Milan purchased a third of the empty houses in the centre of the settlement and initiated the project "Hotel Sextantio". He states that he has to date invested five million euros in the village (Helbert 2005; Preiss 2008; Classe tourist 2014). Today, the hotel consists of 32 rooms that are scattered throughout the village in various properties, concentrated particularly in the area of the old watchtower that dominates the settlement (Figure 4 and 5).

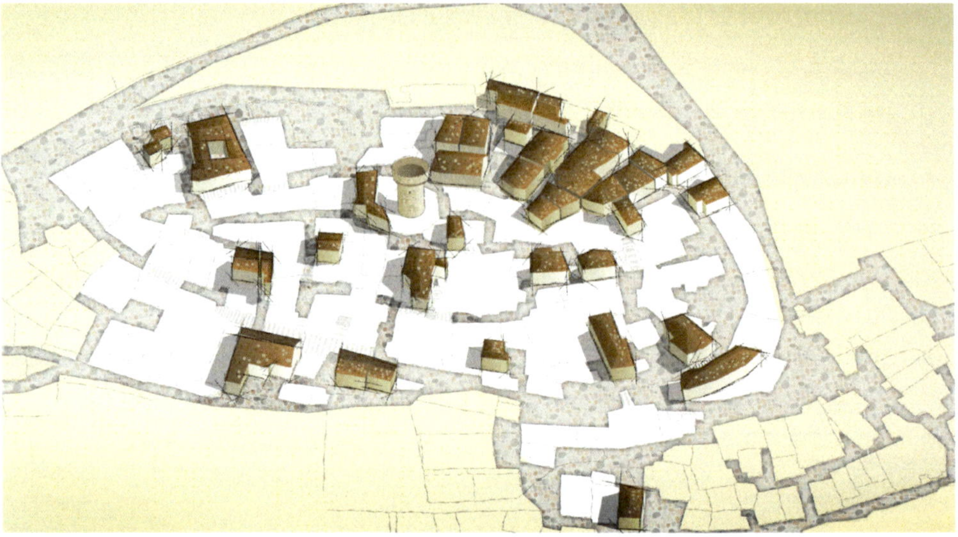

Figure 4: The village centre of Santo Stefano di Sessanio with the property that belongs to the Albergo diffuso

Source: Associazione Sextantio 2015

Figure 5: Historical village centre of Santo Stefano di Sessanio

Source: Comune di Santo Stefano di Sessanio 2015

The basic idea of the project is to preserve the archaic character and evolved structures of the settlement and thus to create a hotel that keeps history alive. The rooms have been restored so that traces of the modest way of life they housed can still be seen (Figure 6). The original floor plan of the individual buildings has been preserved and the furniture and household accessories are original items sourced from the local area. The majority of the modernizations undertaken are of a technical nature and include underfloor heating in the rooms, modern sanitary facilities and WLAN connections. Preiss (2008) describes the result of the restoration as "a kind of beautiful-living minimalism, which attracts tourists from around the world"[3].

However, the user comments on the internet booking platform booking.com (viewed February 2016) do not confirm the notion that guests come from all over the world. Currently about two-thirds of the visitors are actually from Italy, while other guests to Santo Stefano di Sessanio who are active on booking.com originate from Great Britain (15% of the 215 user comments), the Netherlands and Germany. The origin of the users of the booking.com platform may not be representative, but it is unlikely that the actual guest structure is significantly more diverse.

With the support of the local council, the investor has managed to obtain a ban on all new building, so that the original built fabric can be conserved as it is. The properties therefore still appear very simple from the outside (Figure 7) and have only been renovated in the interior. Furthermore, the notion of strengthening the regional value chain is applied to the business. The restoration work is undertaken by local tradespeople and all the textiles are produced by a local weaver. The cultural programme is provided by musicians from the village and the hotel is staffed only by residents of Santo Stefano di Sessanio or the surrounding area.

---

3 „eine Art Schöner Wohnen-Minimalismus, der Touristen aus aller Welt anzieht".

Figure 6: Authentically restored hotel room in Santo Stefano di Sessanio
Source: Associazione Sextantio 2015

Figure 7: Village centre of Santo Stefano di Sessanio
Source: Associazione Sextantio 2015

In addition to the actual accommodation, six shops have been established producing and selling local products. The investor is concerned to preserve traditional handicrafts and the knowledge associated with them. Leisure pursuits on offer therefore include courses on cooking and baking with regional products, weaving and making soap. Two restaurants and a teashop offer meals traditional to Abruzzo.

The hotel can be seen to have jumpstarted activities in the village, as other inhabitants have also opened their own bed-and-breakfast establishments or started selling local specialities in their own shops (Preiss 2008). In this way the remaining local inhabitants also profit from the concept and the entire village is being given a new economic base.

The investor and his two associates have now bought another seven villages or parts of villages with the aim of transforming them according to the "Sextantio principle" (Associazione Sextantio 2015). They too are to be renovated in a fashion that preserves their authenticity to the greatest extent possible, providing guests with the opportunity to experience something of the traditional way of life of Abruzzo while enjoying a comfortable stay.

On the website, the Milan investor records the following appreciation of his own project: "Italy has in its mountains about 2 000 semi-abandoned historic towns and 15 000 where the abandonment was nearly total. In a global market with a trend depriving places of their own identity and absorbing everything that is not clearly protected, [we] should look for safeguard policies in those places of 'marginality' — like in the Appennini Mountains in the South of Italy where abandonment, poverty, and emigration without return, allowed the preservation of past civilisations and the cultures survived until the very recent past" (Associazione Sextantio 2015).

The implementation of the *Albergho diffuso* concept in Santo Stefano di Sessanio has succeeded in giving a sustainable economic base to a village where the population had been depleted by 92% in the previous century. A third of the old village centre could be saved from decay through integration in the *Albergho diffuso*. The new function given to the properties removes them from the sphere of use of the local inhabitants and serves to present a staging of the rural life of bygone times. However, the remaining residents of the village have taken advantage of the situation and have exploited the synergy effects for their own projects. As a result Santo Stefano di Sessanio can be viewed as a model example of a comprehensive and diverse revitalisation based on tourism.

The Toscana Resort Castelfalfi — TUI revitalises a deserted village

In the second case-study village of Castelfalfi, situated in the Tuscan municipality of Montaione 55 kilometres from Pisa, a quite different approach is pursued. The settlement originates from the eleventh century and had about 600 residents in 1950, but by 2007 only two people remained in the village (Tasche 2011). In 2007 TUI bought the almost deserted settlement along with the 12 km^2 of land belonging to it, and in the spring of 2013 opened the "Toscana Resort Castelfalfi" (Ehlers, Tuma 2011a). The developer and building contractor of the resort is a subsidiary of the TUI, the *Tenuta di Castelfalfi S.p.A.*

Figure 8: Layout of the Toscana Resort Castelfalfi
Source: Toscana Resort Castelfalfi 2012

Castelfalfi differs significantly in two ways from Santo Stefano di Sessanio. Firstly, the settlement was completely abandoned when the redevelopment was initiated and, secondly, the project was not undertaken by a private investor but by an internationally active tourism company. The resort is intended to provide fulltime work for 250 people, which would make it a significant employer (Ehlers and Tuma 2011a). The image brochure describes the vision thus: "a sleeping region [...] is being reawakened to new life. Agriculture, handicrafts, the old culture — everything is preserved or, if possible, reintroduced. [...] Soon it all will be there again: the bars, restaurants, boutiques, squares and quiet alleys — just as though it had never gone away"[4] (Toscana Resort Castelfalfi 2012).

The concept of the resort follows two strategies. Holidaymakers are to be received here and thus the necessary catering and hospitality facilities are to be made available. In addition, a large proportion of the renovated and newly built property is to be sold, thus gaining new residents and second-home owners for the village. The development plan includes the following components (Tasche 2011; Tenuta di Castelfalfi 2015; Engel&Völkers Resorts 2015).

For holidaymakers:

- The old tobacco factory located immediately next to the village centre has been redeveloped to house the boutique-hotel "La Tabaccaia" with 31 rooms and suites and an adjoining conference centre. The hotel was opened in 2013.
- The plan also envisions the construction of two other hotels: a four-star spa hotel and the five-star hotel "Di Arte e Cultura". The spa hotel is currently being constructed and is intended to open for the 2016 summer season.

For those interested in property:

- Property that is intended for sale is located in the old centre of the village *"Borgo"* (Figure 8). A total of 41 apartments ranging from 38 m² to 140 m² in size have been constructed within the historical building stock. According to TUI, 90% of these units have already been sold (Tenuta di Castelfalfi 2015). Property prices start at 4 500 euros per square metre. Currently (January 2016), two apartments are available

4 „Eine schlafende Region [...] wird wieder zum Leben erweckt. Landwirtschaft, Handwerk, die alte Kultur — alles bleibt erhalten oder wird, wenn möglich, wieder angesiedelt. [...] Bald schon wird wieder alles da sein: die Bars, Restaurants, Boutiquen, Plätze und stillen Gassen — so als wäre es nie weg gewesen."

through the company Engel&Völkers, priced at 680 000 and 720 000 euros.

- Furthermore, 18 derelict farmhouses scattered around the grounds, the *casali*, are being converted into luxurious country houses (Figure 9, Figure 10). They are sold in their dilapidated state and then renovated within two years according to the individual wishes of the purchaser. The provisions state that the exterior of the houses must be preserved while the interiors can be gutted. Only three of the houses have been completed thus far, while work on another is underway. The house "I Bianchi" with 410 m² of living space and a plot of 2 500 m² is currently available for purchase at a price of 3.2 million euros. Two of the houses have been renovated but not yet sold, these are being let to tourists to generate income.
- Last but not least, the plan envisions the construction of three new so-called villages: San Pietro, Starnino and La Collina. The latter is the only one to have been completed thus far und consists of two-storey dwelling units containing apartments for sale. Seven detached "golf villas" with 240 m² of living space and plots of 2500 m² have also been constructed and are available at a purchase price of 1.7 million euros.

Leisure infrastructure for holidaymakers and property owners:

- In 2010 an 18- and a 9-hole golf course were laid out below the *Borgo* (Figure 8 in the background). The award-winning course attracts a well-off clientele, the group that TUI is targeting with the Castelfalfi project.
- Two restaurants offer Tuscan meals.
- Furthermore, a wellness and beauty area has been established in the immediate vicinity of the apartment houses of the *Borgo*. Named "Le Piscine Café", it includes three swimming pools and a playground.
- The ground floors of the apartment houses in the centre of the village have been converted into retail premises, which predominantly sell goods from Tuscany. The shops are seen as an important component of the revitalisation of the centre of Castelfalfi.

Completion of the entire construction project is scheduled for 2018 (Weiss 2014a).

Figure 9: Country house "I Bianchi"
Source: Engel&Völkers Resorts 2015

Figure 10: Unrestored *casali* in Castelfalfi
Source: Tenuta di Castelfalfi 2015

It can be seen that both the scale of the project and the organizational philosophy of the Toscana Resort Castelfalfi differ fundamentally from the hotel project in Santo Stefano di Sessanio. The TUI subsidiary endeavours to preserve all the exterior walls of the old building stock but the properties are gutted inside. A range of new buildings has also been constructed, substantially altering the original villagescape and its surroundings. Agriculture and the production of wine and olives are intended to continue, but the cultural landscape is being significantly refashioned by the golf course and the three new villages with their apartments and villas, despite the fact that the new buildings are designed in keeping with the local style.

Although the resort clearly revitalises an abandoned village, the original population no longer profits from the measures undertaken. The ADI president therefore does not judge Castelfalfi to be an *Albergo diffuso*, because these developments "need inhabitants, everyday life and an exchange between residents and guests"[5] (Dall'Ara quoted in Baltes 2011). If these requirements are not fulfilled then it is "simply just a tourist village"[6] (Dall'Ara quoted in Baltes 2011). Indeed, it seems reasonable to describe Castelfalfi as an artificial Tuscan holiday world. The concept was evaluated in the magazine "Der Spiegel": "These days [...] Castelfalfi really looks like Germans picture Tuscany, just more luxurious"[7]. Castelfalfi has been described as "a Tuscan Disneyland staged by German managers"[8] (Deckstein 2013), and as a "Playground for the Wealthy" (Ehlers and Tuma 2011b). TUI themselves advertise the complex on the website with the slogan: "Tuscany is a wishful dream and in Castelfalfi it becomes true"[9] (Tenuta di Castelfalfi 2015). The expectations that guests have of Tuscany are thus skilfully picked up upon in Castelfalfi, reproduced in a pure form and stereotypically idealised.

5 „brauche Bewohner, Alltagsleben und den Austausch zwischen Einwohnern und Gästen".

6 „einfach nur ein Touristendorf".

7 „Mittlerweile [...] sieht Castelfalfi tatsächlich so aus, wie Deutsche sich die Toskana erträumen, nur luxuriöser".

8 „von deutschen Managern inszeniertes Toskana-Disneyland".

9 „Die Toskana ist ein Wunschtraum und in Castelfalfi wird er wahr".

Admittedly, TUI is not as successful with its development as originally hoped (Weiss 2014b), as the "open-handed millionaire customer base"[10] (Deckstein 2013) is hesitant to accept the concept. The less wealthy are discouraged by the high prices and exclusivity. Italians have bought about half the apartments in the centre of the village, while the rest are more or less evenly divided between British, North American and Belgian purchasers (Keeve 2013). Inspection of the user comments on the internet booking platform booking.com revealed that a third of visitors to the centrally located hotel La Tabaccaia come from Germany (33% of the user comments were made by Germans; viewed February 2016). Other guests travel from Great Britain (25%), Italy (18%) and the Netherlands (14%). Of course, it remains unclear whether the origin of the users of the booking.com platform corresponds to that of the hotel guests overall.

It is proving particularly difficult to sell the large country houses; so far only two purchasers have been found. The targeted volume of job creation has been correspondingly reduced to 100. Currently only 50 people are employed on permanent contracts. TUI has to date invested 180 million euros in the project, in addition to the purchase price of 250 million euros, and is therefore unlikely to give the project up lightly. Nonetheless, the mid-term aim appears to involve selling the Toskana Resort Castelfalfi to another investor. Overall the track record of this large-scale revitalisation concept is sobering. Even Peter Joussen, a chief executive of TUI, expressed the belief that Castelfalfi "will never make money in its current situation" (quoted in Weiss 2014b).

Rattey: a village in West Pomerania is "reinvented"

There is no known case in Germany that can be compared with the hotel villages in Italy, but there is a rural settlement in West Pomerania that is developing in a similar way. This is the village Rattey, which belongs to the municipality of Schönbeck and is situated 25 kilometres east of Neubrandenburg.

In the centre of the village stands a classical-style country house, built between 1802 and 1806 (Figure 11, Figure 12). It was once the seat of the family von Oertzen, but was sold to the municipality in 1945. In the GDR era the building housed a kitchen for the agricultural cooperative, a cooperative shop, a post office and a kindergarten. In 1990 a lawyer from Bremen named Förster, an expert in insolvency and employment law, bought the country house. At this time only the walls of the original building were still standing. Between 1996 and 1998 Förster organized the restoration of the house and its 20-hectare park in line with

10 „spendable Millionärskundschaft".

the historical records; and in the summer of 1998 opened the "Parkhotel Schloss Rattey" (Förster 2011). Information on booking.com suggests that 80% of the hotel guests come from Germany; other countries of origin include Great Britain, France and the Netherlands (viewed February 2016). However, there were only 22 user comments for Schloss Rattey on booking.com, a basis that does not allow particularly significant conclusions to be drawn.

Figure 11: Exterior view of the country-house hotel
Source: Park Hotel Schloss Rattey 2015

In addition to the country house the complex includes the old stables, the ground floor of which is today used as an event location. This rustic hall is particularly popular with wedding parties for use after the marriage ceremonies that are conducted in the registry office found in the main house. An art gallery displaying pictures from the owner's private collection is located on the upper floor of the stables and is available to view on request.

Winegrowing is a distinctive feature of Rattey, revived by the owner of the country house after he came across drawings of vines in the course of his historical research (Förster 2011). The area benefits from a continental climate with high sunshine duration, so that the region is particularly suited for cultivating the grape varieties Regent, Müller-Thurgau, Phoenix, Ortega and Huxelrebe. When winegrowing began in 1999, a local association of private winegrowers (*Verein der Privatwinzer zu Rattey e.V.*) was founded, which today has 520 members. The vineyards cover an area of about four hectares with space for 15 000 vines. The wine is distributed under the name of "Mecklenburger Landwein", and in 2004 the locations Rattey and Burg Stargard were officially included in the list of German winegrowing areas under the name "Stargarder Land" (Verein der Privatwinzer zu Rattey e.V. 2015). In 2010 the sales outlet "Weinkanne", located in the grounds of the country house, achieved a turnover of about 60 000 euros (Förster 2011). The vineyards have also been used as locations for two of the most prominent garden shows in Germany, the IGA Rostock in 2003 and the BUGA Schwerin in 2009.

In addition to the country house, Förster now also owns 17 other properties in Rattey (see the buildings highlighted in Figure 12). He purchases any house in the village as soon as it becomes vacant (Förster 2011). Several of the houses have been converted into holiday homes that are rented out at upmarket prices. Rattey is thus growing to resemble an *Albergo diffuso*, even if this was not the initial plan but rather developed over time, slowly emerging as an overriding development concept. This trend is what distinguishes this management approach from the many other country houses in Mecklenburg-West Pomerania that are today run as high-priced hotels.

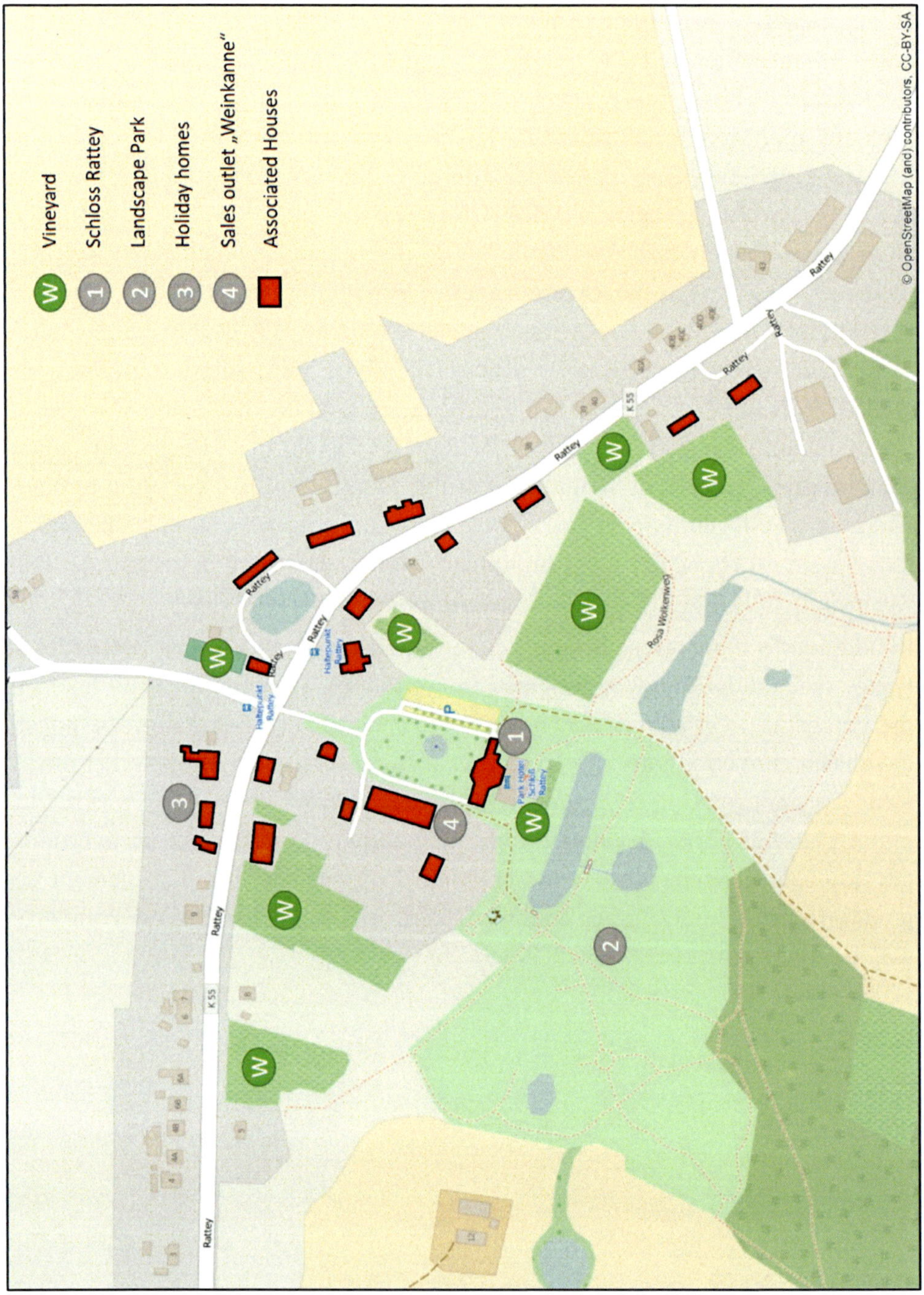

Figure 12: Plan of Schloss Rattey with the landscape park, the vineyards (dark green) and the associated houses (highlighted in red)

Source: Park Hotel Schloss Rattey 2015

The commitment of the investor is felt throughout the village, benefitting the village centre with its late roman church and old smithy in particular. Rattey has 120 residents, many of whom are involved in the winegrowing association. The work and celebrations connected to the vinery invigorate the village community, and indeed the neighbouring settlements. In a structurally weak region characterized by out-migration, the investor has been successful in safeguarding a rural settlement from dereliction, making use of vacant properties, attracting visitors to the rural area of West Pomerania and, last but by no means least, mobilising the village community.

Comparison of the tourism-oriented revitalisation concepts

A comparison of the three case studies reveals the features that Table 1 displays.

The *precondition* for all three projects was the large number of vacant houses that could be used for a hotel or resort project. In Rattey only the country house was initially unoccupied, with other property falling vacant over time, but this was sufficient to jumpstart the revitalisation process. The most important difference between the two Italian initiatives on the one hand and the German project on the other hand, is that the initiative in Rattey was not initially planned at its present scale but developed successively. The most favourable preconditions from the point of view of an investor were found in Castelfalfi, as here the entire village and the properties scattered nearby had been abandoned and were therefore available in their entirety.

The *actors* in all cases are external investors, although individual private investors were involved in Santo Stefano di Sessanio and Rattey while in Castelfalfi investment came from a significant tourism company, TUI. This emphasizes the point made above that the revaluing of the periphery and the introduction of new functions is usually the result of exogenous initiatives.

The *population* of Santo Stefano di Sessanio could be stabilised by the hotel project. In contrast, in Castelfalfi it was the sale of property that first attracted residents back to the settlement, and most purchasers will presumably only use their houses as second homes. The activities of the investor had no direct impact on the number of residents in Rattey. Nonetheless, his commitment has influenced the entire village and provided an attractive image. Of particular significance here is the winegrowers association connected to the hotel, as it has improved community life in the village and also increased contacts to the neighbouring villages, which has created a positive living atmosphere.

The *target group* in Castelfalfi is a wealthy and exclusive clientele, while in Santo Stefano di Sessanio and Rattey discerning guests are targeted, people who prefer high standards and can afford to pay for them. In Castelfalfi a *mixture* of sale and rental is available, although property prices range up to four million euros. In Santo Stefano di Sessanio and Rattey, on the other hand, the rooms and houses are only available for rental so that the focus is on classical holidaymakers and weekenders.

The original *villagescape and cultural landscape* has been particularly successfully preserved in Santo Stefano di Sessanio, which now appears in new splendour. In Rattey the traditional cultural landscape has also profited from the project, indeed the winegrowing landscape was only rediscovered and reconstructed in the course of the restoration. In contrast, the many new buildings and the golf course in Castelfalfi have resulted in a conversion that is principally oriented towards the creation of an artificial holiday world and that attempts to meet the Tuscan wishes and expectations of the visitors. Nevertheless, traditional winegrowing and olive cultivation has been preserved and continues to be pursued.

The historical *built fabric of the dwellings* has been conserved in all three cases. There are, however, differences in the detail. In Castelfalfi only the exterior walls have been preserved and the interiors of the buildings gutted. In contrast, in Santo Stefano di Sessanio upgrading was always based on historical principles, and in Rattey the built fabric of the old houses has also been preserved by the activities of the investor, who always attempts to achieve an authentic restoration and thus contributes to the preservation of the villagescape.

The *economic strength* of all three villages has been improved. In Castelfalfi 100 jobs should be created by 2016. In view of the fact that Castelfalfi was almost derelict, this must be seen as a significant economic recovery. There is no information available about the number of jobs in Santo Stefano di Sessanio. However, the investor only employs people from the local area and has initiated the opening of further tourist-related businesses and shops in the village, leading to the creation of permanent employment above and beyond that provided by the actual hotel. Rattey has also profited from the hotel, the employment it provides and the winegrowing business associated with it.

Table 1: Comparative overview of the three case studies

	Santo Stefano di Sessanio	*Castelfalfi*	*Rattey*
Precondition	Large number of vacant houses that may be used for the hotel/resort project		Vacant country house / estate
Approach	Planned	Planned	Unplanned
Actor	External investor, private individual	External investor, Tourism company	External investor, private individual
Local population	Stabilized	New second-home residents	Strengthened village community
Target group	Wealthy purchasers, discerning guests	Discerning guests	Average to discerning guests
Sale / rental	Sale & rental	Rental	Rental
Villagescape / cultural land-scape	Conservation	Transformation and touristically appealing "conversion"	Reconstruction of the winegrowing landscape
Built fabric of dwellings	Authentic restoration	Preservation of the exterior walls	Authentic restoration
Economic strength	Strengthened by secondary effects and employment	100 jobs in the resort	Strengthened by secondary effects and employment

Turning to the general preconditions necessary for hotel villages, these include:

- Investors who are willing to take risks and who have the vision to recognize the potentials of rural peripheries,
- A preferably wealthy urban population that develops a relevant demand,
- Tourism potential, although the case studies from Abruzzo and West Pomerania show that regions distant from tourist hotspots may also be suitable,
- Available sites for new tourism uses, especially a large number of vacant houses that can be bought as a complex,
- Rural settlements with an appealing villagescape,

Open-minded local residents who support the conversion and restructuring of the village and participate in it accordingly.

All the case studies described have provided important development impulses for peripheral rural settlements. If the advantages and disadvantages of the three concepts connected to the case studies are weighed up, it appears that the *Albergo diffuso* concept is well-suited as a revitalisation strategy for peripheral rural settlements. It represents more than a tourism project. It initiates innovation and creates new opportunities for the local population, although more extensive research is required to assess medium- and long-term effects in particular. Williams sums up the situation from a journalistic perspective: "*Alberghi diffusi* don't just provide travelers a door into a traditional way of life, but are also healthy for the host villages" (Williams 2010). Dall'Ara, the ADI president, elaborates: "the projects also act as a 'driver of development' in their villages, because the managers are encouraged to source all the products used in the village from local producers" (quoted from Williams 2010). The consistent focus on the development of local value chains and the use of endogenous potential represents a significant gain for peripheral rural settlements and provides development options for the future.

The concept of the *Albergo diffuso* has obvious commonalities with community-based tourism. This participative form of tourism has developed since the 1990s and has been implemented in the context of development cooperation, particularly in the global South (Häusler and Strasdas 2003; Blackstock 2005; Lucchetti and Font 2013). Here too the goal is to implement community-oriented projects that use existing resources and strengthen economic, social and cultural structures. Visitor-oriented offerings are developed in small communities distant from mass tourism, working closely together with the residents. An important aim and structuring principle in community-based tourism is participation, so that development prospects and income opportunities are created for the local population (BMZ 2006). The many years of experience and the numerous analyses of community-based tourism projects could provide valuable impulses for the further development of the *Alberghi diffusi* movement. Building upon this as a basis, the risks and mistakes of strategic marketing could be assessed, the participation and qualification of actors and project management assured, and monitoring and sustainability operationalised. In this way the *Alberghi diffusi* concept could be improved and implemented as a convincing approach for intervention in rural peripheries.

Conclusion

The example of the hotel villages has shown how shrinking or even abandoned villages may be revitalised by tourism, particularly in Italy but also, for example, in Germany, France or Greece. The results that can be achieved are diverse and range from the preservation of the cultural landscape and the built fabric of dwellings to an improvement in economic strength to positive effects for the remaining residents. They are also associated with improvements in image and the invigoration of the settlement through tourism.

Despite the many positive effects, however, such a staging of the "rural" can also be critically viewed. Whatever efforts in pursuit of authenticity are made, local traditions are being (re)created in the hotel villages for the purposes of tourism. The case study of the Toscana Resort Castelfalfi shows particularly clearly that a focus on regionality and history can derail and take on the character of an artificial holiday world. In a similar vein, the *Alberghi diffusi* concept aspires not just to integrate remaining inhabitants into the tourism concept as service providers but also as communication partners for guests and as "real" country residents, an aim that certainly seems worthy of discussion. However, in the light of continued processes of regression in peripheral rural regions there is an ongoing search for sustainable concepts that can stabilise and revitalise rural settlements. Hotel villages, despite their acknowledged weaknesses, appear in this context to offer good solutions for dealing with vacant property and developing a new economic basis. The Association of the *Alberghi diffusi* judges their success as follows: "This form of hospitality has showed to be very efficient for the valorisation of towns and villages that are artistically or architecturally peculiar and interesting; through ADs [*Alberghi diffusi*] we can valorise old and dismissed buildings and, at the same time, we can avoid to solve the problems of hospitality only through the construction of new structures" (ADI 2015).

There are very varied ways in which villages can be revitalised as hotels, as the case studies demonstrate. But when visitors are offered a stay in a village where the historical backdrop and traditionally rural atmosphere satisfies a desire for a slower pace of life, peace and (apparent) authenticity, it is possible to describe the *Albergo diffuso* concept as a successful basic model. Particularly convincing is the fact that the majority of the existing *Alberghi diffusi* do not lie in classical tourism regions but are found rather in peripheral areas distant from large centres. Using this concept, the processes of degradation that go hand in hand with the regression of settlements can be halted and, indeed, reversed. The physiognomy and functional diversity of the villages profit, as does the population, the economic base of the settlements and the local value chains.

The potential that can be realised in the rural peripheries of Europe seems not inconsiderable. Due to processes of out-migration from structurally weak rural areas there are significant quantities of vacant property in many places. Although complete abandonments are seldom, sustainable options for the subsequent use of these vacant buildings should be developed. With the *Albergo diffuso* concept, Italy has shown that there is a demand for high-quality offerings. It is, nonetheless, necessary to ascertain whether the finance concepts are viable in the medium term, and to shine a critical spotlight on the economic participation of the local and regional population. Other countries like Spain are currently searching for investors willing to buy and use deserted hamlets and small villages, such as those found in the peripheral regions of Galicia and Asturias. Headlines like "In Spain ghost villages are now going cheap"[11] (Lachmann 2015) and "Spain is giving away its old villages"[12] (Gaulhofer 2015) have attracted attention to this already pressing situation on several occasions in recent years. There is a need to assess the situation in other European countries and to determine whether this type of subsequent use could be promoted in selected locations in the future.

[11] „In Spanien sind Geisterdörfer jetzt günstig zu haben"
[12] „Spanien verschenkt seine alten Dörfer"

References

ADI — Associazione Nazionale Alberghi Diffusi (2015). Alberghi Diffusi. www.alberghidiffusi.it/. Accessed 26 December 2015.

Anonymous (2014). Best of Provence. www.wirtschaftskurier.de/lebensart/artikel/best-of-provence.html. Accessed 26 December 2015.

Associazione Sextantio (2015). Sextantio Albergo Diffuso. www.sextantio.it. Accessed 26 December 2015.

Baltes, J. (2011). Ein ganzes Dorf als Hotel. Touristik aktuell (42), 28.

Bauzeitarchitekten (2015). Wellnesshotel Toscana Resort Castelfalfi mit Studio Florulli. www.bauzeit.com/?noFlash=true&lang=de&page=11&project=1. Accessed 27 December 2015.

BBSR — Bundesinstitut für Bau-, Stadt- und Raumforschung (2010). Landleben — Landlust? Wie Menschen in Kleinstädten und Landgemeinden über ihr Lebensumfeld urteilen. BBSR-Berichte KOMPAKT 10/2010. Bonn.

Blackstock, K. (2005). A critical look at community based tourism. Community Development Journal 40 (1), 39–49.

BMZ — Bundesministerium für wirtschaftliche Zusammenarbeit und Entwicklung (2006). Medienhandbuch Entwicklungspolitik. Berlin.

Bouchez, Y. (2012). Le hameau de Courbefy vendu pour 520 000 euros. www.lemonde.fr/societe/article/2012/05/21/le-hameau-de-courbefy-vendu-pour-520-000-euros_1704982_3224.html. Accessed 26 December 2015.

Classe tourist (2014). Saving Santo Stefano di Sessanio. www.classetouriste.be/saving-santo-stefano-di-sessanio/. Accessed 26 December 2015.

Comune di Santo Stefano di Sessanio (2015). Galeria fotografica. www.comunesantostefanodisessanio.aq.it/c066091/po/gallery.php?g=SWwgcGFlc2FnZ2lv. Accessed 26 December 2015.

Crillon le Brave (2015). Hotel Crillon le Brave Provence www.crillonlebrave.com/de. Accessed 26 December 2015.

Dall'Ara, G. (2007). Albergo diffuso, dall'idea alla realizzazione.. in L'Albergo Diffuso come strumento innovativo per la valorizzazione del potenziale turistico. Quaderni BIC Notes 4 (dicembre 2007) (pp 11–49). Roma: BIC Lazio.

Dall'Ara, G. (2010). Manuale dell'Albergo Diffuso. L'idea, la gestione, il marketing dell'ospitalità diffusa. Milano: Franco Angeli.

Deckstein, D. (2013). Katerstimmung in Kastel Fiffi. Der Spiegel (44), 77–78.

Ehlers, E. (1984a). Bevölkerungswachstum – Nahrungsspielraum – Siedlungsgrenzen der Erde. Studienbücher Geographie. Frankfurt/M.: Diesterweg.

Ehlers, E. (1984b). Die agraren Siedlungsgrenzen der Erde. Gedanken zu ihrer Genese und Typologie am Beispiel des kanadischen Waldlandes. Erdkundliches Wissen 69. Wiesbaden: Steiner.

Ehlers, F. & Tuma, T. (2011a). Die Welt — ein Dorf. www.spiegel.de/spiegel/print/d-78832450.html. Accessed 26 December 2015.

Ehlers, F. & Tuma, T. (2011b). It Takes a Village: TUI Builds Tuscan Playground for the Wealthy. www.spiegel.de/international/europe/it-takes-a-village-tui-builds-tuscan-playground-for-the-wealthy-a-767536.html. Accessed 27 December 2015.

Engel&Völkers Resorts (2015). Toscana Resort Castelfalfi www.engelvoelkers.com/de-de/resorts/toscana-resort-castelfalfi-toskana-italien/. Accessed 27 December 2015.

Förster, K. (2011). Expert interview with the owner of the Schlosshotel in Rattey, Mr. Dr. K. Förster, in July 2011.

Gaulhofer, K. (2015). Spanien verschenkt seine alten Dörfer. www.diepresse.com/home/wirtschaft/international/4825161/Spanien-verschenkt-seine-alten-Dorfer. Accessed 26 December 2015.

Hahne, U. (2011). Neue Ländlichkeit? Landleben im Wandel. Der Bürger im Staat. Themenheft: Raumbilder für das Land — Regionen, Orte und Bauten im ländlichen Raum 61 (1/2), 12–18.

Häusler, N. & Strasdas, W. (2003). Training manual for community based tourism. Addendum to the ecotourism training manual for protected area managers. Zschortau: InWEnt (Capacity Building International).

Helbert, F. (2005). Wir schlafen im Fünf-Sterne-Kuhstall. www.faz.net/aktuell/feuilleton/debatten/santo-stefano-di-sessanio-wir-schlafen-im-fuenf-sterne-kuhstall-1278802.html. Accessed 26 December 2015.

Istituto Nazionale di Statistica (2015). Demografia in Cifre: Popolazione Residente. demo.istat.it/pop2015/index.html. Accessed 26 December 2015.

Kapsaliana Village Hotel (2015). The Kapsaliana Village Hotel. www.kapsalianavillage.gr/de/. Accessed 26 December 2015.

Keeve, V. (2013). Luxusresort in Italien: Das perfekte Stück Toskana. www.tagesspiegel.de/weltspiegel/reise/luxusresort-in-italien-das-perfekte-stueck-toskana/9017876.html. Accessed 26 December 2015.

Lachmann, A. (2015). In Spanien sind Geisterdörfer jetzt günstig zu haben. www.welt.de/finanzen/immobilien/article146050394/In-Spanien-sind-Geisterdoerfer-jetzt-guenstig-zu-haben.html. Accessed 26 December 2015.

Land, R. (2005). Die neue Landwirtschaft und die Dörfer. Presentation in Potsdam-Michendorf, 16./17. March 2005 at the conference „Hightechlandwirtschaft und sterbende Dörfer? Chancen und Probleme der Zivilgesellschaft in den ländlichen Räumen Ostdeutschlands". www.thuenen-institut.de/Publikationen/Land%20Rainer%20 Die%20neue%20Landwirtschaft%20und%20die%20Doerfer %20Maerz%202005.pdf. Accessed 6 January 2013.

Lichtenberger, E. (1979). Die Sukzession von der Agrar- und Freizeitgesellschaft in den Hochgebirgen Europas. In Haimayer, P., Meusburger, P. & Penz, H. (Eds.), Fragen geographischer Forschung (Festschrift für A. Leidlmair) (pp. 401–436), Innsbrucker geographische Studien 5. Universität Innsbruck.

Lucchetti, V. G & Font, X. (2013). Community based tourism: critical success factors. ICRT occasional paper 27. n.p.

de Mallevoüe, D. (2012). Un projet «grandiose» pour le hameau vendu aux enchères. www.lefigaro.fr/actualite-france/2012/07/23/01016-20120723 ARTFIG00282-un-projet-grandiose-pour-le-hameau-vendu-aux-encheres.php. Accessed 26 December 2015.

Nitz, H.-J. (1982). Kulturlandschaftsverfall und Kulturlandschaftsumbau in der Randökumene der westlichen Industriestaaten. Geographische Zeitschrift 70 (3), 162–184.

Nixey, C. (2011). How an entire village can be your hotel for the night. (09.04.2011). The Times.

Park Hotel Schloss Rattey (2015). Schloss Rattey. www.schlossrattey.de/default.aspx. Accessed 26 December 2015.

Polz, K. (2014). Ein Bergdorf als Hotel: Neues Leben in alten Gemäuern. www.pnp.de/nachrichten/ratgeber/reise_und_urlaub/reiseberichte/1390479 _Ein-Bergdorf-als-Hotel-Neues-Leben-in-alten-Gemaeuern.html. Accessed 26 December 2015.

Preiss, M. (2008). Spumante im Hirtenbett. Ein reicher Mailänder hat den verfallenen Abruzzenort Santo Stefano di Sessanio in ein Hoteldorf verwandelt und damit dem Landleben ein Denkmal gesetzt. www.zeit.de/2008/38/Abruzzen. Accessed 26 December 2015.

Reichert-Schick, A. (2013). Abwanderung — Entleerung — Wüstung? Verlauf, Determinanten und Implikationen von Peripherisierungs-, Regressions- und Wüstungsprozessen in ländlichen Siedlungen gestern und heute. Unpublished postdoc thesis (Habilitationsschrift). Trier.

Schmitt, C. (2015). Verwandelt. Im italienischen Friaul verließen die Bewohner ihre Dörfer, Häuser verfielen — bis der Tourismus kam. www.zeit.de/2015/30/friaul-tourimus-verwandelt-italien. Accessed 29 December 2015.

Schüle, C. (2014). Das stille Dorf im Olivenhain. www.zeit.de/2014/43/kreta-kapsaliana-village-hotel. Accessed 26 December 2015.

Tasche, E. (2011). TUI baut Feriendorf für Reiche. www.spiegel.de/reise/aktuell/0,1518,766960,00.html. Accessed 26 December 2015.

Tenuta di Castelfalfi (2015). Toscana Resort Castelfalfi. www.castelfalfi.de/home. Accessed 26 December 2015.

Toscana Resort Castelfalfi (2012). Image Broschüre. www.castelfalfi.de/kontakt/presse/allgemeine-downloads/. Accessed 17 December 2012.

Verein der Privatwinzer zu Rattey e.V. (2015). Verein der Privatwinzer zu Rattey e.V. im Park Hotel Schloss Rattey. www.privatwinzer.de/index.html. Accessed 26 December 2015.

Weiss, R. (2014a). TUI Pushes Italy Holiday-Home Sales in Project-Exit Bid. www.bloomberg.com/news/articles/2014-10-28/tui-pushes-italy-holiday-home-sales-in-project-exit-bid. Accessed 27 December 2015.

Weiss, R. (2014b). TUI tries to exit disastrous italian vacation home project. www.skift.com/2014/10/28/tui-tries-to-exit-disastrous-italian-vacation-home-project. Accessed 27 December 2015.

Williams, G. (2010). Saving Towns by Filling Rooms in Italy. (23.05.2010). The New York Times.

Wilson, E. (2015). Crillon le Brave: A Slice of Provençal Paradise. www.grifcopr.com/blog/crillon-le-brave-a-slice-of-provencal-paradise/. Accessed 26 December 2015.

Rural tourism re-interpreted: how tourist entrepreneurs implement different notions of rurality

Stefan Kordel and Tobias Weidinger

Abstract

This paper addresses what strategies tourist entrepreneurs undertake to implement various notions of rurality by means of the example of rural tourism in the Bavarian Forest. In particular, it is asked how hosts enable guests to consume rurality in light of an overall desire to experience a rural life-world in material and immaterial dimensions. Based on empirical data from 41 expert interviews with tourist entrepreneurs, public authorities involved in tourism and local politicians in more than 20 municipalities in the Bavarian Forest, two types of establishments are analysed in this study, i.e., on-farm establishments and wooden accommodations, which refer to regional building culture. Results show that entrepreneurs meet the demand for idyllic notions and modern equipment, following a binary strategy of conservation and re-interpretation. Besides consumption of ruralities, tourists take also part in the process of constructing rurality, for instance when participating in food production. The juxtaposition of tourist structures that reproduce urban-rural dichotomies on the one hand and are influenced by a hybridisation of the urban and the rural on the other hand, finally encourages diversification of rural tourism sector.

Introducing perspectives on rurality

A variety of approaches in various subdisciplines of social sciences in defining the rural encompass rural as residual to urban, rural as distinctive by means of measurable attributes, e.g., population density, and "rural as contrasted to urban in sociocultural terms" (Cocklin et al. 2002, 3; for some stereotypical differences between urban and rural populations see Champion and Hugo 2004, 8). Currently, demography, spatial planning and especially rural tourism continue to stick to the latter, i.e., a rural-urban dichotomy. Historically, rurality and particularly representations of a rural idyll were developed in contrast to the "unhealthy and socially rough city" (Dirksmeier 2008, 160). Contrasting the rural idyll with the urban nightmare (Yarwood 2005, 24), people emotionally perceive different qualities between the urban and the rural. As a consequence, they search for apparently untouched places and small communities either during tourist stays or when relocating to places in peripheral rural regions (Benson 2013; Stolarick et al. 2010).

Even tourist stakeholders make use of this specific argumentation. In Germany, the Federal Ministry of Economic Affairs and Technology, for instance, suggests commodifying urban-rural dichotomies when marketing rural tourism, stating that:

> *"Rural areas in Germany hold a large tourist potential. Social developments, such as an increasing interest in rural lives among urban populations or a search for authentic and true experiences as well as a growing desire for remoteness and nature offer various chances for developing tourism in rural areas. (...) Rural areas can be positioned as true and modern counter-worlds towards everyday life" (BMWi 2013, 2, 14, translated)*

Highly diverse visitor markets (Carson et al. 2014, 357), a high volatility of consumer preferences (Urry 1991, 52) and, finally, individual perceptions of what is rural (Sharpley and Sharpley 1997, 6) characterise current tourism. Thus, the tourism product in particular, is "largely assembled by, and therefore, unique to the individual tourist" (ibid., 102). As a consequence, the multi-selling of the same physical space is indicated.

This paper analyses what strategies entrepreneurs undertake to implement various notions of rurality. In particular, it is asked how hosts enable guests to consume ruralities and meanwhile co-produce them. Thereby, particular perceptions of the countryside from an urban point of view are elaborated, while contradictions between idealised and realistic notions are depicted. At first, conceptual approaches on rurality are briefly introduced, followed by illustrating manifestations of an overall desire to experience the rural in material and immaterial dimensions in tourism. After that, the development of tourism in the Bavarian Forest is presented, which could only evolve against the backdrop of the peripheral location close to the former Iron Curtain. Then, the methods applied for conducting the case study are described. The main section comprises the discussion of empirical data. How tourist entrepreneurs implement multiple ruralities is illustrated by means of two examples: on-farm tourism and accommodations referring to regional building culture, i.e., wooden lodging. The paper concludes with final remarks on the rurality-tourism nexus.

Encountering ruralities and tourism

Understandings of ruralities for tourism

Transformations of rural spaces from landscapes of production to landscapes of consumption (Hall and Page 2014; Williams and Hall 2000; Butler et al. 1998) resulted in a commodification of rural space for economic purposes, e.g., for tourism or counter-urbanisation (Hall and Page 2014, 229). The product to sell is addressed as rurality, understood as "the social production of a set of meanings" (Mormont 1990, 36) attributed to rural spaces, people, and practices. Since the individual tourist interprets rurality differently, the term 'ruralities' acknowledges multiple meanings.

Thereby, especially tourist stakeholders play with the idea of the rural as anti-urban and thereby assisted in the promotion of particular images of rurality, e.g., the rural idyll (Hall and Page 2014, 242). The latter encompasses the realm of intimacy, materialised through food and intact social lives. Moreover, notions of the rural as a refuge from modernity (Short 1991) associated with hopes of peace, health and security (Hoggart and Paniagua 2001) are widespread semantics ascribed to rural spaces. Although there is no doubt that the concept of the rural idyll remains a relatively fixed and static understanding of rurality (Benson 2013, 521), it seems to be an appropriate concept for understanding an overall desire in Western societies for longing for "security, stasis and nostalgia in an uncertain, changing, and globalising world" (Short 2006, 143). Notwithstanding the fact that the idyll frequently hides social problems and purifies the rural (Silva 2013, 249), it can be considered as a sort of myth that is influential in encouraging people to visit the countryside as tourists (Woods 2005, 13). In other words, an acceptance of cultural symbols related to the rural idyll is observable in rural tourism or specific residential mobilities to rural areas (Hall and Page 2014; Benson 2013; Buller and Hoggart 1994). In sum, each form of place-marketing of rural areas try to sell a peaceful rural idyll "around the concepts of peripherality (…) and the associated distance from modernity and urbanisation" (Hall and Page 2014, 253).

Ruralities and the rural idyll as their specific interpretation, however, are not exclusively defined through qualities and representations. In this respect, Woods (2010) suggests that individuals' practices of rurality have to be taken into account. "This includes attempts to access 'more-than-representational' rural geographies by investigating ways in which rural experiences are felt, sensed, intuited through bodily actions and performances" (ibid., 835).

Besides immaterial qualities and ascriptions, materialities, frequently embodied through human practices, e.g., practices of food consumption or practices of social interaction, should be considered. The role of action of rural dwellers as well as tourists is put at the core, when assuming that rurality is performed to a certain extent (Edensor 2006). Thus, rural space is not only consumed by tourists, but also co-produced by various actors, such as tourists, entrepreneurs and local inhabitants in material and immaterial realms (Jonasson 2012). As a consequence, rural space is transforming, practices are changing and individual meanings alter, although tourists strive for relative stable notions, e.g., the rural idyll (about the meaning of practices in terms of rural tourism see Førde 2009).

Since tourist stakeholders, politicians and popular representations mainly refer to an urban-rural divide, they neglect that traditional definitions of rurality are rendered obsolete for various reasons (e.g., increased human mobility, new uses of rural spaces, McCarthy 2007). Halfacree (2007), for instance, suggests constructions of rural spaces according to a threefold model of rural space and depicts the potential rurality of any place (ibid., 127). To date, ongoing contradictions in scientific debate about rurality are observable. While some stick to the urban-rural dichotomy, others promote an urban-rural continuum (Lane 1994). Yet, there is no doubt that the urban and the rural are influencing each other in manifold ways. Both conceptions are made fruitful in this contribution in order to interpret hybridity of tourist establishments as well as distinctions, implemented in tourist offerings, which mostly draw on urban-rural dichotomies.

Old and new perspectives on rural tourism in Europe

Long-lasting debates about the definition of rural tourism are profoundly illustrated by Hall and Page (2014, 235), who finally suggest that rural tourism should be located in rural areas, built upon special features, e.g., small scale enterprise, open space, traditional in character and connected with local families. While tourism in rural areas is based on rural sceneries, there is no clear definition of rural tourism (ibid., 87). Rein and Schuler (2012, 4–5) suggest to place cultures perceived as rural at the core. Furthermore, rural tourism encompasses personal contact to hosts and experiencing the production and consumption of regional products.

Rural tourism in Europe in general and in Germany in particular can be traced back to the period of industrialisation, when urban upper and middle classes discovered rural areas for recreation purposes especially during the summer months (*Sommerfrische*) (Oppermann 1996, 86). Nowadays, a diversification of rural tourism takes place. With regard to accommodation, a high variety, ranging from campgrounds to B&B to hotels in rural communities and farms, is distinctive (ibid., 88). On-farm tourism, for instance, is agreed to offer uniqueness, authenticity, intimate links with nature (Potocnik-Slavic and Schmitz 2013, 268) and to be a form of tourism that actively integrates the guest into farmers' daily lives (Nilsson 2002; for further categorisation and historical overview see Phillip et al. 2010; Busby and Rendle 2000). This goes in line with what MacCannell (1976, 5) suggested, namely that tourists more and more strive for life-seeing, or, as Bramwell (1994) puts it, wish to integrate with local inhabitants' daily lives. Woods (2014) suggests family farms as emblematic for a rural way of life. Further particularities comprise family relationships, lifestyle and property (Potocnik-Slavic and Schmitz 2013, 266; for a gendered perspective on agritourism see Wright and Annes 2014, 481).

Besides life-seeing, consuming local products and regional cuisine as well as the opportunity to participate in food production are distinctive features of rural tourism (Dubois and Schmitz 2011; Sznajder et al. 2009). Implications of commodifying food on the region's image were at the core of scientific interest (Hall and Page 2014, 247), while establishing emotional links to a certain destination, however, and a subsequent re-visit were sparsely considered.

Following the urban-rural continuum, challenges in defining specifics of rural tourism are significant, e.g., what is defined as rural space, and how to deal with phenomena such as resort-based tourism or theme parks that spill out to rural areas (Lane 1994). Acknowledging hybridity goes in line with including non-farm rural tourism (Oppermann 1996, 89) and offerings associated with alternative perspectives on rurality, e.g., quietness and relaxation or regionality and originality. Therefore, it is analysed, how tourist's desire for authenticity and originality is met by commodification of regional building culture. While the role of Alpine architecture for tourist destinations was discussed for South Tyrol (Innerhofer 2010), studies from entrepreneurs' perspectives in low mountain range areas, where symbolical architectural styles cannot be reproduced to that extent, are lacking. The case study presented here was conducted among tourist stakeholders in the less distinctive low mountain range area of the Bavarian Forest. It is discussed, what strategies entrepreneurs undertake to implement distinct notions of rurality.

Case study and methodological approach

The Bavarian Forest

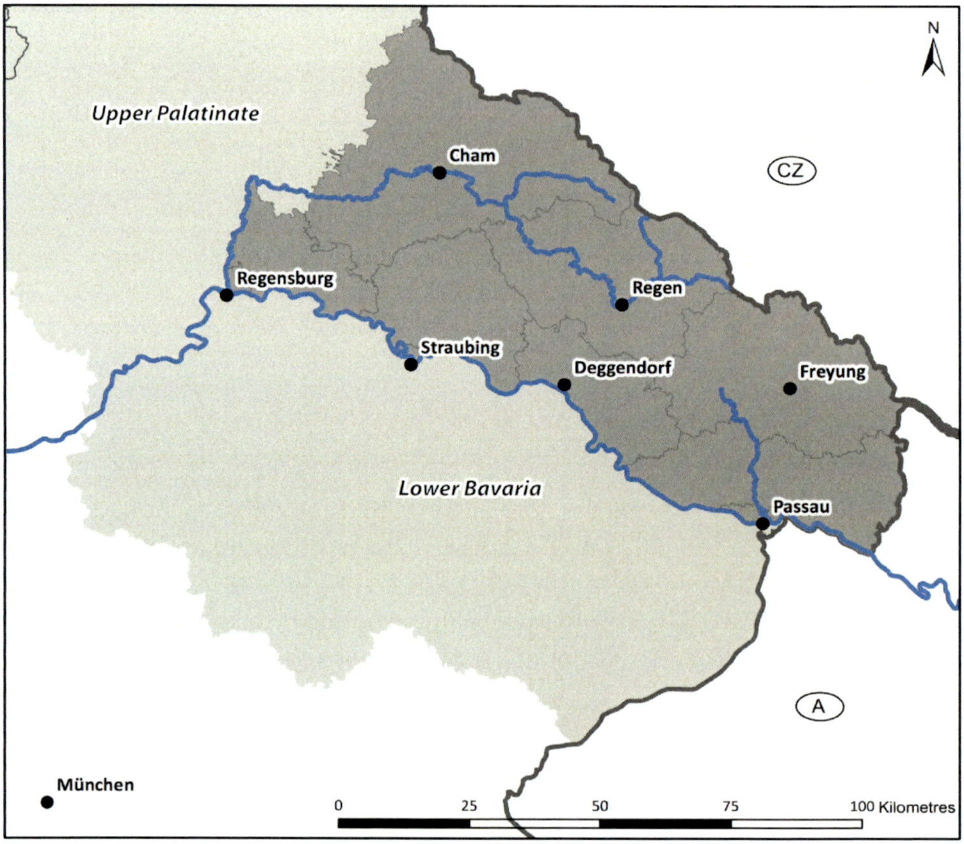

Figure 1: Tourism region Bavarian Forest, Germany

Source: Map based on Federal Agency for Cartography (2015): Administrative Areas 1:250.000, EuroGeographics (2015): Euroglobalmap Cartography: Florian Dworzak

The Bavarian Forest is a low mountain range area in the border triangle of Germany, the Czech Republic and Austria, located in the administrative districts of Lower Bavaria and Upper Palatinate. Large forest areas that were initially considered as resource for timber industry purposes characterise this region. During the 1970s, when the region was put on the margin at the Iron Curtain, the establishment of tourist structures as a strategy for regional development was politically supported by means of designating large forest areas as the first German National Park (Mayer and Woltering 2013). In parallel, mass tourism was encouraged through the construction of large-scale holiday resorts. While initially this led to economic growth by providing alternative sources of employment and income for the local population, the story of success of large-scale resorts has diminished since the 1990s in the course of changing consumer preferences and alternative destinations. Currently, a vast range of municipalities considers resorts as a burden and develop manifold strategies to deal with (Kordel et al. 2015). Even small-scale guesthouses, e.g., family-run establishments, dropped out of the market due to an investment bottleneck (Maier et al. 2013). Simultaneously, a diversification of tourist accommodation structure occurs in the Bavarian Forest, mainly encompassing establishments in higher price segments in order to attract new target groups, e.g., young families and couples as well as people in managerial positions and incentive tourism for firms.

The decreasing number of commercial beds from about 69 971 in 2008 to 58 236 in 2014 captures the consolidation process in tourism sector, despite statistical effects are also significant (BLS 2015, for 2008: accommodations 9 or more beds, including campsites with 3 or more pitches, for 2014: accommodations with 10 or more beds, including campsites with 10 or more pitches). Currently, occupancy rate of accommodations in the Bavarian Forest is as low (32.2% in 2014, BLS 2015) as in other low mountain range areas in Germany, such as the Black Forest 33.5% (own calculation for 2014 after Schwarzwald Tourismus 2015). In 2014, 6.9 million overnight stays were recorded in the Bavarian Forest, mainly covered by domestic tourists (93.9%), with an average duration of stay of 4.0 days (BLS 2015). In line with findings on rural tourism in Baden-Württemberg (Oppermann 1996, 94), a high seasonality of rural tourism is indicated. While agritourism tends to be popular during summer months and school holidays, new establishments, such as wooden lodgings in the Bavarian Forest show good occupancy rates the whole year round due to their uniqueness.

Research methods

Empirical research was undertaken from 2012 to 2014 by means of 41 semi-structured expert interviews with tourist entrepreneurs, public authorities involved in tourism and local politicians in more than 20 municipalities in the Bavarian Forest. Among them, 24 interviews with tourist hosts, either running establishments associated with agritourism or luxury and individualism, were carried out. The interview partners were selected according to two criteria: running agritourism business or an establishment with distinctive features (wooden architecture, family-oriented ownership). The semi-structured interviews aimed at depicting narratives about the development of the tourist company, focusing on innovations as a reaction to guests' perceptions of rurality. The recorded interviews were transcribed verbatim and labelled for analysis. After that, quotations that exemplarily illustrate how entrepreneurs implement ruralities in an outstanding manner, were translated to the English language. For this paper, six interviews were selected. Table 1 provides an overview.

Table 1: Interviewed persons

No.	*Name**	*Function and area of responsibility*	*Type of establishment (if applicable)*
1	Josephine	Member of the local office for rural development	—
2	Hans	Tourist entrepreneur	Agritourism
3	Monika	Tourist entrepreneur, head of organisation for promoting agritourism in the Bavarian Forest region	Agritourism
4	Christine	Representative of agricultural chamber, farmer	—
5	Maria	Tourist entrepreneur	Wooden lodging
6	Angela	Tourist entrepreneur	Wooden lodging

* Due to ethical reasons, names of interview persons were changed.

Strategies for implementing ruralities in current tourist offerings

As stated above, popular images of the countryside reproduce urban-rural dichotomies to a certain extent. How notions of rurality are implemented by tourist entrepreneurs, is analysed by means of two forms of accommodation, i.e., agritourism and lodgings, inspired by wooden architecture. Although varying in motivations to stay, length of stay, origin of guests and socio-demographic structure of guests (Table 2), similar strategies are observable.

Table 2: Comparing agritourism and wooden lodgings

	Agritourism	*Wooden lodgings*
Motivations to stay	Get to know farmers' lives, entertainment	Escape from urban lives, relaxation
Length of stay	Mostly 1 week or more	At least 4–5 days
Origin of guests	Throughout Germany	Surroundings, max. 200km
Socio-demographic structure of guests	Families or grandparents with children	Couples, businessmen

Conservation versus re-interpretation

The first strategies discussed here are situated in a continuum of conservation and re-interpretation. A number of wooden vacation homes, structured as a village located at remote location in the forest represent an outstanding example for the thesis that rural tourism is driven by the desire for a refuge from modernity and urban trappings. Restored wooden farmhouses and chalets have recently become very popular. The owner of an establishment, where traditional buildings of the Bavarian Forest were restored for tourists, explains his philosophy:

> *"This is not only inspired rustically but also includes urban elements. The guest feels good. He perceives this as rural. He would not sleep in a bedroom of a farmer of the year 1910. He wants to have a sauna or infrared cabin after cross-country skiing. Nearly every chalet has this." (Interview Maria)*

Even during vacation in rural areas, tourists prefer luxurious conveniences inside the accommodation. The strategy of re-interpreting traditional building culture shows urban influences to notions of rurality. However, reference to the region's history and everyday lives of the past, e.g., notions of nostalgia are made. The entrepreneur explains her strategy, termed as musealisation in the following quotation:

> *"Musealisation is intended. It is not negative for me. It is nice to see when people sitting on benches realise occasionally that wooden benches are uncomfortable and subsequently reflect that everyday life was hard in former times. No one was sitting on benches for a long time, because there was a lot of work to do. Well, in the Bavarian Forest, everyday life was hard. Well, musealisation is a kind of strategy here." (Interview Maria)*

A conservational attitude is observable when hosts ascribe symbolic meanings to material items in order to make guests conscious about "real life" in former times. Wooden architecture is frequently considered as a signifier for regional building culture, mediating a sense of authenticity and originality (Tschofen 2014). Pechlaner (2014, 49) suggests architecture as visualisation of established cultural discourses. Highlighting local inspirations for interior design of wooden chalets, another entrepreneur points out the challenge to meet the demand of both, the old-fashioned and the modern.

> *"Timber comes from our own forests. We have granite on the terraces. We have a lot of glass. I am keen on glass from Spiegelau [village nearby, which stands symbolically for glass industry in the Bavarian Forest; the authors]. (...) The interior furnishings are modern. The snuggery-like structures are traditional. (...) this is a balancing act between antiquated and modern." (Interview Angela)*

Figure 2: Wooden lodging
Source: Maximilian Knapp

The desire for establishing links to the region through architectural design of accommodations, can be interpreted as a kind of counter-trend towards hybridisation and blurring boundaries between the urban and the rural. Thus, the entrepreneur cited above clearly sticks to the urban-rural dichotomy. While consumption of material notions of rurality, imbued with certain meanings, were at the core so far, the representative of agricultural chamber stresses, how entrepreneurs provide the opportunity to immerse into a rurality back in time by means of specific practices. A revival of everyday practices associated to former times can be experienced by baking bread in a stone oven, for instance.

> *"You immerse into another time, when you experience this. (...) you can experience the smell of bread, you are excited when you taste it. And when the environment is appropriate, it is nice." (Interview Christine)*

Sensory experiences, e.g., smells and tastes, become significant, when tourists practise symbolically meaningful activities. While entrepreneurs involve guests in a kind of food tourism (Hall and Page 2014, 244), guests and hosts co-produce a static understanding of rurality, addressed as a pastoral rurality, which is closely linked to healthy food. One entrepreneur accordingly states:

> *"They get interested in cheese and make cottage cheese at home. They also take honey with them. I could sell every natural food. When I pick strawberries to make jam, the guests accompany me and get excited. This is a dream. Participating, self-made and experiencing. (...) Or they get involved in producing liqueurs. Guests pick blackthorns in buckets so that I can produce liqueurs. And then, we have blackthorn for Christmas. It is simply a matter of take it home." (Interview Monika)*

Involving guests in food production, hosts ascribe meanings to practices and subsequently establish an emotional link between the material item of food and the tourist stay, respectively the region. Selling food as regional products, entrepreneurs enable guests to extend their vacations and consume products adressed as rural signifiers even at their places of origin (Crouch 2006; for a new consumption culture in terms of food, see Miele 2006). Co-production of ruralities becomes also significant when farmers provide opportunities for guests to participate in working lives. A farmer involved in agritourism emphasises tourists' desires for being active, while a second quotation refers to contradictions between tourists' expectations and farmers' working reality.

> *"Working in the stable is very important. We have an ecological farm. During the day, our cows are on the pasture and in the evening, when they return, I collect hay and the children help me. I have special wooden forks for them. (...) When you offer on-farm holidays, the guest wants to have an intact farm. The guest wants to help you." (Interview Hans)*

> *"Nowadays, farms are commonly high-tech. We are currently introducing a milking robot. When the guest comes, he cannot help me. Even the feeding procedure is high-tech. The feed rations have to be exact. It is business thinking." (Interview Christine)*

While Hans specifies how tourists seek active participation on his farm and the role of providing an environment perceived as authentic, Christine stresses that farmers' working lives recently tend towards engineered farming. In order to provide both realities for tourists, the antiquated and the high tech, Hans, as stated above, separated the establishment and introduced a museum for agricultural tools and devices, where tourists can experience the past. While guiding visitors through the museum, his father embeds narratives from everyday life in the Bavarian Forest in the past and thereby provides a local perspective. Involving that specific view is suggested as providing authenticity (Weidinger 2015). In sum, on-farm tourism does not only "commoditise the social and cultural value in farming activity for public consumption" (Wright and Annes 2014, 478; Jackson 1999), yet hosts intend to provide opportunities to co-produce ruralities.

To conclude, it became clear that entrepreneurs running agritourism and wooden lodgings commodify specific material notions of rurality, which are both consumed and imbued with meaning through certain practices. Communicating this static perspective, hosts provide a conservational interpretation of rurality and aim to meet an idyllic rurality. However, re-interpretations of the product rural are implemented, especially when hosts present the "real life" of inhabitants or even co-produce it with tourists. Additionally, dissolving temporal fixations become significant when vacations are dissociated from current stays by means of regional products.

Providing intimacy vs. meeting the desire to interact with local inhabitants
Hosts implement ruralities by means of material culture, references to historicity and regional building culture and therefore make a nostalgic reminiscence. The latter is also observable when entrepreneurs romanticise the peripheral location. The entrepreneur, running the restored farmhouses accordingly states:

> *"For metropolitans, it is like being on the moon. (...) We have no streetlights. It is completely dark at night. Where else do you have this in a city?" (Interview Maria)*

This strategy makes use of the urban-rural dichotomy by emphasising atmospheric, i.e., immaterial notions ascribed to rural areas. Intimacy plays a major role for spending holidays in such accommodations. According to the entrepreneur, the establishment aims to recruit guests who are very much involved in business and want to opt out for a while. While providing a relaxing atmosphere is a major concern, offerings related to active tourism are relatively rare. However, hosts offer activities such as chopping wood that are somehow addressed as being close to nature. Enabling practices that symbolically represent a certain rurality, entrepreneurs allow a flight to a kind of counter-world to the globalised and mobile businessman's life world. Nevertheless, refuge from urban lives does not intend capping all kind of social interactions, but allow the guest to decide whether and to what extent he or she wants to interact with locals. Opportunities for interactions with local inhabitants, for instance, are provided outside the holiday accommodation, e.g., in taverns.

> *"Tourists get a list of taverns in the village. When they once have been in the tavern and met the landlord afterwards, they like to chat with them, or even with locals, they once met in the tavern. When guests go for a walk in the forest and meet locals, they talk to them. That is a kind of unspoiled rural. Guests like that!" (Interview Angela)*

Not addressed as contrary to the desire to be on ones' own, hosts provide opportunities to meet local inhabitants. The latter is also fostered in the accommodation itself, when farmers enable guests to access farmers' private lives.

> *"Guests only enter my apartment when I want it. Some come and say, let's have a cup of coffee. Either I let them in or we go outside. I would not do this with everyone. I am not obliged to do so." (Interview Monika)*

The wish to enter privacy and family life is frequently considered as core value of rural societies, is especially popular among tourists. The statement above, however, contends that access to private spheres is limited to regular guests and depends on the host's attitude. As carried out in a study on agritourism in Austria and Norway (Daugstad and Kirchengast 2013), physical (separate entrance) or a temporal compartmentalisation (opening hours) are indicated in tourist offerings, located in private surroundings.

There is no doubt that focusing on interdependencies between materialities and practices that are perceived as authentic, not only among tourists, but also among local inhabitants (Weidinger 2015) challenge tourist entrepreneurs' established strategies in rural areas. Additionally, tourists strive for consuming an idealised rural reality, while entrepreneurs, especially those running on farm lodging, aim to deconstruct this by providing an insight to farmers' daily lives. Further contradictions encompass the past and present continuum or the desire for intimacy and strategies to enable interactions with locals. Hosts establish individual strategies to meet that needs and thus commodify rural space in manifold ways. As indicated in the following quotation, specialisation and commodification of endogenous potentials, either material or immaterial plays an important role for economic success. A member of the local office for rural development states:

> *"The more remote the establishment is situated, the more it has to specialise in something. (...) Each establishment has to emphasise this. This would be an important key to success." (Interview Josephine)*

Concluding remarks

Against the backdrop of a renaissance of rurality in current Western societies and an increasing commodification of rural space, either for tourist purposes or as an attractive site of living, this paper asked, what strategies tourist entrepreneurs undertake to implement various notions of rurality. The two sorts of establishments analysed in this study, on-farm establishments and wooden accommodations, intend to market at least a duality of past and present. Moreover, they stand for a juxtaposition of how entrepreneurs implement the demand of tourists to consume and co-produce multiple notions of rurality and subsequently encourage diversification of the rural tourism sector. Despite locations are different, marketable offers (Table 3), encompassing encounters with animals and farmers and consumption/production of food on the one hand, and experiences of wilderness and consumption of amenities on the other, clearly relate to notions, commonly addressed as the rural idyll. References are given to nature, romanticism and nostalgia (Bell 2006). It is worth reminding authenticity and access to real life of others as key assets for successful rural tourism, yet it is worthwhile to highlight interdependencies of stakeholders, i.e., entrepreneurs, tourists and local population. Especially the latter are considered as predestined for interpreting authenticity and finally ruralities (Weidinger 2015).

Deriving from the assumption that rural tourism is situated in rural areas and entrepreneurs commodify notions of authentic and original cultures, a mere focus on consumption is not sufficient, as also production and co-production become increasingly important. Ascribing symbolical meanings to material and immaterial culture or introducing emotional notions, entrepreneurs actively intervene in constructions of rurality, either constructing idylls or more realistic views, close to the working life. Tourists, however, also take part in the construction process, for instance when they participate in food production and thus co-construct the consumed product. Finally, establishing emotional links to material items, perceived as authentic, can be addressed as a resource for further visits.

Table 3: Ruralities in agritourism and wooden lodgings

	Agritourism	*Wooden lodgings*
Location	Intact small-scale family farms	Luxurious accommodations in peripheral locations with reference to regional building culture
Marketable offers	Encounters with animals and farmers Consumption and production of food	Experiences of wilderness Consumption of amenities
Notions of rurality	Pastoral idyll and 'realistic' farmer's lives Urban-rural dichotomy predominates	Intimacy and tranquillity Urban-rural dichotomy predominates, 'urban' elements are integrated

Considering notions of rurality, this contribution contends that urban-rural dichotomies are still at the core in marketing rural tourism. However, the examples elaborated here show that this dichotomy is blurring, as "rural areas are in no way immune to transformations" (Uusitalo and Assmuth 2013, 71). To sum up, the immersion of urban ways of life to rural areas and vice versa is observable in entrepreneurs' strategies.

Research on rural tourism should reflect societal trends that were investigated elsewhere, e.g., counter-urbanisation, rural gentrification, urban farming, as it is obvious that complex processes recently created a global countryside (Woods 2013). Acknowledging the variety of ruralities, implemented by tourist entrepreneurs, offers a potential for peripheral rural areas, which are not equipped with distinctive features, either natural or cultural ones. The creativity of entrepreneurs and the willingness to combine traditional and modern notions of rurality are preconditions for successful tourist establishments in peripheral areas.

References

Bell, D. (2006). Variations on the rural idyll. In Cloke, P., Marsden, T. & Mooney, P. H. (Eds.), The handbook of rural studies (pp. 149–160). London: Sage.

Benson, M. (2013). Living the "Real" Dream in la France profonde? Lifestyle Migration, Social Distinction, and the Authenticities of Everyday Life. Anthropological Quarterly 68 (2), 501–526.

BLS — Bayerisches Landesamt für Statistik (2015). Eckdaten der Tourismusregionen in Bayern. www.statistik.bayern.de/statistik/tourismus. Accessed 23 November 2015.

BMWi — Bundesministerium für Wirtschaft und Technologie (2013). Tourismusperspektiven in ländlichen Räumen. Handlungsempfehlungen zur Förderung des Tourismus in ländlichen Räumen. Berlin.

Bramwell, B. (1994). Rural tourism and sustainable rural tourism. Journal of Sustainable Tourism 2 (1–2), 1–6.

Buller, H. & Hoggart, K. (1994). International Counterurbanization: British Migrants in Rural France. Aldershot: Avebury.

Busby, G. & Rendle, S. (2000). The transition from tourism on farms to farm tourism. Tourism Management 21 (6), 635–642.

Butler, R., Hall, C. M. & Jenkins, J. (Eds.) (1998). Tourism and Recreation in Rural Areas. Chichester: Wiley.

Carson, D. A., Carson, D. B. & Lundmark, L. (2014). Tourism and Mobilities in Sparsely Populated Areas: Towards a Framework and Research Agenda. Scandinavian Journal of Hospitality and Tourism 14 (4), 353–366.

Champion, T. & Hugo, G. (2004): Introduction: Moving Beyond the Urban-Rural Dichotomy. In Champion, T. & Hugo, G. (Eds.), New forms of urbanization: beyond the urban-rural dichotomy (pp. 3–24). Aldershot: Ashgate.

Cocklin, C., Bowler, I. & Bryant, C. (2002). Introduction. In Bowler, I., Bryant, C. & Cocklin, C. (Eds.), The Sustainability of Rural Systems: geographical Interpretations (pp. 1–12). Dordrecht, Boston, London: Kluwer Academic Publishers.

Crouch, D. (2006). Tourism, consumption and rurality. In Cloke, P., Marsden. T. & Mooney, P. H. (Eds.), The handbook of rural studies (pp. 355–364). London: Sage.

Daugstad, K. & Kirchengast, C. (2013). Authenticity and the pseudo-backstage of agri-tourism. Annals of Tourism Research 43, 170–191.

Dirksmeier, P. (2008). Strife in the rural idyll? The relationship between autochthons and in-migrants in scenic regions of South Bavaria. Erdkunde 62 (2), 159–171.

Dubois, C. & Schmitz, S. (2011). The way to agritourist development from constraints to opportunities (19th Annual Colloquium of the International Geographical Union Commission on the Sustainability of Rural Systems). Galway: National University of Ireland.

Edensor, T. (2006). Performing rurality. In Cloke, P., Marsden, T. & Mooney, P. H. (Eds.), The handbook of rural studies (pp. 355–364). London: Sage.

Førde, A. (2009). Creating 'The Land of the Big Fish': A Study of Rural Tourism Innovation. In Nyseth, T. & Viken, A. (Eds.), Place reinvention. Northern perspectives (pp. 93–110). Farnham: Ashgate.

Halfacree, K. (2007). Trial by space for a 'radical rural': Introducing alternative localities, representations and lives. Journal of Rural Studies 23 (2), 125–141.

Hall, C. M. & Page, S. (2014). The geography of tourism and recreation. Environment, place and space. London: Routledge.

Hoggart, K. & Paniagua, A. (2001). What rural restructuring? Journal of Rural Studies 17 (1), 41–62.

Innerhofer, E. (2010). Regionale Kernkompetenzen — die Rolle von Architektur und Design. In Pechlaner, H. & Schön, S. (Eds.), Regionale Baukultur als Erfolgsfaktor im Tourismus. Nachhaltige Vermarktung von Destinationen (pp. 23–38). Berlin: Erich Schmidt Verlag.

Jackson, P. (1999). Commodity cultures: the traffic in things. Transactions of the Institute of British Geographers 24 (1), 95–108.

Jonasson, M. (2012). Co-producing and co-performing attractive rural living in popular media. Rural Society 22 (1), 17–30.

Kordel, S., Weidinger, T. & Kocher, M. (2015). Wandel des Tourismus in ländlichen Räumen. Die Destination Bayerischer Wald erfindet sich neu. Praxis Geographie 11, 38–43.

Lane, B. (1994). What is Rural Tourism. In Bramwell, B. & Lane, B. (Eds.), Rural Tourism and Sustainable Rural Development (pp. 102–111). Clevedon: Channel View Publications.

MacCannell, D. (1976). The Tourist. A new theory of the leisure class. London: Macmillan.

Maier, J., Weber, J. & Weizenegger, S. (2013): Analyse und Evaluierung ausgewählter tourismuspolitischer Leitziele in Bayern 2011/12 anhand von Fallstudien im Allgäu, Bayerischen Wald und Fichtelgebirge. In Job, H. & Mayer, M. (Eds.), Tourismus und Regionalentwicklung in Bayern (pp. 42–68). Hannover: ARL.

Mayer, M. & Woltering, M. (2013). Tourismusförderung und Regionalentwicklung im Nationalpark Bayerischer Wald. In Gamerith, W., Anhuf, D. & Struck, E. (Eds.), Passau und seine Nachbarregionen. Orte, Ereignisse, Verbindungen — ein geographischer Wegweiser (pp. 349–359). Regensburg: Friedrich Pustet.

McCarthy, J. (2007): Rural geography: globalizing the countryside. Progress in Human Geography 32 (1), 1–9.

Miele, M. (2006). Consumption culture: the case of food. In Cloke, P., Marsden, T. & Mooney, P. H. (Eds.), The handbook of rural studies (pp. 344–354). London: Sage.

Mormont, M. (1990). 'Who is rural?' or, how to be rural: towards a sociology of the rural. In Marsden, T., Lowe, P. & Whatmore, S. (Eds.), Rural restructuring (pp. 21–44). London: Fulton.

Nilsson, P. A. (2002). Staying on farms: An Ideological Background. Annals of Tourism Research 29 (1), 7–24.

Oppermann, M. (1996). Rural Tourism in Southern Germany. Annals of Tourism Research 23 (1), 86–102.

Pechlaner, H. & Innerhofer, E. (2014). Architektur und Tourismus: Perspektiven für Identitäts- und Erlebnisräume. In Kunst Meran (Ed.), Journal zur Ausstellung Alpen Architektur Tourismus am Beispiel Südtirol (pp. 48–49). Bozen.

Phillip, S., Hunter, C. & Blackstock, K. (2010). A typology for defining agritourism. Tourism Management 31 (6), 754–758.

Potocnik-Slavic, I. & Schmitz, S. (2013). Farm Tourism Across Europe. European Countryside 5 (4), 265–274.

Rein, H. & Schuler, A. (Eds.) (2012). Tourismus im ländlichen Raum. Wiesbaden: Gabler.

Schwarzwald Tourismus (2015). Geschäftsbericht 2014. Freiburg.

Sharpley, R. & Sharpley, J. (1997). Rural Tourism. An Introduction. London: International Thomson Business Press.

Short, B. (2006). Idyllic ruralities. In Cloke, P., Marsden, T. & Mooney, P. H. (Eds.), The handbook of rural studies (pp. 133–148). London: Sage.

Short, J. (1991). Imagined Country: Society, Culture and Environment. London: Routledge.

Silva, L. (2013). Concluding Remarks on Perceived and Lived Ruralities and the Future of Rural Europe. In Silva, L. & Figueiredo, E. (Eds.), Shaping Rural Areas in Europe. Perceptions and Outcomes on the Present and the Future (pp. 247–254). Dordrecht: Springer.

Stolarick, K., Denstedt, M., Donald, B., & Spencer, M. G. (2010). Creativity, tourism and economic development in a rural context: The case of Prince Edward County. Journal of Rural and Community Development, 5 (1–2), 238–254.

Sznajder, M., Przezbórska, L. & Scrimgeour, F. (2009). Agritourism. Wallingford (UK), Cambridge (MA): CABI.

Tschofen, B. (2014). Tourismusarchitektur als Atmosphärenarbeit. In Kunst Meran (Ed.), Journal zur Ausstellung Alpen Architektur Tourismus am Beispiel Südtirol (pp. 48–49). Bozen.

Urry, J. (1991). The tourist gaze. London: Sage.

Uusitalo, E. & Assmuth, L. (2013). Having, Loosing, Being in the Periphery: Interpretations of Locality in the National Landscape of Koli, Eastern Finland. In Silva, L. & Figueiredo, E. (Eds.), Shaping Rural Areas in Europe. Perceptions and Outcomes on the Present and the Future (pp. 57–73). Dordrecht: Springer.

Weidinger, T. (2015). Encountering local inhabitants' perspectives in terms of authenticity: The example of rural tourism in Southern Germany. Dos Algarves 25, 4–24.

Williams, A. & Hall, M. (2000). Tourism and migration: new relationships between production and consumption. Tourism Geographies 2 (1), 5–27.

Woods, M. (2005). Rural geography: Processes, responses and experiences in rural restructuring. London: Sage.

Woods, M. (2010). Performing rurality and practising rural geography. Progress in Human Geography 34 (6), 835–846.

Woods, M. (2013). Rural Development, Globalization and European Regional Policy: Perspectives from the DERREG Project. Geographia Polonia 86 (2), 99–109.

Woods, M. (2014). Family Farming in the Global Countryside. Anthropological Notebooks 20 (3), 31–48.

Wright, W. & Annes, A. (2014). Farm Women and Agritourism: Representing a New Rurality. Sociologia Ruralis 54 (4), 477–499.

Yarwood, R. (2005). Beyond the Rural Idyll: Images, Countryside change and geography. Geography 90 (1), 19–31.

Intra-European lifestyle migration of German (pre)retirees to rural Spain: a contribution to rural development in times of the financial and real estate crisis? The migrants' point of view

Tobias Weidinger and Stefan Kordel

Abstract

This article concerns lifestyle and leisure-oriented migration of German (pre)retirees aged 50+ to the hinterland of the Spanish Costa Blanca. Against the backdrop of the current financial and real estate crisis it is discussed, how migrants, addressed as relatively affluent, contribute to rural development in terms of economic and social practices. Deriving from migrants' biographies, analyses take an individual perspective for a deeper understanding of practices due to their desired way of living. Results show that the migration decision, which is inspired by idealised notions of the countryside, i.e., a rural idyll, and repeatedly negotiated through meaningful practices in post-migration lives. In light of the crisis, they both persist on and alter practices with regard to consumption or civil commitment in order to support local population and thus uphold idyllic notions of rurality. Thereby, everyday practices also reflect their relatively affluent status, which, from their perspective, contribute to rural development.

Intra-European North-South migration: An old phenomenon in a changed context

Since the EU treaty allowed citizens to choose a site of living within the EU, the scope of migration profoundly changed. Since then, especially intra-European migration increased and scientific debates focused on various types of migration, while individual motivations as well as various implications at destinations were at the core of analyses. While studies mainly focused on migration to prosperous metropolitan regions, rural and peripheral areas in Southern Europe have been sparsely considered, except from case studies, for instance, on working migration to rural Greece (Kasimis and Papadopoulos 2005), Spain (Camarero et al. 2012) and Portugal (Fonseca 2008). Other studies discussed international migration to rural areas for lifestyle-related motives and the search for the rural idyll in particular (Benson 2014; Eimermann et al. 2012). Migration processes of the specific age group of (pre)retirees were extensively illustrated for Mediterranean countries (for an overview see Casado Díaz et al. 2004; King et al. 2000). The latter group represents well-off people, for whom "the choice of a place to live is, in principle, almost entirely disconnected from income considerations" (Polèse 2009, 153–54). While hegemonic societal discourses in Europe mostly consider the elderly as a burden in light of demographic change, there is evidence about the economic effect for regions (ibid., 163–64; Vollet et al. 2005) as well as positive attitudes towards in-migration of (pre)retirees (Weidinger and Kordel 2015).

Given the context of the current financial and real estate crisis in Spain started in 2007 and its manifold consequences, this article focuses on German (pre)retirees in rural Spain and their contribution to rural development from the individuals' points of view. While few studies discussed the nexus of labour migration and crisis (Domínguez-Mujica et al. 2014) or family-oriented counterurbanisation and crisis (Gkartzios 2013), people participating in lifestyle and leisure-oriented moves to rural and peripheral areas were not analysed due to this specific setting.

This contribution aims to bring together a biographical-narrative approach on rural in-migration of (pre)retirees and rural development research. Migration biographies of lifestyle migrants, mostly residing for a long time at places at the Spanish Costa Blanca, which, by them, are individually perceived as rural, are analysed with regard to social practices throughout the crisis. To be more concrete, it is asked, what contribution do German (pre)retirees make with regard to rural development in Spain in times of the crisis from a self-reflexive perspective.

The article firstly refers to previous research on individual motivations of (pre)retirees moving to rural areas and takes into account specific practices on-site. Thereby, conceptual aspects concerning this and effects on rural development ascertained from intra-regional (pre)retirement migration are depicted. Secondly, the site of study, i.e., the northern Costa Blanca, Spain, and the applied methods are introduced. Thirdly, results are presented according to the research question set up above. Migrants' contributions to rural development are analysed by considering preferences for sites of living and consequences for real estate sector, consumption practices and civil commitment. Finally, the conclusion highlights the added value of lifestyle migration approach for analysing rural in-migration from an individual perspective.

State of research

Lifestyle migration with special focus on (pre)retirees

In current Western societies, a shift in migration motives from economic considerations towards issues of quality of life can be observed. Migration processes following the hedonic imperative, e.g., the strive for happiness, comfort or beautiful environments (Polèse 2009, 153) are conceptually embedded in a wide range of frameworks, set up by sociologists, ethnographers, anthropologists and human geographers. Considered as consumption led-migration, the amenity migration approach focuses on pull factors of destinations and thus highlights place-related issues, such as environmental quality or cultural richness (Moss and Glorioso 2014; Moss 2006). However, this perspective neglects individual needs that are associated to identity making processes, which are impressively mirrored in migration biographies. Lifestyle migration instead focuses on relatively affluent individuals moving to places that signify, for the migrant, a better quality of life (Benson and Osbaldiston 2014; Janoschka and Haas 2013; Benson and O'Reilly 2009a). Within this approach, representations of a better life and how such images have been altered or reproduced in post-migration lives in manifold ways, were analysed (Benson and O'Reilly 2009b, 610). Thus, another distinctive feature is the assumption that migration is seen as an ongoing search rather than a final act, which results in a repeated negotiation in post-migration life.

Both, drivers to relocate and post-migration lives are highly interlinked to other experiences of mobilities, e.g., travel (Kordel 2015b; for detailed analysis of tourism-migration nexus see Rodríguez 2001). Due to relatively high pensions and a huge amount of spare time, lifestyle or leisure-oriented moves are especially common among the specific age group of (pre)retirees (Kordel 2015a; Duncombe et al. 2003; also discussed as affluent mobility by Schriewer and Encinas 2007). Moreover, (pre)retirees' self-conceptions of ageing process changed towards a self-determined understanding, where distinctive consumption is considered as status symbol (Janoschka and Haas 2011). Intra-European movements are facilitated by EU legislations (Ackers and Dwyer 2000), cheap and easy travel (Rodríguez-Pose and Ketterer 2012) and specific structural conditions at destinations, e.g., the housing market. Popular target regions for North and Central Europeans aged 50+ were initially consistent to areas of mass tourism in Europe (Casado Díaz et al. 2004). Accordingly, coastal areas predominated, while more peripheral and rural areas have recently come into the fore (Benson 2014; Eimermann et al. 2012; for discussions about counterurbanisation see Buller and Hoggart 1994). Motives to migrate to rural communities inherit "a clear distinction between a hurried and distressed life before migration and the life sought after migration, with the promise of a liberating future away from the daily constraints of urbanisation" (Gaspar 2015, 17). Accordingly, in-migrants strive for a certain idealised vision of the countryside based on "middle-class myths of a lost rurality in northern Europe" (King et al. 2000, 33). Benson and O'Reilly (2009b, 620) describe these migrants as rural idyll seekers, for whom social harmony, continuity, and community spirit are important values in post-migration lives. Accordingly, the peculiarity of interactions with local inhabitants represents a meaningful signifier for lifestyle migrants and subsequently a part of the desired way to live.

While studies in the U.S. (Brown and Glasgow 2008), Latin America (Matarrita-Cascante and Stocks 2013), Europe (Steinicke et al. 2012) and Australia (Davies 2014), carried out from the amenity migration angle, already dealt with questions of rural development through relatively affluent newcomers, e.g., retirees, the huge amount of case studies on lifestyle migration are lacking this perspective (except Benson 2015 and Weidinger and Kordel 2015).

In-migration of relatively affluent (pre)retirees and implications for rural development

As migration of older people to rural areas is reshaping both the physical, economic and social character of places (Davies 2014), it is discussed controversially, as both a boon and a burden (Haas 1990). Various studies, mainly carried out by gerontologists and economists in the U.S., focused on these implications, stressing the demand of in-migrants on the real estate market and the construction sector (Vollet et al. 2005). As a consequence, abandoned properties can be restored and used as retirement retreats (Buller and Hoggart 1994). However, this may result in rising real estate prices in situ, which can displace young families (Schmied 2005). Subsequently, land use conflicts may arise between retirees' urban and leisure-oriented land use and economic activities at destinations, e.g., farming and industry (Zasada et al. 2010; Janoschka 2009a; Janoschka 2009b; Vollet et al. 2005).

As studies, carried out in the U.S., show, retired migrants are not only relatively affluent compared to the autochthonous elderly population (Bennett 1993), they also tend to have high disposable incomes due to their pensions and savings, which are frequently addressed as immune to economic downturns (Brown and Glasgow 2008). Most of this money earned elsewhere is spent in the receiving community — apart from local taxes on food, appliances, clothing, entertainment and health services (Polèse 2009; Vollet et al. 2005), but also for furniture, cars, clothing or toys for grandchildren (Bennett 1993). Additionally, visiting friends and relatives (VFR) tourism provides a further basis for consumption expenditures. Thereby, local economy could overcome seasonal tourism cycle, create additional demand for both tourists, local inhabitants and newcomers (Sunil et al. 2007) and increase the demand for additional employment. A new phase of growth and socio-economic transformation resulted in labelling (pre)retirees as "an ideal economic stimuli" for less-populated areas (Isserman

1994). Nevertheless, the in-migration of (pre)retirees could lead to rising living costs. Negative consequences could also occur, ever, if migrants' health deteriorates as this may be accompanied with an increased demand and rising costs for medical facilities (Bennett 1993).

Newcomers can also be an important social resource for communities as they are mostly considered to be relatively better educated and have allocable spare time. Thus, they are more likely to be involved in civic engagement, both socially and politically (Brown et al. 2011). In-migrants may also donate for local purposes (Schmied 2005) and stimulate the arts and cultural scene (Brown et al. 2011). As a consequence, various stakeholders, for instance civic organisations or local governments, can benefit from the knowledge of newcomers (ibid.; Bennett 1993). However, many senior citizens only become involved politically, if they feel treated negatively by political decisions in the past or fear to be treated negatively in the future (Janoschka 2009a).

While consequences of in-migration of (pre)retirees on rural development were analysed by means of statistical measures and from the perspective of various local stakeholders, e.g., economy, social actors or local politics, the understanding of practices from individuals' migration biographies is central to this contribution. To put it in other words, the migration decision is repeatedly negotiated in post-migration lives through meaningful everyday practices. Since migrants' relative privilege is reflected within their everyday practices, they can become aware of their affluence and reflect upon their practices concomitant with economic and social consequences (Benson 2014, 61). Thus, lifestyle migration can provide an explanatory approach for implications of in-migration of relatively affluent (pre)retirees on local inhabitants, hierarchies and cultures. To sum up, this paper aims to bring together qualitative research from individuals' perspective and rural development. In light of the construction of a better way of life in a rural setting, it is especially relevant how migrants perceive both individual motivations and practices, which reaffirm the existence in Spain and individual explanations for development.

Methodological approach

Site of the study

In the course of the proliferation of mass tourism in the 1960s and 1970s, Spanish Mediterranean coastal areas were discovered by national and international real estate promoters, who met a demand for holiday homes in a tourist setting. As a consequence, a massive construction boom in terms of infrastructure and real estate (roads, hotels and holiday homes) began. Manifestations of the latter are spatially scattered urbanisations that are newly established urbanised residential complexes without any link to the historic core centre (Mantecón et al. 2009; Jurdao 1979). Specific legislations in Spain in the 1980s encouraged foreigners to purchase properties and to settle down as tourists, entrepreneurs or retirement migrants in order to compensate seasonality in the tourism sector (O'Reilly 2000, 480). This development could accelerate and immerse towards the hinterland of coastal areas (Figure 1) and resulted in an urban and leisure-oriented land use displacing traditional agricultural cultivation of fruit, vine, almonds, olives and citrus fruits (Zasada et al. 2010, 137)[1].

Figure 1: View of the Vall de Pop / Jalón Valle
Source: Tobias Weidinger 2015

In line with the framework of tourism life-cycle, retirement migration to Southern Europe grows alongside tourism from the 1970s (Rodríguez 2001; Foster and Murphy 1991). Former regular guests decided to use their former holiday homes as permanent retirement retreats.

---

[1] Furthermore, the government implemented neoliberal land regulations in the Land Act of 1998, which facilitated the dedication of land for building activities to a considerable extent (García 2010, 971).

To date, the province of Alicante with its 300+ days of sunshine and an annual average temperature of 19° Celsius received a massive influx of Northern and Central European (pre)retirees[2]. Accordingly, Alicante represents the province with the highest proportion of foreign residents all over Spain (26% in 2014, according to INE 2015). The number of registered German residents, for example, rose from 3 853 (1986), to 13 676 (1998) to 22 269 (2014), thereof 16 269 aged 50+ (INE 2015; Valero Escandell 1989). Thus, at least 23% of all German (pre)retirees registered in Spain live in this province. In the last two decades, especially the hinterland of the Marina Alta, i.e., the north eastern part of the province, gained more and more importance as a place of residence for German (pre)retirees[3].

The Marina Alta is characterised by a large landscape diversity, including sandy beaches and rugged coastlines with small bays as well as low mountain ranges in close distance to the coastline. Spurs of the Baetic Cordillera can reach altitudes of up to 1 000 m a.s.l.

The Vall de Pop, which was branded as Jalón Valley by the real estate industry, is considered as one of the "hot spots" of rural in-migration. While agriculture predominates on the valley bottom, the urbanisations mentioned above were built on the hillsides (Figure 1). The village of Lliber, for instance, represents the municipality with the second highest share of foreigners in the entire country (70.5%, Table 1), whereby lifestyle migrants from Northern and Central Europe, i.e., mainly British and German (pre)retirees are the largest group. In some municipalities also working migrants from Eastern Europe, Latin America and Africa were attracted for employment in agriculture or the construction sector (Collantes et al. 2014; Bayona-i-Carrasco and Gil-Alonso 2013; Morén-Alegret 2008).

---

2 Beside the Costa Blanca and its hinterland, the Balearic Islands (Friedrich and Kaiser 2001), the Canary Islands (Breuer 2005) and the Costa del Sol (Kordel 2015) represent other numerically important destinations for German (pre)retirees, who want to retire abroad. It is important to know that statistical figures underestimate the actual number of residents, as individuals often do not register at their new site of living because of different reasons (Schriewer and Encinas 2007).

3 According to Gibler and Taltavull (2010) the younger and more affluent (pre)retirees settle in the north eastern part of the province, whereas price-sensitive house-owners mainly choose the Southern part of the Province.

Table 1: Population development in Llíber (Vall de Pop)

	1999	*2004*	*2009*	*2014*
Inhabitants	524	837	1057	1080
Share of foreign residents	44.7	58.1	67.4	70.5
British residents	141	265	463	499
German residents	42	104	119	116
Share of population aged 50+	58.8	54.6	56.4	62.4
Share of population aged 50+ (national average)	32.8	33.1	33.8	36.7

Source: INE 2015

The financial and real estate crisis, deriving from the burst of the real estate bubble resulted in remarkable chain effects and changes in the Spanish economy: nearly 2.3 million people lost their jobs and public debts increased (López and Rodríguez 2011; García 2010, 968). Consequently, massive cut-backs in welfare state and dispossessions of persons who were not longer able to service their loans lead to social protest movements among all realms of society, e.g., "Juventud Sin Futuro" (JSF), "Indignados" or "movimiento 15-M" (Abellán et al. 2012). However, García pointed out that "it is possible to observe some city (and regional) variations in the creation of and reactions to the crisis" (ibid.; for detailed analysis see Charnock et al. 2014). Until the first quarter of 2008, the unemployment rate in the province of Alicante increased from 11.0% (1/2008) to 29.5% (4/2013), continuously representing a figure that is comparably higher than the Spanish average (9.6%, 1/2008 and 25.7%, 1/2013, all figures from INE 2015) and also fluctuating during the course of the year due to seasonality in tourism sector. As foreign workers are disproportionately affected by dismissions, subsequent emigration processes came to the fore (Domínguez-Mujica et al. 2014). For the specific group of Northern European retirees, Huete et al. (2013) indicate a certain degree of emigration from Spain. Yet, when explaining out-migration with implications of the financial and real estate crisis, the authors neglect specific registration practices on the one hand, e.g., de-registering for avoiding taxations but keep on staying in Spain illegally. On the other hand, age-related constraints, decline of health or the loss of a spouse (Betty and Hall 2015) continue to exist as reasons for emigration even during the crisis.

Methods

Drawing on the assumption of lifestyle migration research that individuals repeatedly negotiate better lives at destinations, the migration process has to be addressed as part of one's biography (Halfacree and Boyle 1993, 337; Halfacree and Rivera 2012), which goes beyond pure residential change. Approaching individuals' post-migration lives by reconstructing biographies deepens the "understanding the way individuals and households experience the place of living" (Stenbacka 2012, 60). Particularly long-time experiences at a certain place of living make individuals sensitive in evaluating transformations at destinations.

The research project titled with "Crisis! What Crisis? Europe! What Europe?"[4] focused on how lifestyle migrants from Northern and Central Europe perceive consequences of the current financial and real estate crisis in Spain and subsequently maintain or alter everyday practices. In the course of the study, 39 qualitative interviews with German and British (pre)retirees aged 54 to 73 aimed to depict narratives about the perception and the reaction on the crisis, forms of political and social commitment having derived from the crisis. For this purpose, interviews encompassed the following topics: reasons for moving, the choice of current site of living, involvement in the local community, social life, representations of town and countryside. In case respondents did not refer to aspects of the crisis during the interview, they were asked to elaborate their perceptions and their particular role at the end of each interview.

German and British (pre)retirees were selected as they represent the largest or second largest foreign nationality in the majority of the municipalities at the northern Costa Blanca (county Marina Alta) — both at the coastline and the hinterland. In line with the relational approach of Rivera Escribano (2007), who suggests symbolic and perceived distance from town as valuable when evaluating rurality, the hinterland cannot be separated from the coastal area by means of metric distance (contrary to what Huete et al. 2013 suggested, i.e., a ten kilometre line from the coast). Accordingly, only migrants, who defined themselves as living on the countryside were considered for analysis. To be more concrete, narratives of distinction towards the coastal areas were addressed as distinguish-

4 The project was initiated by Dr. Stefan Kordel and Dr. Andrés Gerique at the Institute of Geography at the University of Erlangen-Nuremberg in 2013.

ing criterion[5]. Biographical-narrative interviews were conducted in December 2013 and March 2015, while respondents were selected randomised in public space during markets and festivities, in cafés and bars or at the doorstep. The average duration of the interviews was about one hour. All interviewees were mainly at good mental and physical health.

Eight biographical-narrative interviews with Germans are central to this contribution (Table 2). The interviewees are either retired or live solely on their savings earned in previous working life; respectively pursuing current activities in the arts or the tourism sector (lifestyle entrepreneurs, Stone and Stubbs 2007). Respondents either live permanently in Spain or even stayed there for a long period of time that ranges from eight to 43 years. The latter represents a peculiarity, which distinguish German lifestyle migrants in the hinterland from British citizens, who have arrived relatively late (Huete et al. 2013). Three types of housing predominate among respondents: old fincas or farms on the campo, i.e., large properties partially used for agricultural cultivation, hillside urbanisations and houses situated in the village centre. Except from one, all dwellings were bought and not rented (Table 2).

5 The words "countryside", "Dorf", "dörflich", "Hinterland", "Land", "ländlich" and "rural" acted as codes.

Table 2: Sample of interviews presented in this contribution

Pseudonym	*Age*	*Status*	*Family status*	*Type of dwelling*	*Place of residence*	*Duration of stay and period of residence in Spain*
Joseph	68	retired	liaised	finca on the campo	Teulada	permanently, since 2003
Peter & Claudia	70/68	both retired	married	urbanisation	Orba	six months/ year, since 1972
Tom & Sarah	73/61	both retired	married	house in the village	Tormos	permanently, since 2005
Barbara	55	fine artist	married	farm on the campo	Jalón	permanently, since 1984
Maria	64	fine artist	widowed	house in the village	Sanet y Negrals	permanently, since 2000
Judith	60	retired, renting a holiday apartment	married	finca on the campo	Benissa	permanently, since 2005
Kerstin	64	retired	married	urbanisation	Jalón	permanently, since 1995
Franziska	54	single mother	divorced	house in the village	Alcalalí	permanently, since 2007

Source: Authors' compilation

Prior to the biographical-narrative interviews, 24 guideline-based, problem-centred interviews with local stakeholders, both Spaniards and foreigners, involved in tourism, real estate and housing, social and political affairs, were conducted in April and December 2013 in order to get a more comprehensive view of local implications of the financial and real estate crisis. For all qualitative interviews, data processing was undertaken as follows: firstly, all interviews were transcribed literally and then translated into English; secondly, data were treated by narrative analysis (Lieblich et al. 1998). In order to guarantee anonymity of interviewed persons, pseudonyms were used.

Contribution to a cushioned landing? Self-reflexive ascriptions of German (pre)retirees in times of the crisis

The presentation and discussion of empirical results is organised in three sections: At first, migrants' contributions to rural development are analysed with regard to preferences for sites of living and consequences for the real estate sector, followed by considering their consumption practices and finally approaching perspectives on civil commitment.

Preferences for sites of living and consequences for the real estate sector

While various studies on European citizens at Spanish coastal areas emphasised living preferences in urbanisations close to tourist facilities (Friedrich and Kaiser 2001; King et al. 2000), a couple, who migrated to the Costa Blanca hinterland, drew another picture when reflecting on motivations. After searching for a new domicile in Spain and Greece, they settled in the small village of Tormos (317 inhabitants in 2014).

> *Tom (addressed to his wife): "Would you like to answer or shall I?"*
>
> *Sarah: "It's simple. We want to escape tourism. We do not want to be part of the tourist stuff. And in the hinterland it's cosier. Costs are lower than in Dénia (author's remark: city at the coast line)."*
>
> *Tom: "If we want to have trouble, anytime, we are in the thick of it within a 20 minutes' drive [...]. If I want to have my silence, I settle down in the hinterland. It's completely different regarding the sense of community. Everyone knows everyone here."*

Deriving from the assumption that individuals evaluate the site of living due to their personal situation in relation to other places (Kordel 2016), in-migrants are weighing pros and cons, for instance, with regard to economic aspects. Lower costs for housing and living make it possible to fulfil one's own dream of the good life, e.g., the restoration of a traditional building in a village or the construction of a finca (Kordel 2015b). The desire for dissociating themselves from ordinary tourists is achieved by means of refusing tourist practices on the one hand and by spatial segregation on the other, e.g., selection of a residential area in the hinterland (see the findings on German second home owners in Tuscany, Seidl 2009). Notwithstanding the latter, they choose locations, which are easily accessible to the coast in order to be able to take advantage of various offerings. Besides this, other consumption-related motivations come to the fore, encompassing both tranquillity, authenticity and a sense of community, which are assessed as desirable for a better way of life.

The quest for a specific way of living associated to rural areas among German (pre)retirees, manifested through relocation to the hinterland, had massive implications for the village population and its development, as stated by a German editor of a newspaper:

> *"The whole society benefitted from the construction sector with its string of consequences. You have to make yourself aware of! [...] The gardener, the furniture shop, the plumber, the carpenter — all are connected to the construction sector. [...] It's a village population here. They are craftsmen and farmer. Most of them are common people. All of them jumped upon the bandwagon. It failed, but they still keep going." (male, German)*

Having observed positive implications for local economy resulting from inmigration of relatively affluent people, the editor acknowledges a development towards a mono-functional structure of the labour market based on the construction sector and services linked to it. Even in the course of the crisis the way of doing business is not questioned. German (pre)retirees residing in the hinterland for many years observed this development on the long run. Accordingly, they evaluate the connection between the real estate sector and implications of the crisis.

> *Franziska: "It's not just since the last five years. It started over 20 years ago and was a very creeping process. Deriving from the construction boom — another urbanisation and yet another. It was foreseeable that the bubble will burst one day, especially when you have 1 000 prospective buyers and 100 000 houses. [...] In a manner of speaking, nobody has a job here. But there are no poor people. Not even among the Spanish. [...] This is because everyone owns a great deal of land here. [...] The hills, where urbanisations are located now, also belonged to the people. In the '70s/'80s/'90s the property was sold for a tremendous amount of money, which is now in their bank accounts. And they subsist on it."*

Sales of land and the construction sector could contribute to create wealth and prosperity among local population in the short run. German (pre)retirees are aware of their role and critically acknowledge that local economies stick to the construction sector and related services for relatively affluent migrants in the long run. For this kind of orientation of the local economy, i.e., capturing wealth that is gained outside the region (Polèse 2009, 164), Davezies (2008) introduced the term "residential economies". In terms of developing rural economies, a certain path-dependency is indicated.

Consumption practices

According to their desired way of life in the rural, German (pre)retirees show everyday practices which aim to establish a close relationship between them and members of the local community. Acting locally, migrants want to identify themselves with the host community (Janoschka 2009a, 135). As noted in the context of second home dwellers, Manzo and Perkins (2006, 339) emphasized that individuals who are involved in their neighbourhood are more willing to spend time and money on it.

> *Sarah: "At the very beginning we were both thinking that we are here for just a visit. Well, that we are guests. Meanwhile I don't view this the same way. We spend our pension here. We pay our taxes. We are residents here. We pay our property taxes. Every year we go to the tax office to do our income-tax return. Now I'm a member of the Spanish society who actually spends more than a Spaniard."*

The respondent is conscious about her economic power. Relative affluence through regular, disposable income compared to local population is associated with positive implications for fiscal revenue of municipalities and local economy, i.e., spending of pensions on-site. Besides fiscal dimensions, German (pre)retirees evaluate their consumption expenditures.

> *Joseph: "If foreigners were not demanding that much, crisis would be even worse. As far as I know the figures, the majority of economic profit is generated from European money, i.e., from Central Europe. Restaurants and everything that is attached with it; shopping opportunities, starting with Lidl, Aldi through to high-priced consumption goods. Spaniards do not have the money, but foreigners are purchasing on a large scale. In my opinion, this is a worthy cause for Spaniards."*

Highlighting the role of relatively affluent individuals from Northern and Central Europe in the specific context of the current crisis, Joseph once more refers to economic power. Especially foreigners do not reduce expenditures, but maintain a consumption-oriented way of life. Foreigner's demand for goods and services may contribute to keep jobs on-site and diminishes consequences of the crisis for Spaniards.

> *Claudia: "If tourists or those, who live here permanently, didn't purchase, it would be even worse for them. Therefore, I always try to buy locally to support the local people. When we have a problem with electricity, we always go down to the village and say: Come on, do us a favour! And they appreciate that. That is possibly because we are more rural than perhaps in Dénia."*

A binary purpose for consuming locally is addressed in the quotation above: firstly, migrants argue to support local economies and secondly benefit from proximity and regular interactions to local craftsmen as they get problems solved easily. For the majority, these practices are distinctive from those commonly ascribed to people living at coastal areas. The desire for involvement in local economies reflects the desired way of life addressed as rural.

Crisis-related effects especially for the village population, threatens (pre)retirees' rural and social idyll. In order to mitigate those negative effects and to resolve the mismatch between represented and lived rurality (Halfacree 2014, 98), (pre)retirees alter practices with regard to consumption, i.e., buying more locally.

Civil commitment

While social involvement is considered as compensation for absent family and kins as well as changing social roles, previously associated to working life, (pre)retirees tend to re-define their social interactions in post-migration lives (Haas 2013). Thus, meaningful activities and practices such as neighbourly help, participation in clubs and associations or charitable commitment structure their everyday life, provide sociability and contribute to one's self-development in terms of continuity (ibid. 1385–86). In the Vall de Pop, a voluntary organisation (Jalón Valley Help) was founded by British residents as early as 1992 with the objective to "help those in need within the Vall de Pop and to support local charities and organisations" (homepage, Jalón Valley Help 2015). In times of the crisis, a local councilwoman for social affairs explains a raising number of poor people who become increasingly less protected by family and kin. She also recognises changing practices of lifestyle migrants:

> *"It's not like before, perhaps, when one family has a difficult particular situation. We all have to deal with it. And you find that people are working together. [...] There's a lot more solidarity amongst the foreign residents. The foreigners are much more conscious. And the German Club has also given, in the last 3 years at Christmas time, a lot of money to Caritas. Caritas sometimes pay people's electricity bills, when they have no money, but mainly is food. I find that's the Europeans, who are much more willing to help. Councillor for social affairs in municipality." (female, British)*

Two explanations for civic involvement exist: lifestyle migrants render their economic status visible for compatriots or others and thereby produce reputation (Haas 2013). Moreover, volunteering is addressed as a mean to get in contact with Spanish population and finally an effort for integration (ibid.; Janoschka and Haas 2011, 287). As stated above, however, migrants justify involvement in charitable organisations by referring to the crisis at present. While Haas (2013) pointed out that these practices predominate among British citizens due to cultural notions, such practices play a minor role among German (pre)retirees. Instead, they criticise British compatriots for their desire to get attention by the media. Those migrants maintaining loose contacts to Spanish neighbours or acquaintances are familiar with the current situation of family members and prefer individual and personal assistance in manifold ways.

> *Judith: "Spaniards would not accept, when you approach them and say: Things are really going bad for you. [...] I hand a 100 euro over to you. You can't do that. They would never accept. As well as they would never seek your aid. But when you are aware of their situation and you are a caring person, you can help by hiring someone and no longer hire a Polish worker, who always did a great and cheap job as an illicit worker. You turn to your neighbour instead, when you know, that he can do this job, too. Well, you willingly pay slightly more if you hire the Spaniard. [...] I have a cleaning lady, whom I employ, because she is absolutely in need of money. And I also employ her spouse, because they are pretty good Spanish neighbours."*

Precondition for practices, such as hiring people from the village in need for help are established social contacts, which are mainly based on the desired way of living in a rural municipality (Short and Stockdale 1999). Accordingly, Judith is distancing herself from those migrants, who have limited contact with Spanish people. Heaving learned cultural codes and habitus in order to become part of the rural community (Sardinha 2015, 44), the respondent is able to act in the appropriate way, i.e., she knows how to support them in times of the crisis. It becomes obvious that Judith changes her everyday practices, again, in order to preserve a rural idyll. However, by taking on the role as employers, migrants perform privilege, and, as Benson (2014, 62) argues for US citizens in Panama, they "enter into a system of class relations". This working alliance finally counteracts the desired egalitarian relationship between migrants and locals.

Conclusion

This article concerned lifestyle and leisure-oriented migration to areas perceived as rural. Thereby, rural and peripheral areas were not defined by means of statistical measures, but according to a relational and social-constructivist understanding of space. Ascribing meanings to the hinterland of the Costa Blanca, Spain, they aim to fulfil a better way of life. Against the backdrop of their relatively affluent status on the one and the current financial and real estate crisis on the other hand, it was asked, how German lifestyle movers perceive to contribute to rural development. Thereby, the contribution could add value to existing literature in two dimensions:

Firstly, in order to grasp implications of in-migration for rural development in the specific context of the crisis, it is essential to understand the migration process as ongoing search rather than a final act. Thus, analyses of post-migration lives from individuals' perspectives provide a basis for discussing economic and social implications for local inhabitants, hierarchies and cultures. It was shown that German (pre)retirees evaluate their contribution to rural development in times of the crisis positively in terms of consequences for the real estate sector, consumption expenditures and social welfare, e.g., charity activities and neighbourly help. Moreover, they maintain and intensify certain practices in order to mitigate the consequences of the crisis, while changing others.

According to the claim set up by Gaspar (2015, 26) "to identify the distinctive or common features of [rural in-migration] when compared with the wider trends of sun-seeking retirement migration", perception of the crisis and practices responding to it can be considered as slightly different from the urbanised coast. While living spatially distant from Spanish society predominates in urbanisations and anonymity is indicated, face to face assistance is common among rural dwellers. In order to preserve the specific rural way of life, which initially attracted them, migrants strive for various cooperation between local population and them in times of crisis. Thus, locals become co-constructors for migrants' better way of lives (Weidinger and Kordel 2015, 38).

Secondly, steering towards individuals' migration biographies is considered as a methodological surplus. Particularly in light of long-time experiences at a new chosen site of living, individuals are able to evaluate transformations having occurred over a long period of time. Results show that social practices are interlinked to the crisis to some extent. Thus, a deeper understanding of migrants' practices, carried out by means of qualitative research, could be valuable for applied questions on rural development.

Acknowledgments

We are grateful to the German Academic Exchange Service (DAAD) for partially funding this research in the course of the "University dialogue with Southern Europe" programme (Hochschuldialog Südeuropa). Furthermore, we thank our students at the Institute of Geography, who assisted in conducting interviews at the Costa Blanca in spring and winter 2013 as part of a student's project.

References

Abellán, J., Sequera, J. & Janoschka, M. (2012). Occupying the #Hotelmadrid: A Laboratory for Urban Resistance. Social Movement Studies: Journal of Social, Cultural and Political Protest 11 (3–4), 320–326.

Ackers, L. & Dwyer, P. (2004). Fixed laws, fluid lives: the citizenship status of post-retirement migrants in the European Union. Ageing and Society 24 (3), 451–475.

Bayona-i-Carrasco, J. & Gil-Alonso, F. (2013). Is Foreign Immigration the Solution to Rural Depopulation? The Case of Catalonia (1996–2009). Sociologia Ruralis 53 (1), 26–51.

Bennett, D. G. (1993). Retirement Migration and Economic Development in High-Amenity, Nonmetropolitan Areas. The Journal of Applied Gerontology 12 (4), 466–481.

Benson, M. (2014). Negotiating Privilege in and through Lifestyle Migration. In Benson, M. & Osbaldiston, N. (Eds.), Understanding Lifestyle Migration. Theoretical Approaches to Migration and the Quest for a Better Way of Life (pp. 47–68). Basingstoke: Palgrave Macmillan.

Benson, M. (2015). Class, Race, Privilege: Structuring the Lifestyle Migrant Experience in Boquete, Panama. Journal of Latin American Geography 14 (1), 19–37.

Benson, M. & O'Reilly, K. (2009a) (Eds.). Lifestyle Migration. Expectations, Aspirations and Experiences. Farnham: Ashgate.

Benson, M. & O'Reilly, K. (2009b). Migration and the search for a better way of life: A critical exploration of lifestyle migration. Sociological Review 57 (4), 608–625.

Benson, M. & Osbaldiston, N. (2014) (Eds.). Understanding Lifestyle Migration. Theoretical Approaches to Migration and the Quest for a Better Way of Life. Basingstoke: Palgrave Macmillan.

Betty, C. & Hall, K. (2015). The myth of no return? Why retired British migrants in Spain return to the UK. In Torkington, K., David, I. & Sardinha, J. (Eds.), Practising the Good Life. Lifestyle Migration in Practices (pp. 123–136). Newcastle upon Tyne: Cambridge Scholars Publishing.

Breuer, T. (2005). Retirement Migration or rather Second-Home Tourism? German Senior Citizen on the Canary Islands. Die Erde 136 (3), 313–333.

Brown, D. L., Bolender, B. C., Kulcsár, L. J., Glasgow, N. & Sanders, S. (2011). Intercounty Variability of Net Migration at Older Ages as a Path-Dependent Process. Rural Sociology 76 (1), 44–73.

Brown, D. L. & Glasgow, N. (2008): Rural Retirement Migration. Berlin: Springer.

Buller, H. & Hoggart, K. (1994). International Counterurbanization: British Migrants in Rural France. Avebury: Aldershot.

Camarero, L., Sampedro, R. & Oliva, J. (2012). Foreigners, neighbours, immigrants: translocal mobilities in rural areas in Spain. In Hedberg, C. & do Carmo, R. M. (Eds.), Translocal Ruralism: Mobility and Connectivity in European Rural Spaces (pp. 143–162). Dordrecht: Springer.

Casado-Diaz, M. A., Kaiser, C. & Warnes, A. M. (2004). Northern European retired residents in nine southern European areas: characteristics, motivations and adjustment. Ageing & Society 24 (3), 353–381.

Charnock, G., Purcell, T. & Ribera-Fumaz, R. (2014). The Limits to Capital in Spain. Crisis and Revolt in the European South. Basingstoke: Palgrave Macmillan.

Collantes, F., Pinilla, V., Sáez, L. A. & Silvestre, J. (2014). Reducing Depopulation in Rural Spain: The Impact of Immigration. Population Space and Place 20 (7), 606–621.

Davezies (2008). La République et ses territories: la circulation invisible des richesses. Paris: Éditions du Seuil.

Davies, A. (2014). Urban to Rural Elderly Migration: Renewing and Reinventing Australia's Small Rural Towns. In Dufty-Jones, R. & Connell, J. (Eds.), Rural change in Australia (pp. 43–55). Farnham: Ashgate.

Domínguez-Mujica, J., Guerra-Talavera, R. & Parreño-Castellano, J. M. (2014). Migration at a Time of Global Economic Crisis: The Situation in Spain. International Migration 52 (5), 113–127.

Duncombe, W., Robbins, M. & Wolf, D. A. (2003). Place Characteristics and Residential Location Choice Among the Retirement-Age Population. Journal of Gerontology 58 (4), 244–252.

Eimermann, M., Lundmark, M. & Müller, D. (2012). Exploring Dutch Migration to Rural Sweden: International Counterurbanisation in the EU. Tiidschrift voor economische en sociale geografie 103 (3), 330–346.

Fonseca, M. L. (2008). New waves of immigration to small towns and rural areas in Portugal. Population, Space and Place 14 (6), 525–535.

Foster, D. M. & Murphy, P. (1991). Resort cycle revisited: The retirement connection. Annals of Tourism Research 18 (4), 553–567.

Friedrich, K. & Kaiser, C. (2001). Rentnersiedlungen auf Mallorca? Möglichkeiten und Grenzen der Übertragbarkeit des nordamerikanischen Konzepts auf den Europäischen Sunbelt. Europa Regional 9 (4), 204–211.

García, M. (2010). The Breakdown of the Spanish Urban Growth Model: Social and Territorial Effects of the Global Crisis. International Journal of Urban and Regional Research 34 (4), 967–980.

Gaspar, S. (2015). In Search of the rural idyll: Lifestyle migrants across the European Union. In Torkington, K., David, I. & Sardinha, J. (Eds.), Practising the Good Life. Lifestyle Migration in Practices (pp. 14–32). Newcastle upon Tyne: Cambridge Scholars Publishing.

Gibler, K. M. & Taltavull, P. (2010). Using preferences for international retiree housing market segmentation. Journal of Property Research 27 (3), 221–237.

Gkartzios, M. (2013). 'Leaving Athens': Narratives of counterurbanisation in times of crisis. Journal of Rural Studies 32, 158–167.

Haas, H. (2013). Volunteering in retirement migration: meanings and functions of charitable activities for older British residents in Spain. Ageing and Society 33 (8), 1374–1400.

Haas, W. H. (1990). Retirement migration: Boon or burden? Journal of Applied Gerontology 9 (4), 387–392.

Halfacree, K. H. (2014). Jumping Up from the Armchair: Beyond the Idyll in Counterurbanisation. In Benson, M. & Osbaldiston, N. (Eds.), Understanding Lifestyle Migration. Theoretical Approaches to Migration and the Quest for a Better Way of Life (pp. 92–115). Basingstoke: Palgrave Macmillan.

Halfacree, K. H. & Boyle, P. J. (1993). The challenge facing migration research: the case for a biographical approach. Progress in Human Geography 17 (3), 333–348.

Halfacree, K. H. & Rivera, M. J. (2012). Moving to the Countryside ... and Staying: Lives beyond Representations. Sociologia Ruralis 52 (1), 92–114.

Huete, R., Mantecón, A. & Estévez, J. (2013). Challenges in Lifestyle Migration Research. Reflections and Findings about the Spanish Crisis. Mobilities 8 (3), 331–348.

INE — Instituto Nacional de Estadística (2015). INEbase. www.ine.es. Accessed 15 Mai 2015.

Isserman, A. M. (1994). State economic development policy and practice in the United States: A survey article. International Regional Science Review 16 (2), 49–100.

Jalon Valley Help (2015). Homepage. www.jalonvalleyhelp.com. Accessed 15 May 2015.

Janoschka, M. (2009a). Konstruktion europäischer Identitäten in räumlich-politischen Konflikten. Sozialgeographische Bibliothek 11. Stuttgart: Franz Steiner.

Janoschka, M. (2009b). The Contested Spaces of Lifestyle Mobilities. Regime Analysis as a Tool to Study Political Claims in Latin American Retirement Destinations. Die Erde 140 (3), 251–274.

Janoschka, M. & Haas, H. (2011). Stadtentwicklung, Lokalpolitik und Ehrenamt als Experimentierfelder des alltäglich gelebten Europas. Zur Partizipation in transnationalen Migrationsgesellschaften. In Johler, R., Matter, M. & Zinn-Thomas, S. (Eds.), Mobilitäten. Europa in Bewegung als Herausforderung kulturanalytischer Forschung (pp. 295–305). Münster: Waxmann.

Janoschka, M. & Haas, H. (2013) (Eds.). Contested Spatialities, Lifestyle Migration and Residential Tourism. London/New York: Routledge.

Jurdao, F. (1979). España en venta: compra de suelos por extranjeros y colonización de campesinos en la Costa del Sol. Madrid: Ayusa.

Kasimis, C. & Papadopoulos, A. G. (2005). The multifunctional role of migrants in the Greek countryside: implications for the rural economy and society. Journal of Ethnic and Migration Studies 31 (1), 99–127.

King, R., Warnes, T. & Williams, A. (2000). Sunset Lives. British Retirement Migration to the Mediterranean. Oxford: Berg.

Kordel, S. (2016). The production of spaces of the 'good life' — the case of lifestyle migrants in Spain. Leisure Studies 35 (2), 129–140.

Kordel, S. (2015a). Striving for the "good life" — home-making among senior citizens on the move. An analysis of German (pre)retirees in Spain and Germany in a continuum of tourism and migration. Doctoral thesis. University of Erlangen.

Kordel, S. (2015b). Being a tourist — being at home: Reconstructing tourist experiences and negotiating home in retirement migrants' daily lives. In Torkington, K., David, I. & Sardinha, J. (Eds.), Practising the Good Life. Lifestyle Migration in Practices (pp. 105–122). Newcastle upon Tyne: Cambridge Scholars Publishing.

Lieblich, A., Tuval-Mashiach, R. & Zilber, T. (1998). Narrative research. Reading, analysis and interpretation. Thousand Oaks: Sage.

López, I. & Rodríguez, E. (2011). Das spanische Modell. Zeitschrift für kritische Sozialwissenschaft 42 (1), 113–144.

Mantecón, A., Huete, R. & Mazón, T. (2009). Las urbanizaciones "europeas". Una investigación sobre las nuevas sociedades duales en el Mediterráneo. Scripta Nova. Revista Electrónica de Geografía y Ciencias Sociales XIII (301). Barcelona: Universidad de Barcelona. www.ub.es/geocrit/sn/sn-301.htm.

Manzo, L. C. & Perkins, D. D. (2006). Finding Common Ground: The Importance of Place Attachment to Community Participation and Planning. Journal of Planning Literature 20 (4), 335–350.

Matarrita-Cascante, D. & Stocks, G. (2013). Amenity migration to the global south: Implications for community development. Geoforum 49, 91–102.

Morén-Alegret, R. (2008). Ruralphilia and Urbophobia versus Urbophilia and Ruralphobia? Lessons from Immigrant Integration Processes in Spain. Population, Space and Place 14 (6), 537–552.

Moss, L. A. G. (2006) (Ed.). The Amenity Migrants. Seeking and Sustaining mountains and their cultures. Oxfordshire: CABI.

Moss, L. A. G. & Glorioso, R. (Eds.) (2014). Global Amenity Migration: Transforming Rural Culture, Economy and Landscape. Kaslo: New Ecology Press.

O'Reilly, K. (2000). 'New Europe, Old Boundaries': British Migrants in Spain. Journal of Social Welfare and Family Law 22 (4), 479–491.

Polèse, M. (2009). The Wealth and Poverty of Regions. Chicago, London: The University of Chicago Press.

Rivera Escribano, M. J. (2007). Migration to rural Navarre: Questioning the experience of counterurbanisation. Tijdschrift voor Economische en Sociale Geografie 98 (1), 32–41.

Rodríguez, V. (2001). Tourism as a recruiting post for retirement migration. Tourism Geographies: An International Journal of Tourism Space, Place and Environment 3 (1), 52–63.

Rodríguez-Pose, A. & Ketterer, T. D. (2012). Do local amenities affect appeal of regions for migrants. Journal of Regional Science 52 (4), 535–561.

Sardinha, J. (2015). Idyllic Seekers and Liminal Beings: Lifestyle Migrants in Central Portugal. In Torkington, K., David, I. & Sardinha, J. (Eds.), Practising the Good Life. Lifestyle Migration in Practices (pp. 33–51). Newcastle upon Tyne: Cambridge Scholars Publishing.

Schmied, D. (2005). Incomers and Locals in the European Countryside. In Schmied, D. (Ed.), Winning and losing: The changing geography of Europe's rural areas (pp. 141–166). Aldershot: Ashgate.

Schriewer, K. & Encinas, I. (2007). Being Misleading About Where One Resides. European Affluence Mobility and Registration Patterns. Ethnologia Europea Journal of European Ethnology 37 (1–2), 98–106.

Seidl, D. (2009). „Wir machen hier unser Italien…". Multilokalität deutscher Ferienhausbesitzer. Münster: Waxmann.

Short, J. & Stockdale, A. (1999). English migrants in the Scottish countryside: Opportunities for rural Scotland. Scottish Geographical Journal,115 (3), 177–192.

Steinicke, E., Čede, P. & Löffler, R. (2012). In-migration as a new process in demographic problem areas of the Alps. Ghost towns vs. amenity settlements in the Alpine boarder area between Italy and Slovenia. Erdkunde 66 (4), 329–344.

Stenbacka, S. (2012). "The Rural" Intervening in the Lives of Internal and International Migrants: Migrants, Biographies and Translocal Practices. In Hedberg, C. & do Carmo, R. M. (Eds.), Translocal Ruralism: Mobility and Connectivity in European Rural Spaces (pp. 55–72). Dordrecht: Springer.

Stone, I. & Stubbs, C. (2007). Enterprising expatriates: lifestyle migration and entrepreneurship in rural southern Europe. Entrepreneurship and Regional Development 19 (5), 433–450.

Sunil, T.S., Rojas, V. & Bradley, D. E. (2007). United States' international retirement migration: the reasons for retiring to the environs of Lake Chapala, Mexico. Ageing and Society 27 (4), 489–510.

Valero Escandell, J. R. (1989). La inmigración en la provincial de Alicante. Doctoral Thesis. Alicante.

Vollet, D., Callois, J.-M. & Roussel, V. (2005). Impact of Retirees on Rural Development: Some Observations from the South of France. The Journal of Regional Analysis & Policy 35 (2), 54–68.

Weidinger, T. & Kordel, S. (2015). German spa towns as retirement destinations: How tourists negotiate(d) relocation and locals assess in-migration. Dve Domovini / Two Homelands 42, 37–52.

Zasada, I., Alves, S., Müller, F.C., Piorr, A., Berges, R. & Bell, S. (2010). International retirement migration in the Alicante region, Spain: process, spatial pattern and environmental impacts. Journal of Environmental Planning and Management 53 (1), 125–141.

Art for the village: On the linkages between public and participatory art and village development in the rural periphery of Germany

Theresa Dietl

Abstract

In the face of demographic and structural change, spatial planning is in search of appropriate perspectives and instruments for the development of peripheral rural areas. The current uncertainty opens space for experiments such as artistic projects for village development. This leads to the question: What aspects can public and participatory artistic projects contribute to village development planning? A case study on "Kunst fürs Dorf — Dörfer für Kunst" (Art for the Village — Villages for Art) is presented, a German national competition and a project initiated by Deutsche Stiftung Kulturlandschaft that took place in 2013. The paper focuses on one of the three participating villages: Blankensee, which is located in the rural periphery of Mecklenburg-West Pomerania. Through the analysis of qualitative data (mainly interviews with actors involved), characteristics of the project are identified that might be inspirational for participatory village development processes in other places. Thereby, the focus is not put on the outcome but on the framework, as set by the foundation, and the processes as initiated by artist Barbara Caveng.

Introduction

"Art? That is not my business" one might think as a geographer or spatial planner. In this paper, nonetheless, it will be argued that art as an "unfettered form of creativity" (Rendell 2006, 3) has something to say that is relevant for geographers or spatial planners. Especially in the genres of public and participatory art, some elements link well to the current struggle and debate about village development planning in the face of demographic and structural change. Correspondingly, this paper is guided by the question: What aspects can public and participatory artistic projects contribute to village development processes?

To develop the argument, one example will be looked at in detail, an artistic project that took place in the frame of the German national competition "Kunst fürs Dorf — Dörfer für Kunst" (Art for the Village — Villages for Art) in 2013. The project was initiated and funded by the foundation Deutsche Stiftung Kulturlandschaft. There, three artists were selected to spend six months each in one rural community and develop together with the inhabitants an artistic project that made the place subject of discussion. In this paper, I will focus on one of these projects: "Mi kricht hier keener mehr wech" (nobody's gonna get me outta here) by Berlin-based visual artist Barbara Caveng.

It took place in the municipality of Blankensee which is located in the rural periphery of Mecklenburg-West Pomerania, right on the Polish border (Figure 1). Since its incorporation in 2004, it consists of the two neighbouring villages Blankensee and Pampow.

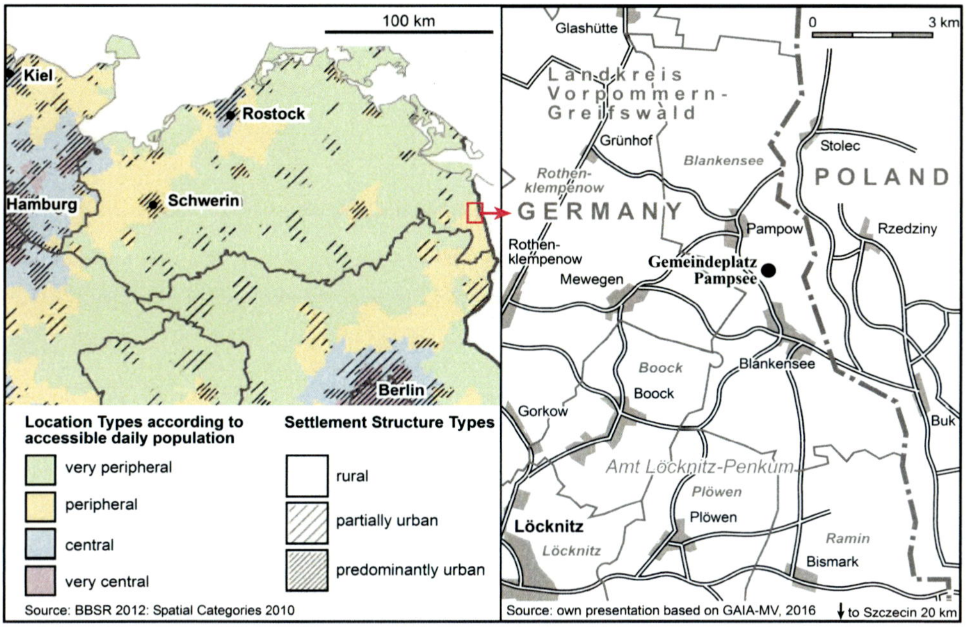

Figure 1: Location of the municipality of Blankensee

Source: Spatial Categories 2010 by BBSR 2012 and Theresa Dietl 2015, based on openstreetmap.org

The paper is divided into five sections. In section one, the background of the research, the reasons for the selection of the case study and the methodological approach are outlined. In section two, the challenges many peripheral villages face in the context of demographic and structural change are summarised. Furthermore, some attempts from the perspective of village development planning to react to them are outlined. In chapter three, the artwork that was developed in Blankensee is described and the art form it belongs to is characterized. In section four, my empirical findings from Kunst fürs Dorf in Blankensee are brought together with current approaches to village development planning, focussing on structural aspects — the project's framework as set by the funding institution — and on elements of the process on site, initiated by the artist. Finally, the findings are summarised and future prospects given.

Background and research methodology

Artistic activities in villages are manifold, ranging from local choirs, theatre groups or workshops of (local) artists to big festivals or artistic residencies (Kegler 2014; Henkel 2012). The focus in this paper is on an artistic project from the field of participatory and public art. Both art forms are interconnected. Participatory art[1] has been characterized by Claire Bishop, one of the most influential art historians in the field, as the following: "To put it simply: the artist is conceived less as an individual producer of discrete objects than as a collaborator and producer of situations; the work of art as a finite, portable, commodifiable product is reconceived as an ongoing or long-term project with an unclear beginning and end; while the audience, previously conceived as a 'viewer' or 'beholder', is now repositioned as a co-producer or participant" (Bishop 2012, 2, emphases in the original). While participatory art puts a focus, for example, on the participation of non-artists, public art emphasizes the public concern of the art project and its presentation outside the walls of a gallery or museum. Susan Lacy, who coined the term "new genre public art" in the 1990s, speaks about it as "visual art that uses both traditional and non-traditional media to communicate and interact with a broad and diversified audience about issues directly relevant to their lives [...] based on engagement" (Lacy 1995, 19).

In art theory, there is a need for in-depth research on processes of participatory and public art projects (Kester 2015). Moreover, looking at the German context and at such artistic activities in villages, research is still in its infancy. Just recently, Brita Polzer (2013) published a first collection of contemporary artistic activities in villages in German-speaking countries. Significantly more studies exist on artistic projects in cities in comparison to those in villages. However, they offer little insight for this paper since they usually focus on the outcomes of such activities rather than the processes; asking what do these projects contribute to urban development rather than what processes did lead to the contributions (for an overview of debates on "urban art" see, e.g., Hildebrandt 2012). So far, specifically the relationship between public and participatory artistic projects and village development planning has only been touched on in rather short articles. According to Simone Tippach-Schneider (2010), who focuses on the 2009 series of Kunst fürs Dorf, the analysed projects fostered dialogue

[1] Bishop states that numerous terms for "post-studio practices" exist, such as community art, socially engaged art, participatory art, collaborative art, contextual art, social practise (Bishop 2012, 1). Since this paper does not focus on art history, it is not relevant to get deeper into the distinctions among them.

among the inhabitants about their village's future and encouraged initiative and engagement after the projects. Simone Helmle and Claudia Neu (2010) summarise in their article a half-day session at a sociological conference where artistic projects in villages were presented. They highlight, for example, that such projects can raise awareness and create possibilities for communication and encounters. Also, central actors, who might not be determined a priori, are needed and a project's success is influenced by the capacity of villagers and artists to find their own way within an open process. As yet, the only in-depth study that exists offers a sociological point of view on Kunst fürs Dorf in its 2009 series. Marlen Schröder (2011) determines the encouragement of civil engagement by the artistic projects and the dependency of a project's success on the openness of the participants.

Given the variety of artistic projects taking place in rural areas, why have I chosen Kunst fürs Dorf as the case study to research the link of art and village development planning? First, because Kunst fürs Dorf (for the first time in its third series in 2013[2]) was a German wide and known project, where about 100 villages and 140 artists applied for. Second, because it was run by a foundation for rural development, the Deutsche Stiftung Kulturlandschaft, and not — as usually the case — by an art or cultural institution. Thus, the objective of Kunst fürs Dorf was not only to foster the production of artworks in villages but also to have in mind a possible contribution of art to the development of rural areas. And third, because Kunst fürs Dorf operates with a progressive art term by focussing on participatory and public art.

The six-month process of Kunst fürs Dorf was an intense and complex reality. The findings presented here are the preliminary outcome of an exploratory field study in the context of a PhD thesis. This article is based on the analysis of three forms of empirical data: relevant documents on the project published by the foundation; 17 semi-structured interviews with relevant actors, in particular the artist, inhabitants of the villages, the curator and staff of the foundation; and participatory observations in the villages during the run of the project. In order to get into the collected data and identify a relevant story that lies in it, I have used situational analysis by Adele Clarke (2005), which is rooted in grounded theory. Through situational mapping as a method of analysis, a deeper understanding of a given situation is aimed at, grasping the variety of elements in it.

2 Kunst fürs Dorf took also place in 2009 in Mecklenburg-West Pomerania and in 2011 in Lower Saxony.

Clarke also points out that there “[n]ot only are no tabula rasa researchers, but also we usually come with a lot of baggage. Such ideas and perceptions become intellectual wallpaper...” (Clarke 2005, 85). The ideas in this paper are thus also influenced by my own working experience in spatial planning and with participatory projects. As common in qualitative data analysis, the various forms of data were coded; and continued memoing took place.

Challenges of village development and the perspectives of planning

As widely discussed and researched in recent years, villages currently undergo strong structural and demographic changes. Agriculture has lost its role as the key economic sector in rural areas. Consequently, many villages, especially in the rural periphery, are affected by a chain of changes that influence the quality of life: a lack of job opportunities, people leaving the places to find work elsewhere, a declining birth rate, an increasing number of elderly people, the closing of shops, bars, schools and other facilities, the growing vacancy of houses, a lack of sufficient public transportation lines, etc. (e.g., Henkel 2012; Berlin-Institut 2011).

The municipality of Blankensee exemplary reveals these structural and demographic changes. During the period of the GDR, about 200 people were employed by the local agricultural production cooperative (LPG). After the German reunification, agriculture as an employer lost its importance. Nowadays, not more than 25 people work in the meanwhile privatised agricultural company. Many young people have left the place to find work elsewhere, e.g., in neighbouring towns such as Pasewalk or Löcknitz, or in Hamburg or Berlin; a high percentage of elderly people remains in the villages. Some people from other areas in Germany have moved to Blankensee, some of them driven by favourable land prizes. Blankensee and Pampow have about 300 inhabitants each. From the previous three pubs and restaurants in Pampow and the two in Blankensee only one remains open today. All local shops, as well as the kindergarten and school were closed. The frequency of the bus transportation lines to the places around declined significantly. Nevertheless, a variety of cultural and sport activities take place in the villages. The whole year around, active associations organise a number of events. During the summer months, bike tourist stop in Blankensee on their way north to the Baltic Sea. And as the villages are situated close to the Polish border — not more than 20 kilometres away from Poland’s seventh-largest city, Szczecin — there is hope that young Polish families will move to the community, rebuild old houses and counteract the dramatic aging of the population. Some families have done so already; and in 2013 the only baby in Blankensee and Pampow was born to a Polish mother.

Given this background, spatial planners and professionals in related fields are in search for new perspectives, strategies and instruments for the development of peripheral rural areas (Oswalt 2013). There is an uncertainty of where the journey leads to. This becomes evident, for instance, in the many projects or competitions that are run in villages in Germany today, funded by numerous governmental bodies, institutions or organisations[3]. These projects no longer only focus on hard factors, such as the construction of new roads or the renovation of old houses that shall bring development, but more and more also regard soft factors. These are, for instance, the strengthening of village communities and identification with the place, cooperations among and between villages as well as the fostering of local initiatives (Hernández and Harteisen 2013; Danielzyk and Knieling 2011; Mose 2005).

Not only are these keywords from research on rural development, but also have they become politically relevant. They were used, for example, at the national "Zukunftsforum Ländliche Entwicklung" organised by the German Ministry of Food and Agriculture in 2014. The Federal Minister of Food and Agriculture, Christian Schmidt, appeals to all possible actors to contribute to an integrated rural development: "I warmly invite you: take up the practical advices and help to create the rural development. Your assistance and engagement is crucial for our rural areas to remain attractive" (Schmidt in BMEL 2014, 4; my translation[4]). This call exemplarily shows that local citizens are considered increasingly important for the development of rural areas. It is the task of spatial planning to take specific actions in order to make locals aware of the development of their environment, help them communicate their concerns and perspectives and get them involved. This also requires a redefinition of the role of the spatial planner, becoming more and more an "initiator, moderator or manager for

---

3 One can mention here, for example, the competition "Menschen und Erfolge. Aktiv für ländliche Infrastruktur" run by the Federal Ministry for the Environment, Nature, Conservation, Building and Nuclear Safety (BMUB); or small scale initiatives financed by LEADER.

4 Original quote in German: „Ich lade Sie herzlich ein: Greifen Sie die praktischen Hinweise auf und gestalten Sie selbst die ländliche Entwicklung mit. Ihre Mitarbeit und Ihr Engagement sind entscheidend, damit unsere ländlichen Räume attraktiv bleiben."

specific development processes" (Danielzyk 2005, 465; my translation[5]). But other actors can also make a contribution to that — such as artists.

The artwork of Blankensee

In October 2013, artist Barbara Caveng left the villages Blankensee and Pampow after an intense period of six month living and working in the community. Two final art objects remained in the villages and became property of the community. First, the "KunstKiosk" (art kiosk; Figure 2), an old trailer that had been refurbished and furnished. It was operating as a meeting point, serving coffee and freshly baked cake to locals and bike tourists that passed the place on the national bike route during the summer months. On the trailer's side walls, quotes from locals gave insights into the villages' history through personal stories. Colourful umbrellas, made from used fabrics collected in the community, also functioned as signs for the histories of its former users. The second object, a 14 meter long "Holzmiete" (pile of fire wood; Figure 3) was located in the fields between both villages on the newly created community square of "Kunstgemeinde Pampsee" (art community Pampsee — a merger of the two villages' names). It was inaugurated as part of the project in June 2013 (Figure 4). The Holzmiete was designed and constructed from local wood, and the sentence "Mi kricht hier keener mehr wech" was inscribed in tall letters, also a quote from one of the citizens, that thus became the title of the whole project.

The "Kunstkiosk" and the "Holzmiete" are the physical products, the material outcome of Kunst fürs Dorf. But as in the genres participatory and public art, the final material outcome is just one aspect of the artwork. Of the same importance is what has happened during the process. In this way, artist Caveng said about her work in Blankensee:

> *"The artwork is the structure that happens here. The artwork is the invisible art. That is for sure. Nonetheless, it is good if something is visualised, for instance for the people that now form this artwork as a social plastic, for them to be able to remember it — and talk about it through a sign like this [the pile of fire wood]." (2013)*[6]

---

5 „Planung muss sich stärker als Initiator, Moderator und Manager von konkreten Entwicklungsprozessen verstehen."

6 All interviews with actors involved in Kunst fürs Dorf were conducted in German and translated into English for the purpose of this paper. I would like to thank all interviewees for their contribution.

Figure 2: "KunstKiosk" in Pampow
Source: Photo taken by Theresa Dietl, August 2013

Figure 3: "Holzmiete" on the community square
Source: Photo taken by Theresa Dietl, April 2014

Figure 4: Inauguration of the community square of the "Kunstgemeinde Pampsee"
Source: Photo taken by Theresa Dietl, June 2013

Kunst fürs Dorf linked to new approaches to village development planning

As seen above, participatory and public art are connected to new approaches of planning for peripheral rural areas through keywords such as communication, interaction, issues relevant to people's lives, or engagement (Bishop 2012; Lacy 1995). How they are reflected and connected to each other in practice will be outlined in the following chapter by looking at the example of Kunst fürs Dorf. Since "[h]ow a public artwork comes into existence is a very important part of its meaning" (Finkelpearl 2001, XII), the framework in which Kunst fürs Dorf took place will be analysed.

The framework: Kunst fürs Dorf by Deutsche Stiftung Kulturlandschaft

Kunst fürs Dorf was neither embedded in a specific spatial planning context in the community of Blankensee nor was it set in motion as a village development tool. Nonetheless, already in the call for application, the funding institution linked the programme to the challenges many peripheral villages face today, emphasising social aspects such as "loss of identity", "loss of occasions and locations for interpersonal relations and communication", or "dwindling of social cohesion, which is essential for the future viability of village settlements"[7] (Deutsche Stiftung Kulturlandschaft 2012). The goal of the foundation, as stated in the project's call, was not to find specific solutions for these particular problems. But by starting from naming these challenges, it can be argued, that the artistic initiatives were considered a means for contributing to their solving. Furthermore, the stated aims were very much connected to planning issues, since Kunst fürs Dorf should induce:

- "Manifold encounters and an intensive exchange between the locals and 'their' artist"
- "Ideally a broad and sustained public opinion making and participation process"
- "Change the public perception of the rural"
- "Suggestions for a constructive discussion about the rural reality"
- "Ideally further ideas and approaches for action that in the future might lead to an improved situation of local life" (Deutsche Stiftung Kulturlandschaft 2012).

Besides the influence the project should have on the perception of rural areas (which will not be the focus in this paper), above mentioned keywords were clearly expressed here: encounter, exchange, participation, action. How these were transformed from words to actions guided by the project's framework will be outlined now.

One of the basic elements of Kunst fürs Dorf was that the foundation required from the artists to move to the villages for the duration of the project.[8] The critique spatial planners from time to time face, that they are hardly present on site and know the situation in a given environment mainly from maps and documents, was thereby eliminated. Temporarily living in the villages became a key factor for Kunst fürs Dorf because in this way, the artist became a local citizen herself — with all her being. She built personal relations with the people and learnt about the situation in Blankensee and Pampow directly from them as well as through her own observations. This gave her the opportunity to develop a site-specific project, taking into account what mattered in the community.

---

7 Original quote in German.

8 Even if this requirement is rarely laid down in artistic projects, it is not unique to Kunst fürs Dorf. It was applied, for instance, in the frame of "Kunst im Dorf" (Art in the Village) in the German community Hohenstein in 2002 (Gemeinde Hohenstein n.d.).

The second requirement assisted this as well: the applying artists were not allowed to suggest a concept for the artwork with their application but instead had to develop the concept on site. This was an unusual requirement since in many programmes, artists are selected exactly because of the specific concept they suggest for a given area. But for Deutsche Stiftung Kulturlandschaft, it was central that the artists got inspired by the prevailing situation in the villages.

This leads to another relevant element: the foundation was ready to assume risk in that it could not anticipate what "product" it would get at the end of the six-month period. It was expected that also and especially because of the experimental character — of letting things go — new insights could be developed. One employee of the foundation highlighted this as follows after the end of the project:

> *"I think that the ... engagement of the foundation in such an area is probably easier than when ... public money goes to a project like this. Because then, you probably have to say from the beginning what result you specifically hope for. And you have to prove afterwards that it actually had occurred. And that is, I think, especially what then again hinders the people to really become active in the village. Because then they cannot freely get into the process. But they get a project where it is quite clear from the beginning what will happen." (2014)*

Many projects where locals shall get involved work exactly this way: the final outcome is already defined at the beginning of a project. Indicators for what would mark the success of the project are enunciated at the start. At the end of the project — many times linked to the payment of the final rate of the project's fee — it is measured if the indicators have been implemented, for instance if 50 or 100 people have gotten involved.[9] For Kunst fürs Dorf it worked differently. The artist fee of 20 000 euro (including material) was not connected to an ex ante defined project goal or to specific indicators that had to be fulfilled. The foundation gave the project out of its hand and into the hands of the artists. Its staff did not intervene during the duration of the project.

Because the process on site totally depended on the artist and the villagers, the foundation undertook special efforts with the selection of the involved actors. The design of the selection process is thus the forth important element that has been identified as a possible contribution of this artistic project to peripheral village development planning. In the first round, two selections took place at the

[9] See, for example, projects run in the publicly financed programme "Soziale Stadt" in German cities.

same time: one for the villages and one for the artists. A jury of experts concerned with rural development selected the three final villages, while a jury of experts from the art field selected six artists. In a second round, the six preselected artists presented themselves and their work to representatives from the three winner villages, who then chose "their artist". This was a key element because power was handed over to the villagers. In village development planning, many times decisions are taken elsewhere (Henkel 2012) and made by people (e.g., planners) that are not related to and selected by locals. Aiming for strengthening locals as actors, as required for a successful peripheral village development (Neu 2013; Mose 2005) involving them in the selection of their collaborators can be instrumental. However, this poses the question of how villagers (as any other non-professional actors) are able to evaluate the quality of applicants. In Kunst fürs Dorf, this has been solved in that the hard skills of the applying artists had been checked for by the expert jury in the first round. The second reason for the importance of the selection of the artist by the villagers is that the interpersonal relation between the artist and the locals became a key factor during the run of the project. In planning projects, this factor is generally not taken into account. It is rarely considered whether or not the chemistry is right between the planner (or any other project responsible) and the locals that shall get involved. Instead, the focus is put, for example, on qualification. One citizen of Blankensee, who was a member of the selection committee, stated exemplarily what many said:

> *"We had conceded the favour to choose our artist. And then we have picked from what was for offer a little bit the one who, I would say, who you could feel ... the one who was most likely able to do something together with the people. And then it was totally irrelevant what discipline this was." (2013)*

The importance of the human relations as stated here leads to the analysis of process oriented aspects of Kunst fürs Dorf.

The process: Barbara Caveng's artistic project in Blankensee and Pampow

The artistic project in Blankensee and Pampow was based on the citizens' local concerns and knowledge. Using local knowledge — involving the citizens as "local experts" — links to the demand for identifying local specificities and potentials for an adequate peripheral village development (Mose 2005), given the locals' importance as a key source for this identification. Nonetheless, harvesting and building on local knowledge is still not often turned into practise. This might be the case because it is time-consuming and might require personal approaches (for the chances and limits see, e.g., BMVBS and BBSR 2010, 20). Caveng framed such a personal and time-consuming approach: during the first weeks of her stay, she visited almost 30 households for coffee or tea and conducted interviews. In this way, she got insights into the villagers' everyday lives and personal stories. From these visits, she identified the topic of her artistic project. Initially, she intended to focus on the relationship between the German and Polish inhabitants, which was also the first idea of the locals. But during the interviews, Caveng became aware that the relation between Blankensee and Pampow was more relevant. Even if both villages belong to one municipality, little had been done in order to link these two closer together since their incorporation in 2004. By making this topic subject of the artistic project — one that concerns many villages all over the country — vivid discussion in the community rose on the question whether or not the relation between the two was problematic. As stated in the interviews, for many villagers, this was not a subject they were conscious about before. In this way, the project brought a local issue to the surface which citizens had to tackle.

Given Caveng's intention to strengthen the relationship between Blankensee and Pampow, she developed formats of collaborative work for the inhabitants of both villages: the "KunstKiosk" (with all elements involved as mentioned above) and the "Holzmiete" on the "Gemeindeplatz". Within about five weeks of intensive work, the KunstKiosk was refurbished and furnished in May and June 2013. In the end of June, it was inaugurated. About 40 villagers[10] participated in different kinds of work for the preparation of the kiosk and its run throughout the summer. At irregular time intervals, meetings where next steps were discussed and work instructions were given took place in smaller groups and under direction of the artist. Informal arrangements were also reached between the artist and the locals

---

10 All numbers of participants according to a panel produced by the artist for the exhibition "HEIMISCH" in the "Galerie im Saalbau" in Berlin 2014.

and among the locals themselves. Since the end of May, the "Sonnenschirmgruppe" (umbrella group; Figure 5) met regularly. Five to ten women and one man gathered weekly to design and produce umbrellas and other equipment for the KunstKiosk. The construction of the Holzmiete on the Gemeindeplatz started only six weeks before the end of the artist's stay. An intensive working period in August and September took place. About 20 people participated in the so called "Holzbrigade" (wood brigade). They assisted in cutting wood, building wooden frames for tall letters and piling up the wood at the Gemeindeplatz in a time-consuming manner for many days. During the working process as well as with the final products, opportunities and places for communication and encounter were created. Given the fact that many peripheral villages lack such opportunities and places (Neu 2013; Berlin-Institut 2011), the project contributed to its solving. This outcome was also identified for other public and participatory artistic projects (Polzer 2013; Helmle and Neu 2010).

Figure 5: Meeting of the "umbrella group" — production of umbrellas from old fabric
Source: Photo taken by Theresa Dietl, August 2013

Caveng initiated participation opportunities according to the interests and skills available in the community. In this way, locals of very different ages and backgrounds got involved, for instance, by painting a trailer, doing carpentry work, refurbishing furniture, running the kiosk, baking cake, donating fabric, making music, sewing umbrellas, speaking to people, chopping down trees, piling wood, distributing flyer, etc. No external assistance was needed, a criterion applying to many artistic activities in rural areas (Polzer 2013). In the beginning, some locals were sceptical on how they could contribute to the project since they did not consider themselves good painters or drawers. But by discovering that the artistic project in their municipality was not about creating a "traditional" work of art, one woman said exemplarily:

> *"Well, we are just no artists. I mean, I'm glad if I can get a flower painted on. ... I will get involved with the works that someone can do who is not artistically skilled." (2013)*

Especially for people that were little engaged before, it was an important experience to be able to contribute to the project. By continuously stressing that it was only through the joint efforts that the project came to exist, the artist gave respect and recognition to the locals. Reflecting on a conversation with a visitor, Caveng said:

> *"'Why do you always say 'we?' [I was asked.] And then I said: excuse me, but this is We. It is not me who is doing all this work. Then you would not be here at the kiosk at the moment." (2013)*

Also the citizens were aware of their importance for the realisation of the artistic project, as exemplarily expressed in an interview with a local woman, referring to a meeting in May 2013:

> *"In the last meeting, all who really want to be involved registered. And then, we all registered, almost everybody that was present that day. Well, as I said, she shall not manage all this here by herself. She will get appropriate support by us. And yes, what we can do as amateurs, we will do to support her energetically." (2013)*

At this "KunstbürgerTreffen" (meeting of the art citizens), Caveng introduced her idea about the community square and the art kiosk. Also, she suggested to name the two villages in the art project's context "Pampsee" and formed working groups for specific tasks. About 60 people from both villages participated in that meeting, which corresponds to approximately 10% of the municipality's population. In total, about 110 villagers were involved in the project. There were also villagers that spurned the project for different reasons, and others that were indifferent to it. Additionally, since most activities took place during the day, working people could not get involved as much as, for instance, pensioners or the unemployed. Joining Kunst fürs Dorf was not obligatory but an offer to which people could respond freely — some accepted it, others refused.[11]

The artist had the leading role throughout the whole project. She offered possibilities for collaboration but never gave her key position away. Her position was accepted by the villagers, as stated in many interviews, due to her expertise in running participatory artistic projects. But after the end of Kunst fürs Dorf, it was a challenge for the villagers to take the artist's position and distribute her roles. Caveng held numerous functions: she was not only the artist, the creative and aesthetic head of the project, but also the initiator of action, the coordinator, moderator, manager or motivator. Many of these functions were mentioned above as new roles spatial planners have to take up (Danielzyk 2005). As shown for Kunst fürs Dorf, also artists can fulfil a number of these new roles needed in peripheral village development planning. At times, Caveng was overloaded with the tasks that came with these numerous roles, also given the fact that as a citizen of Blankensee and Pampow she was involved in personal stories and struggles, lacking the opportunity for a retreat. Not only her but also the villagers expressed that the high level of engagement could only be maintained for a limited period of time. For a more permanent activity, the intensity would have to be reduced.

The involvement of a considerable number of people in the project was, above all, caused by the personal relations Caveng built with the people. Reflecting on her achievement to take many different villagers on board, she said:

> *"I have to say really, I think what is relevant for being on a par with the people, [is that] I really have an interest in these people." (2013)*

Many were personally invited to participate, e.g., during her house visits in the beginning of her stay. The artist did not start the project by asking for the assistance of the locals. Instead, she started with small steps that created relations. This stands in contrast to many spatial planning projects where it is rarely focused on building personal relations prior to the actual work. Also, for participatory artistic activities, it is not always the case that artists take time to establish such relations, which sometimes lead to astonishment when people

---

[11] It has often been criticised that the success of a public and participatory artistic project can be measured by the number of people that participated. Effects do not correspond to the number, but instead, they often occure on an individual level and only after an unpredictable time (Bishop 2012; Helmle and Neu 2010; Finkelpearl 2001).

refuse to participate.[12] As the first act of her project, Caveng asked the people of Blankensee and Pampow to help her furnish her temporary apartment. In a lively ceremony, people from almost 20 households brought furniture, dishes etc.. In this way, first personal contacts were established and relations (also symbolic) were created in that the one would be living with the objects of the others.

Kunst fürs Dorf was a challenge for the locals. Almost none of them had participated in an artistic project before. The artist was — at least in the beginning — an extraordinary figure in the municipality, and some of her ideas seemed too edgy to be turned into reality. Scepticism concerning new approaches and ideas was not unusual in Blankensee and Pampow, as one interviewee expressed:

> *"... you have the mentality here: it's not working! If you do something here, immediately you will be shown all reasons why something won't work." (2013)*

Scepticism is common in many places, also in peripheral villages facing change (Oswalt 2013). However, being open to new ways of responding to these challenges, including self-responsibility and risk (Neu 2013), is crucial for peripheral village development (Oswalt 2013). The contribution of Kunst fürs Dorf was seen in this aspect, as stated by one interviewee:

> *"In yesterday's discussion [of many villagers on issues concerning Blankensee], you have seen ... 'We need more money! They have to give us more money!' No one comes up with the idea that you are responsible yourself for your money. When you do this [Kunst fürs Dorf] for fun and it sticks in the minds of one or two, well, maybe this is also possible ... Or when he just gets the impulse for something different, something that can work as well, well then it would be a progress." (2013)*

Kunst fürs Dorf provided an opportunity for practicing openness towards new ideas and approaches. This was possible through the gradual development of the project on site and continuous feedback and exchange with the locals. The artist stretched the limits and asked for things that were not considered feasible by the locals at first. One example for this is the construction of the wooden sculpture as the final sequence of the project. Caveng summed up her experience about that as follows:

---

12 Based on my own observations of a number of projects in this field, since research on such projects is rare and information and documentation is often provided by the actors themselves (see, e.g., projects portraited in Polzer 2013).

> *"When I said, I want to do this [build the wooden sculpture], everyone asked: how do you want to do that? Well, A, from the material aspect, and B, from the people? ... Some things are difficult. But now, looking back at the procedure ... within one week, the material, the logistics ... and a very positive response from everyone that I show this work. This is always a good sign that the decision [to build the wooden sculpture] was the right decision." (2013)*

In this way, Kunst fürs Dorf worked as a catalyst for opening up the community for more experiments to come.

Conclusion

In Blankensee and Pampow as well as in many other peripheral villages in Europe, things change due to structural and demographic transformations. The population declines and gets older, the infrastructure decreases, houses remain empty etc. This paper aimed to explore possible contributions of one exemplarily public and participatory artistic project, Kunst fürs Dorf, to peripheral village development planning. Besides creating places and opportunities for communication and encounter for the locals, Kunst fürs Dorf did not offer particular solutions for specific problems peripheral villages face. The project neither made people move back to Blankensee or Pampow, nor brought it new jobs or led to the re-opening of the supermarket. Given this, one citizen of Blankensee said frustratedly:"Well, two births last year, one Polish probably this year. Any more questions? Considering this, art cannot achieve anything" (2013).

Nevertheless, aspects of Kunst fürs Dorf were identified in this paper that can contribute to peripheral village development. These aspects refer to processes rather than to specific effects. To sum up the findings, concerning the project's framework, it was significant that the artist lived in the village for the duration of the project and developed the project on site, informed by local knowledge and based on the given situation. This was made possible by the foundation's openness towards experiments and an unpredictable outcome. The selection process, in which the villagers had chosen "their" artist based on to the chemistry they felt towards her, was also crucial. Concerning the process of the artistic project on site, it was highlighted that the project was a joint work by the artist and the villagers, where importance was given to the citizens as local experts. Moreover, the personal relations between the artist and the villagers has seemingly contributed to both, the participation rate of the local people as well as the matching process of individual skills and involvement in the project.

Given the fact that it is claimed necessary by scientists and politicians that local citizens get involved in the development of their villages (BMEL 2014; Hernández and Harteisen 2013; Danielzyk and Knieling 2011; Mose 2005), I consider these identified aspects from an artistic project inspirational for non-artistic projects in the context of village development in peripheral rural areas. Kunst fürs Dorf did not contribute through finding solutions for specific problems or being directly involved in village development projects. It rather gave hints for participatory spatial planning on how to design participatory processes and get locals involved in the development of their environment as well as on how to foster self-responsibility and risk (Neu 2013). Perhaps, some of these criteria can be identified in non-artistic projects as well; perhaps, it is not specifically due to "art" that the mentioned contributions were made. Whether art could achieve something different than non-art is not a question in this paper since answering it would require a comparative empirical study. But it can be argued that art as an "unfettered form of creativity" (Rendell 2006, 3) allows freedom that is not possible in rather traditional ways of spatial planning (Altrock and Beeck 2013). In contrast to spatial planners, artists have the freedom to design a project's process freely, e.g., to investigate a significant amount of time in building personal relations with the villagers. The artistic freedom thus can lead to testing ways and approaches out of the ordinary which might give new answers to the perspective in which rural peripheral areas can develop.

There are many questions related to Kunst fürs Dorf that could have been asked as well. From the perspective of art theory, for example, a major question one can raise is whether art could or should be instrumentalised, i.e., in order to assist with village development planning. From the perspective of planning, one can ask how projects like Kunst fürs Dorf could migrate from a short-term intervention to a long-term programme without losing its flexibility and experimental character. Also, one can put the focus on the outcome of Kunst fürs Dorf rather than on the process and could evaluate what has remained of the project after a given time.

A short insight into this last aspect shall be given, asking what has remained of Kunst fürs Dorf two years after its implementation. The project in Blankensee and Pampow was not over with the end of Caveng's stay in the community. Since then, the citizens have built on what they had developed during the time with the artist. For instance, citizens assisted Caveng in the preparation of two exhibitions in Berlin and Dortmund in 2015, the KunstKiosk was functioning as a meeting place for locals and bike tourist in 2014 and 2015, a group of citizens from both villages still meet monthly for creative production, and other cultural activities, outreaching to other communities, have been developed.[13] And despite the fear that it would be demolished shortly after its construction, the Holzmiete still marks the community square as a memento for a successful experiment where self-responsibility was practiced and risk taken.

[13] For details see www.kunstgemeinde-pampsee.net/die-kunstgemeinde-pampsee/ (Last access 29.05.2015).

References

Altrock, U. & Beeck, S. (2013). Stadtentwicklung und die Praxis von „Interventionen". RaumPlanung 167 (3), 21–26.

Berlin-Institut für Bevölkerung und Entwicklung (Ed.). (2011). Die Zukunft der Dörfer. Zwischen Stabilität und demografischem Niedergang. Berlin.

Bishop, C. (2012). Artificial Hells: Participatory Art and the Politics of Spectatorship. London and New York: Verso.

Bundesministerium für Ernährung und Landwirtschaft (BMEL) (Ed.) (2014). Ländliche Entwicklung aktiv gestalten. Leitfaden. Berlin.

Bundesministerium für Verkehr, Bau und Stadtentwicklung (BMVBS) & Bundesinstitut für Bau-, Stadt- und Raumforschung (BBSR) (Eds.) (2010). Miteinander Stadt Entwickeln. Stadtentwicklung, Zivilgesellschaft und bürgerschaftliches Engagement. Berlin.

Clarke, A. (2005). Situational Analysis: Grounded Theory After the Postmodern Turn. Thousand Oaks, London. New Delhi: Sage.

Danielzyk, R. & Knieling, J. (2011). Informelle Planungsansätze. In Akademie für Raumforschung und Landesplanung (ARL) (Ed.), Grundriss der Raumordnung und Raumentwicklung (pp. 473–498). Hannover: ARL.

Danielzyk, R. (2005). Informelle Planung. In Akademie für Raumforschung und Landesplanung (ARL) (Ed.), Handwörterbuch der Raumordnung (pp. 465–469). Hannover: ARL.

Deutsche Stiftung Kulturlandschaft (2012). Kunst fürs Dorf — Dörfer für Kunst: Ausschreibung 2013. KÜNSTLER. www.burg-halle.de/uploads/tx_aoeburg/kunst_fuers_dorf__2013_ausschreibung_kuenstler.pdf; DÖRFER. www.leadernetzwerk-sachsen-anhalt.de/fileadmin/user_upload/Aktuelles/Doerfer_fuer_Kunst_2013_Ausschreibung_Doerfer.pdf. Accessed 23 May 2015.

Gemeinde Hohenstein (n.d.). Kunst im Dorf. Ein Modellprojekt im Entwicklungsprogramm Ländlicher Raum. www.gemeinde-hohenstein.de/gemeindeleben/kulturelles/kunst+im+dorf/kunst+im+dorf.html. Accessed 23 May 2015.

Oswalt, P. (2013). Der ländliche Raum ist kein Baum: Von den zentralen Orten zur Cloud. In Faber, K & Oswalt, P. (Eds.), Raumpioniere in ländlichen Regionen: Neue Wege der Daseinsvorsorge. Edition Bauhaus 35 (pp. 6–16). Leipzig: Spector Books.

Finkelpearl, T. (2001). Dialogues in Public Art. Cambridge (Massachusetts) and London: MIT Press.

Helmle, S. & Neu, C. (2010). Leitsystem Kunst. Kulturpolitische Mitteilungen 131 (IV/2010), 68–69.

Henkel, G. (2012). Das Dorf: Landleben in Deutschland — gestern und heute. Stuttgart: Theiss.

Hernández, A. A. & Harteisen, U. (2013). Förderprogramm LEADER — Impulsgeber für innovative Landentwicklung in Europa. Ländlicher Raum, no. 01/2013, 22–26.

Hildebrandt, P. M. (2012). Urbane Kunst. In Eckardt, F. (Ed.), Handbuch Stadtsoziologie (pp. 721–744). Wiesbaden: VS Verlag für Sozialwissenschaften.

Kegler, B. (2014). Ganz nah dran: Der ländliche Raum zwischen Breitenkultur und Soziokultur. In Schneider, W. (Ed.), Weißbuch Breitenkultur. Kulturpolitische Kartografie eines gesellschaftlichen Phänomens am Beispiel des Landes Niedersachsen. Hildesheimer Universitätsschriften 29 (pp. 57–79). Hildesheim: Universitätsverlag Hildesheim.

Kester, G. (2015). Editorial. FIELD: A Journal of Socially-Engaged Art Criticism. www.field-journal.com/about. Accessed 15 November 2015.

Lacy, S. (Ed.) (1995). Mapping the Terrain: New Genre Public Art. Seattle: Bay Press.

Mose, I. (2005). Ländliche Räume. In Akademie für Raumforschung und Landesplanung (ARL) (Ed.), Handwörterbuch der Raumordnung (pp. 573–579). Hannover: ARL.

Neu, C. (2013). Mehr Lebensqualität für weniger Menschen: Herausforderungen für eine neue Daseinsvorsorge im peripheren ländlichen Raum. In Faber, K & Oswalt, P. (Eds.), Raumpioniere in ländlichen Regionen: Neue Wege der Daseinsvorsorge. Edition Bauhaus 35 (pp. 17–25). Leipzig: Spector Books.

Polzer, B. (Ed.) (2013). Kunst und Dorf: Künstlerische Aktivitäten in der Provinz. Zürich: Scheidegger & Spiess.

Rendell, J. (2006). Art and Architecture: A Place Between. New York: I.B.Tauris.

Schröder, M. (2011). Kunst fürs Dorf — Dörfer für Kunst: Innovative Wege zur Aktivierung bürgerschaftlichen Engagements in entlegenen Räumen. Unpublished Magister thesis. Universität Rostock.

Tippach-Schneider, S. (2010). Leitsystem zum Neuen. Ein Projekt der REINIGUNGSGESELLSCHAFT im ländlichen Raum. Kulturpolitische Mitteilungen 129 (II/2010), 72–73.

4 Renewable energies, climate change and sustainable mobility

Aligning local adaptive land use management in coastal regions with European policies on climate adaptation and rural development

Leena Karrasch and Thomas Klenke

Abstract

In European coastal regions natural and anthropogenic change processes, such as changes in climate or ecosystems, demographic distributions or economic potentials, appeal for adaptive and ecosystem-based strategies to manage human activities. With respect to an integrated European policy context, the coastal regions are of significant ecological and socio-economic interest.

Case study research focusing on participatory adaptive land use planning and assessment in the community of Krummhörn (periphery southern North Sea coastal region, north-west Germany) provides an example of an adaptive governance approach that promotes ecosystem-based management, increases the regional adaptive capacity and close knowledge-gaps for better-informed decision-making. We consider to what extent participatory adaptive land use planning and assessment enables decision-makers to align sustainable land use management of flood prone coastal areas with common goals of European climate and rural development strategies.

Introduction

Since low lying European coastal areas will especially be threatened by possible negative impacts of climate change (ESPON 2011; Schmidt et al. 2014), any sustainable development in coastal regions has to successfully implement climate adaptation and mitigation measures (IPCC 2014; Wilbanks et al. 2007; Wong et al. 2014). In particular, rising sea level, change in temperature and increased frequency of extreme weather conditions, including heavy rain and storms, are growing challenges for coastal peripheral region in particular (Gaslikova et al. 2013; Grabeman and Weisse 2008; Jacob et al. 2008; Nicholls et al. 2007). Beside the effects of climate change, socio-economic change processes will influence the future of the coastal areas, especially regions in rural peripheries. At present, many rural areas in Germany observe a decline and shift in population, changes in economic incentives and land use as well (OECD 2007). All the changes interfere. Localised and embedded into spatial settings, these changes challenge rural development as managed by spatial planning.

Consequently, public debates on all political levels have taken place to handle the challenges and to simultaneously improve regional development. Especially on EU level, there is a claim for participatory approaches to plan climate adaptation actions, support rural development policy, and take land-sea interactions into account (COM 2013, Regulation (EU) 1305/2013, Directive 2014/89/EU).

This study pursues the question, how participatory adaptive land use planning and assessment (PALUPA) contributes to implement common goals of EU policies, namely the EU strategy on adaptation to climate change (COM 2013), the Directive on establishing a framework for maritime spatial planning (Directive 2014/89/EU) and the Regulation on support for rural development (Regulation (EU) 1305/2013), into adaptive land use management of flood-prone coastal areas. Land use management scenarios have been developed together with regional decision-makers in a PALUPA process to raise the awareness and to design a more sustainable land use management to cope with the future challenges driven by climate and socio-economical change processes. To analyse the connections between PALUPA and EU climate adaptation and regional development policies, the content of three relevant EU policy documents (COM 2013, Regulation (EU) 1305/2013, Directive 2014/89/EU) has been screened for common objectives in the scope of undertaking sustainable management of natural resources, climate adaptation and mitigation actions, closing knowledge gaps for better-informed decision-making, as well as fostering capacity building and participatory governance approaches. In this context, we show how the results of PALUPA are aligned with the selected EU policies and how the common objectives might be integrated into regional planning activities.

Case study area

The case study area Krummhörn is a peripheral coastal community located within the southern North Sea region in the north-western part of Germany (Figure 1). This low lying rural region is endangered by terrestrial fluvial floods as well as storm surge events, since one third of the area is located below sea level. The current flood defence, a massive sea wall (dike), protects the hinterland from storm surges. To make current land use possible, the hinterland needs to be constantly drained. Expecting an increase in extreme precipitation and sea level rise, the current drainage system is not able to cope with increasing water run-off.

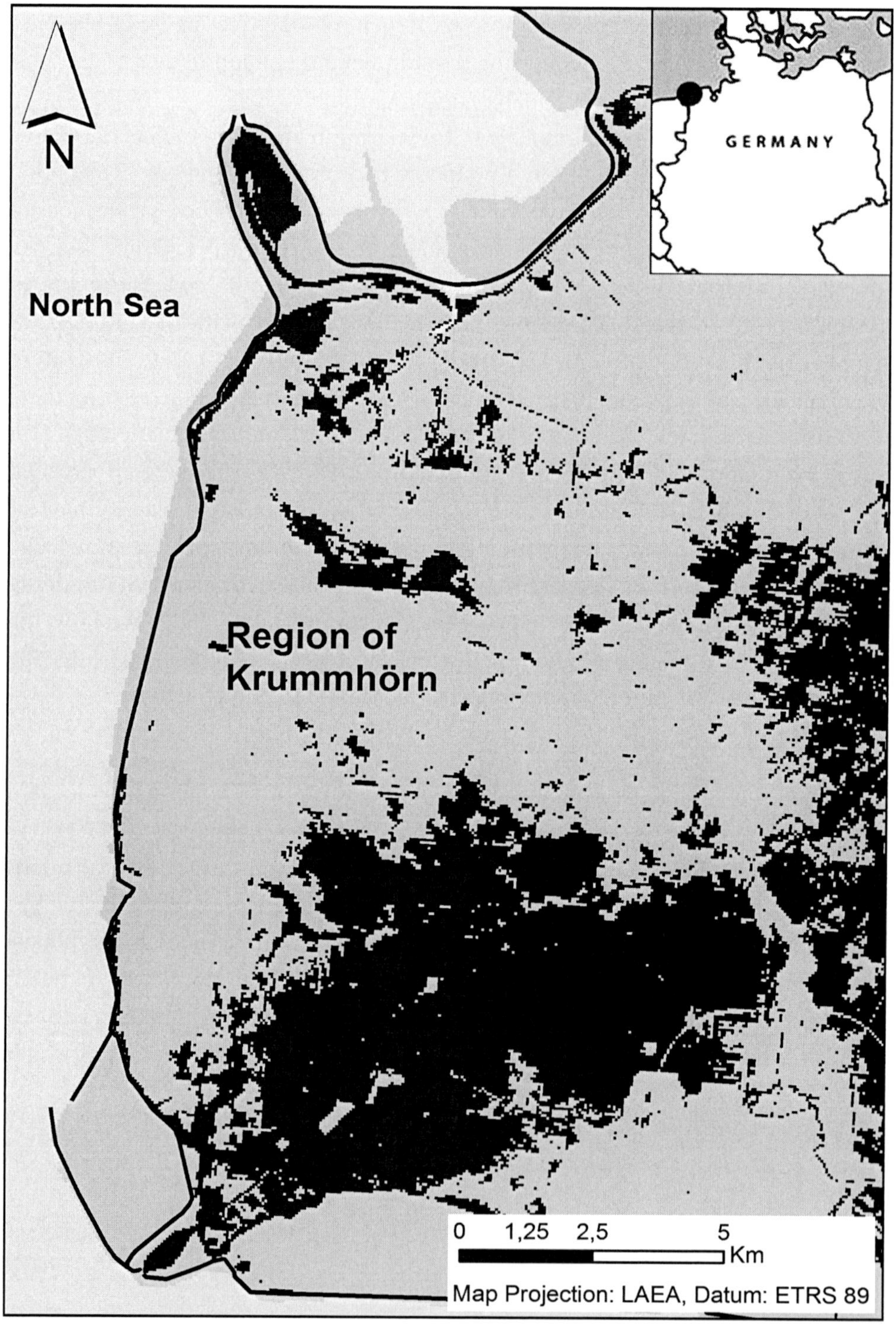

Figure 1: Case study area coastal community of Krummhörn, located at the north-western region of Germany. All dark grey spots are areas below sea level

Participatory adaptive land use planning and assessment

During a 4 year collaboration (from 2011 to 2015), regional decision-makers and knowledge-brokers co-developed a participatory adaptive land use planning and assessment (PALUPA) methodology. Independent from ongoing political decision-making processes, PALUPA aimed to elucidate and develop recommendations for possible land management adaptation strategies.

Decision-makers have been involved from the beginning of the PALUPA process. They were engaged to select and define the borders of the case study area. Additionally, during the research process, different qualitative methods for data acquisition have been applied (Karrasch et al. 2014). Possible adaptation measures on both an individual and group level have been discussed in semi-structured interviews (Bogner et al. 2009; Flick 2006), regional forums, and focus groups to foster group discussions and consensus building (EEA 2001; Hare and Pahl-Wostl 2002; Pahl-Wostl 2002). The decision-maker group was representative for the community since all relevant actors who represent the demands of the case study area and have the possibility to take and implement decisions have been included. They provided specific local knowledge, in theory as well as in practice.

In order to generate, transfer and integrate results, researchers served as knowledge-brokers to communicate relevant scientific approaches (for instance the concept of ecosystem services) to the stakeholders in a comprehensible manner (Karrasch et al. 2014; Ward et al. 2009). The knowledge-brokers also guided the whole PALUPA process.

To take the complex problems inherent to human-environmental systems of the case study region up, land management scenarios were used to integrate science and decision-making. By introducing different scenarios (see below), the knowledge-brokers provided examples how to deal with future challenges and develop sufficient adaptation and mitigation options. Additionally, the knowledge-brokers introduced the concept of an ecosystem-based management. Following the definition of UNEP/GPA (2006), ecosystem-based management considers the socio-economic systems of a community not independently but as integral part of ecosystems. It is "concerned with the processes of change within living systems and sustaining the services that healthy ecosystems produce. Ecosystem-based management is therefore designed and executed as an adaptive, learning-based process that applies the principles of the scientific method to the processes of management" (UNEP/GPA 2006, 5). The knowledge-brokers applied the ecosystem services approach (MA 2003) to emphasise the benefits people obtain from ecosystems.

Table 1: Local and regional stakeholders involved in the participatory integrated planning and assessment

Sector	*Decision-maker*	*Institution*	*Level*
Water management	Engineer	Water Board Emden	Autonomous public corporation (local)
	Director	Water Board Emden	Autonomous public corporation (local)
	Director	Dike Board Krummhörn	Autonomous public corporation (local)
Nature conservation	Administrator	National Park authority, Lower Saxony	State authority (regional)
	Administrator	Biosphere Reserve authority, Lower Saxony	State authority (regional)
	Administrator	Lower Nature conservation authority	Autonomous public corporation (regional)
Agriculture	Director	Farmers Association East Frisia	Registered association (local, regional)
	Chairman	Farmers Association East Frisia	Registered association (local)
	Administrator	Chamber of Agriculture, Lower Saxony	Autonomous public corporation (regional)
Policy	Major	Community of Krummhörn	State authority (local, regional)
	Administrator	County of Aurich	Public corporation (regional)
	Coordinator	Sate Estate Agency, Norden	State authority (local, regional)
	Government administrator	Community of Krummhörn	Public corporation (regional)
Tourism	Director	Tourism Association East Frisa	Independent limited liability company (local, regional)

Source: Authors' compilation

PALUPA aims to actively involve decision-makers belonging to different sectors, their knowledge, values and perceptions to support decision-making processes (Hare and Pahl-Wostl 2002; Salter et al. 2010). In terms of adaptive governance processes, the knowledge-brokers took the leadership and therefore a key function for trust building, conflict management, networking, and knowledge generation (Folke et al. 2005). Adaptive governance processes reduce uncertainty in terms of future decisions and provide decision-makers possible solutions concerning changes in land use management (Nyberg 2009; Rijke et al. 2012).

Scenario process: local participation in land use management

Land management scenarios were used to anticipate spatio-temporal changes in the case study area. They assemble possible adaptation and mitigation options. Each scenario fosters a particular change in land use. Based on this change, there is also a change in key ecosystem services provided by a certain land use (Table 2).

The first scenario (trend) indicates no changes in land use management compared to present practice. The vast open landscape is characterized by intensively and extensively use of grassland and arable land. Possible negative impacts of climate change lead to adverse effects — due to rising sea level the time water can be naturally discharged by gravity into the sea decreases. In combination with increasing precipitation, the water levels increase leading to inland floods. Consequently, to maintain land use, costly pumping activities to drain the region have to be increased. The second scenario (water management) illustrates huge water retention (polder) areas to store water to avoid floods in times of extreme events. The retained freshwater in the polders may be used during dry periods for irrigation measures or recreational purposes. Naturally growing reed stands will be harvested and might be used for green energy generation. The third scenario (carbon sequestration) is similar to the water management scenario, but instead of harvesting the reeds, the reed stands will contribute to sequester atmospheric carbon dioxide and to develop a more nature conservation land management. These three initial scenarios have been provided by the knowledge-brokers to communicate different ideas how future land management might be designed.

Table 2: Main features of scenarios considered in the participatory integrated planning and assessment

	Trend scenario	*Water management scenario*	*Carbon sequestration scenario*	*Actor-based scenario*
Change in land use	No change in land use, business as usual. Agriculture, dairy farming, tourism and nature conservation are the main land uses. These uses are depending on a strict water management which aims to keep a certain water level in the whole region and to avoid floods from the sea and the hinterland.	Water storage in polder areas. The large polder areas are located in the low lying parts of the community. Harvesting of reed stands. Decrease in arable land and grassland.	Water storage in polder areas. The large polder areas are located in the low lying parts of the community. Peat formation in wet marshes for sequestering atmospheric CO_2. Decrease in arable land and grassland.	Maintenance of present land use, combination of different land uses (mosaic structure). Small polder areas in very low-laying areas.
Used key ecosystem services	Food production Forage production Recreation and tourism	Biomass for energy Freshwater Hazard regulation by water retention Prevention of saltwater intrusion.	Reduction of greenhouse gases Hazard regulation by water retention Prevention of saltwater intrusion.	Food production Forage production Hazard regulation by water retention Prevention of saltwater intrusion. Recreation and tourism Community identification
Adaptation or mitigation to climate change	Increasing pumping capacity to secure the drainage of the region.	Restrict floods to the polder area. Use of retained fresh water in dry periods for irrigation measures. Prevention of subsurface saltwater intrusion. Use of biomass to generate green energy.	Restrict floods to the polder area covered with rich reed stands. Prevention of subsurface saltwater intrusion. Development towards natural former vegetation.	Restrict floods to the small polder areas. Use of retained fresh water in dry periods for irrigation measures. Prevention of subsurface saltwater intrusion.

Notes: Provisioning services secure the income of the farmers and are also the first income source of the community. An increase in the regulating services provides a higher safety standard. Cultural ecosystem services are emphasized by the stakeholders. The maintenance of the regional characteristics and aesthetic value of the landscape is of major importance for the stakeholders, not only for tourism as second major income source but also for their well-being. The stakeholders prefer a mosaic-structure of many different land uses.

Source: Authors' compilation

Taking these three initial scenarios into account, the decision-makers developed a fourth scenario. In this actor-based scenario, the main land uses (agriculture, tourism and nature conservation) as well as the drainage system shall be maintained. The main change in land use regarding climate adaptation measures is the establishment of polder areas for water retention in case of extreme weather events. In contrast to the initial options, the stakeholders preferred rather small polders in low lying areas without settlements. These polders should serve for freshwater retention, prevention of saltwater intrusion and recreational purposes and not be used for active peat formation (as proposed as mitigation option in the carbon sequestration scenario).

The water management and carbon sequestration scenarios present extreme forms of changes in land use. At the beginning of PALUPA, the stakeholders were not in favour of designing a water retention polder. One explanation is that for centuries the people in this low lying coastal area follow a hazard reduction approach, by "keep floods away from urban areas" (Oosterberg et al. 2005, 23). During the PALUPA, the decision-makers started to re-think possibilities for water management. Even though the mitigation option of sequestering atmospheric carbon dioxide in wet marshes was not preferred, adaptation in terms of the water retention polder was conceivable. In terms of regional adaptive capacity, the decision-makers′ willingness to develop a water retention polder showed a change in behaviour, based on knowledge generation and awareness rising. The benefits taken from this change was mainly the reduced risk of and vulnerability to floods. This also contributed to a multifunctional coastal defence zone, combining different existing and future measures to tackle the impacts of climate change. Thus, regarding possible adaptation and mitigation measures in terms of the actor-based scenario, an increase in the regional adaptive capacity was observable.

EU climate adaptation and coastal rural development policy

On the upper policy level, three European policies on climate adaptation and rural development policies, namely the EU strategy on adaptation to climate change (COM 2013), the EU Directive on establishing a framework for maritime spatial planning (Directive 2014/89/EU) and the EU regulation on support for rural development by the European Agricultural Fund for Rural Development (EAFRD) and repealing Council Regulation (EC) No 1698/2005 (Regulation (EU) 1305/2013) are particularly relevant with respect to cross-sectoral spatial management. The common goals are listed in Table 3.

Table 3: Main features and common goals of relevant EU climate adaptation and coastal rural development policy documents

	COM 2013	Directive 2014/89/EU	Regulation (EU) 1305/2013
Common goals	Strengthening Europe´s resilience to the impacts of climate change	Framework for maritime spatial planning, taking into account land-sea interactions	Supporting rural development
Sustainable management of natural resources (ecosystem-based management)	Green infrastructure, ecosystem-based approaches	Articles 1 and 5	Articles 3, 4, 5, 8, 17, 13, 18, 53 and 55
Climate adaptation and mitigation actions	Actions 1, 3, 6, 7 and 8	Marine ecosystem services deliver benefits in terms of climate change mitigation and adaptation	Articles 5, 15, 8, 15 and 59
Closing knowledge gaps for better-informed decision making	Actions 4 and 5	Article 10	Articles 5, 14, 53 and 55
Capacity building, participatory governance approaches	Action 2	Articles 4, 9, 11 and 12	Articles 5, 7, 16, 35, 52, 53, 54 and 56

Source: Authors' compilation

Even though especially the binding parts of the EU strategies and regulations are expected to have a significant influence on coastal management on a local to regional scale, the direct up-take by the local decision-makers is reserved and rather low. During the participatory integrated planning and assessment process the knowledge-brokers became aware that the involved decision-makers were not familiar with the details of the European climate and rural development policies. The decision-makers considered the content of these policies being vague and rather difficult to understand although the general aims were favoured. The decision-makers were not aware of the implications these policies can generate on different policy levels. Additionally, resources were considered missing on EU or national institutional level to implement the objectives and start collaboration processes. This was particularly the case for non-binding communication of the EU.

Aligning local adaptive land use management with EU climate adaptation and rural development policies

Joint foresight planning of regional decision-makers in a PALUPA process guided by knowledge-brokers provides an option to link land use management in a peripheral municipal area and European climate adaptation and rural development policies (Figure 2). Even though EU policies were not directly addressed in the PALUPA process, clear linkages between its outcome and policy objectives emerged.

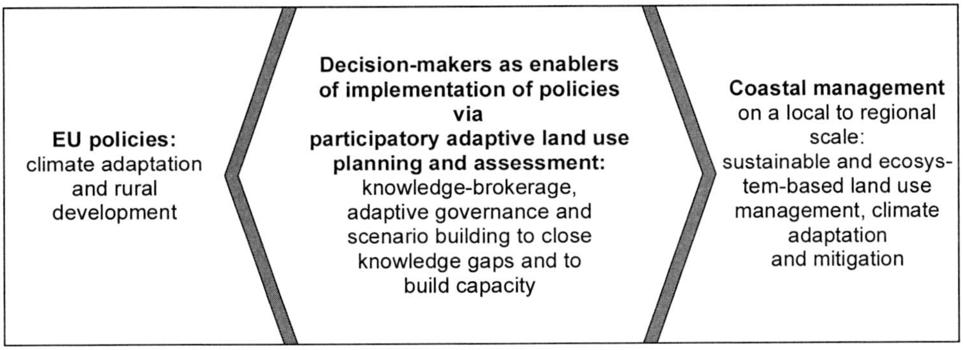

Figure 2: Relationship between EU policies (framework and upper strategic level) and coastal management (level of action taking)

Source: Authors' considerations

The PALUPA process takes the needs and world-views of the decision-makers into account. The decision-makers are open to consider external suggestions and alternative land use developments, which have been provided by the knowledge-brokers in form of the different scenarios. In the view of the decision-makers, the content of land management scenarios is better understandable and easier to realise than the content of EU policies, which have in their opinion a more theoretical character.

Considering the first common goal, a sustainable management of natural resources and including an ecosystem-based management, the knowledge-brokers used the concept of ecosystem services (MA 2003) to raise the stakeholders' awareness of fostering an ecosystem-based management. The translation of the ecosystem services approach into understandable units (for a detailed description see Karrasch et al. 2014) increased the awareness to enhance an ecosystem-based land use management in the case study area.

The co-development of scenarios (Rotmans et al. 2000; Salter et al. 2010) enabled to communicate and develop possible adaptation strategies of the case study region, contributing to the second common goal of taking climate adaptation and mitigation actions. Scenarios showing possible futures of the case study region strengthened as visual support the whole adaptive governance process and lead to an improved understanding of the complex interrelationships of socio-ecological systems (Thompson et al. 2012).

The third goal, closing knowledge-gaps for better-informed decision-making, is addressed by the role of the knowledge-brokers in the PALUPA process. A close relationship between the knowledge-brokers and decision-makers promotes the transformation of information into knowledge (Reed et al. 2014). The knowledge-brokers communicated the benefits of different adaptation measures for the decision-makers, ensured that the process was transparent and comprehensive, and gave feedback showing the results of every step taken in the process. Furthermore, the ongoing knowledge exchange between the knowledge-brokers and decision-makers as shown in this investigation "has the potential to significantly enhance the impact of environmental management research, policy and practice" (Reed et al. 2014, 338).

Regarding the forth common goal, the EU policies require active stakeholder involvement and capacity building. In the case study area, the heterogeneous group of decision-makers represents cross-sectoral, trans- and interdisciplinary integration on local and regional levels. The inclusion of all relevant decision-makers in the PALUPA process increased the groups′ acceptance and consensus building with regard to the development of new strategies (Schmidt et al. 2014). Additionally, decision-making processes are enhanced by participation of different stakeholders (Casado-Asensio and Steurer 2014; Pahl-Wostl 2002; Reed 2008, 2014; Ridder and Pahl-Wostl 2005; Pidgeon and Fishoff 2011). Regional adaptive capacity implies the decision-makers′ motivation to take advantages and benefits from changes rather than focusing on negative impacts. An increase in regional adaptive capacity is also valuable for decision-makers, since it promotes awareness raising, knowledge generation and development of new competences, and change in behaviour towards the development and implementation of adaptation strategies (Brooks and Adger 2005; Juhola and Kruse 2015). Furthermore, the provision of possible adaptation measures in terms of adaptive governance provides problem oriented solutions, which help to increase the regional adaptive capacity and the decision-makers′ understanding of benefits of change. The collaboration and networking of the decision-maker group has been increased. This is an important task for the whole adaptive governance process (Folke et al. 2005). An increase in regional adaptive capacity enables decision-makers to reduce risks and vulnerabilities associated with climate change impacts and socio-economic changes (Adger et al. 2007; Smit and Wandel 2006). As described in Casado-Asensio and Steurer (2014), climate and rural development policies are moving from command-and-control regulations towards more comprehensive and integrated strategies, focusing on sustainability and governance approaches, which are common goals of the three considered policies.

The results of this research will be used to generate recommendations for a sustainable development of the case study region and the development of a climate adaptation strategy which can be implemented in official regional planning. For the future, the development of transferable solutions for adaptation in other European coastal regions including cross-border solutions for sound collaboration between different countries will be a necessary step to strengthen horizontal integration by including and consulting trans-national stakeholders, in particularly because peripheral coastal regions of the southern North Sea face similar challenges. The requirements for transferability to other European low-lying coastal areas are the development of best-practice examples and generalisation of experiences and learning processes. By tackling problems and finding sound solutions for the case study area, it has to be ensured that these results are available for other similar projects.

Conclusion

The conducted PALUPA process in a peripheral low-lying coastal area focuses on fostering an ecosystem-based management to increase regional adaptive capacity and can be classified as an adaptive governance process dealing with the problems of climate and socio-economically induced change in rural development. By using different land management scenarios, PALUPA makes decision-makers aware of the challenges inherent to climate change and socio-economic change processes. Key elements of PALUPA are cross-sectoral participation, collaboration, and the work on a common goal to increase engagement of local and regional decision-makers. The case study shows that the generated results can be aligned with common EU policies. European climate and rural development policies aim to realise sustainable and adaptive management of coastal areas and emphasize the need for ecosystem-based climate adaptation and mitigation strategies. In order to foster a better-informed decision-making and to close knowledge gaps, they claim to support capacity building, participatory processes, trans- and interdisciplinary cooperation and cross-border collaborations. The alignment of PALUPA with EU policies is only possible because the policies do not counteract the needs and interests of the decision-makers, and a practical implementation of the actor-based scenario fostering a more sustainable and adaptive land use management can lead to the implementation of common policy objectives.

The investigation shows that local initiatives on future land use management, rather coincidently than intended, realise goals of the European climate and regional development policies. Here, the applied approach might serve as a "blueprint" to integrate EU policies into adaptive land use management of flood prone coastal areas (themed Europeanising coastal regions) in a more systematic manner. Transferable approaches based on transdisciplinary methods will strengthen the role of local stakeholders as enablers of an adequate implementtation of EU policies.

Acknowledgement

This study was part of the collaborative research project COMTESS (Sustainable coastal land management: Trade-offs in ecosystem services). The authors would like to express their gratitude for the support they have received through contract No. 01LL0911 of the German Federal Ministry of Education and Research, part of the funding measure "Sustainable Land Management". We are thankful for the time and effort of the local and regional stakeholders and the active participation they spent during the research process. We would also like to thank Christian Aden for his assistance in designing Figure 1.

References

Adger, W. N., S. Agrawala, M. M. Q. Mirza, Conde, C., O'Brien, K., Pulhin, J., Pulwarty, J., Smit, B. & Takahashi, K. (2007). Assessment of adaptation practices, options, constraints and capacity. In Parry, M. L., Canziani, O. F., Palutikof, J. P., van der Linden, P. J. & Hanson, C. E. (Eds.), Climate Change 2007: Impacts, Adaptation and Vulnerability. Contribution of Working Group II to the Fourth Assessment Report of the Intergovernmental Panel on Climate Change (pp. 717–743). Cambridge: Cambridge University Press.

Bogner, A., Littig, B. & Menz, W. (Eds.) (2009). Interviewing Experts. Research Methods Series. Basingstoke: Palgrave Macmillan.

Brooks, N. & Adger, W. N. (2005). Assessing and Enhancing Adaptive Capacity. In Lim, B. & Spanger-Siegfried, E. (Eds.). Adaptation policy frameworks for climate change: Developing strategies, policies and measures (pp. 165–181). Cambridge: Cambridge University Press.

Casado-Asensio, J. & Steurer, R. (2014). Integrated strategies on sustainable development, climate change mitigation and adaptation in Western Europe: communication rather than coordination. Journal of Public Policy 34 (3), 437–473.

COM (2013). Communication from the Commission to the European Parliament, the Council, the European Economic and Social Committee and the Committee of the Regions. An EU Strategy on adaptation to climate change. COM(2013)216. Brussels, 16 April 2013.

Directive 2014/89/EU of the European Parliament and of the Council of 23 July 2014 establishing a framework for maritime spatial planning, OJ L 257, 28.08.2014, 135–145.

EEA (2001). Participatory integrated assessment methods: An assessment of their usefulness to the European Environmental Agency. Technical Report 64, Copenhagen.

ESPON (2011). ESPON Climate: Climate Change and Territorial Effects on Regions and Local Economies. Dortmund: ESPON & IRPUD.

Flick, U. (2006). An introduction to qualitative research. 3rd edition. London: Sage.

Folke, C., Hahn, T., Olsson, P. & Norberg, J. (2005). Adaptive governance of social-ecological systems. Annual Review of Environment and Resources 30, 441–473.

Gaslikova, L., Grabemann, I. & Groll, N. (2013). Changes in North Sea storm surge conditions for four transient future climate realizations. Natural Hazards 66 (3), 1501–1518.

Grabemann, I. & Weisse, R. (2008). Climate change impact on extreme wave conditions in the North Sea: An ensemble study. Ocean Dynamics 58 (3–4), 199–212.

Hare, M. & Pahl-Wostl, C. (2002). Stakeholder categorisation in participatory integrated assessment processes. Integrated Assessment 3 (1), 50–62.

IPCC (2014). Climate Change 2014: Synthesis Report. Contribution of Working Groups I, II and III to the Fifth Assessment Report of the Intergovernmental Panel on Climate Change [Core Writing Team Pachauri, R. K. & Meyer, L. A. (Eds.)]. Geneva: IPCC.

Jacob, D., Göttel, H., Kotlarski, S., Lorenz, P. & Sieck, K. (2008). Klimaauswirkungen und Anpassung in Deutschland — Phase 1: Erstellung regionaler Klimaszenarien für Deutschland. Climate Change 11/08. Dessau: Umweltbundesamt.

Juhola, S. & Kruse, S. (2015). A framework for analysing regional adaptive capacity assessments: Challenges for methodology and policy making. Mitigation and Adaption Strategies for Global Change 20 (1), 99–120.

Karrasch, L., Klenke, T. & Woltjer, J. (2014). Linking the ecosystem services approach to social preferences and needs in integrated coastal land use management — A planning approach. Land Use Policy 38, 522–532.

MA — Millennium Ecosystem Assessment (2003). Ecosystems and Human Well-Being. A Framework for Assessment. Washington: Island Press.

Nicholls, R. J., Wong, P. P., Burkett, V. R., Codignotto, J. O., Hay, J. E., McLean, R. F., Ragoonaden, S. & Woodroffe, C. C. (2007). Coastal systems and low-lying areas. In Parry, M. L., Canziani, O. F., Palutikof, J. P., van der Linden, P. J. & Hanson, C. E. (Eds.), Climate Change 2007: Impacts, Adaptation and Vulnerability. Contribution of Working Group II to the Fourth Assessment Report of the Intergovernmental Panel on Climate Change (pp. 315–356). Cambridge: Cambridge University Press.

Nyberg, J. B. (2009). Adaptive Management: Strategies for coping with change and uncertainty. In Bawa, K. S. & Seidler, R. (Eds.), Dimensions of Sustainable Development 2 (pp. 1–32). Oxford: EOLSS Publishers.

OECD (2007). OECD rural policy reviews: Germany. Paris: OECD Publishing.

Oosterberg, W., Van Drimmelen, C. & van der Vlist, M. (2005). Strategies to harmonize urbanization and flood risk management in deltas. 45th Congress of the European Regional Science Association: "Land Use and Water Management in a Sustainable Network Society", 23–27 August 2005, Amsterdam, The Netherlands.

Pahl-Wostl, C. (2002). Participative and stakeholder-based policy design, evaluation and modelling Processes. Integrated Assessment 3 (1), 3–14.

Pidgeon, N. & Fischhoff, B. (2011). The role of social and decision sciences in communicating uncertain climate risks. Nature Climate Change 1, 35–41.

Reed, M. S. (2008). Stakeholder participation for environmental management: a literature review. Biological Conservation 141 (10), 2417–2431.

Reed, M. S., Stringer, L. C., Fazey, I., Evely, A. C. & Kruijsen, J. H. J. (2014). Five principles for the practice of knowledge exchange in environmental management. Journal of Environmental Management 146, 337–345.

Regulation (EU) No 1305/2013 of the European Parliament and of the Council of 17 December 2013 on support for rural development by the European Agricultural Fund for Rural Development (EAFRD) and repealing Council Regulation (EC) No 1698/2005, OJ L 347, 20.12.2013, 487–548.

Rijke, J., Brown, R., Zevenbergen, C., Ashley, R., Farrelly, M., Morison, P. & van Herk, S. (2012). Fit-for-purpose governance: A framework to make adaptive governance operational. Environmental Science and Policy 22, 763–784.

Ridder, D & Pahl-Wostl, C. (2005). Participatory Integrated Assessment in local level planning. Regional Environmental Change 5 (4), 188–196.

Rotmans, J., van Asselt, M., Anastasi, C., Greeuw, S., Mellors, J., Peters, S., Rothman, D. & Rijkens, N. (2000). Visions for a sustainable Europe. Futures 32 (9–10), 809–831.

Salter, J., Robinson, J. & Wiek, A. (2010). Participatory methods of integrated assessment — a review. Wiley Interdisciplinary Reviews: Climate Change 1 (5), 697–717.

Schmidt, A., Striegnitz, M. & Kuhn, K. (2014). Integrating regional perceptions into climate change adaptation: a transdisciplinary case study from Germany's North Sea Coast. Regional Environmental Change 14 (6), 2105–2114.

Smit, B. & Wandel, J. (2006). Adaptation, adaptive capacity and vulnerability. Global Environmental Change 16 (3), 282–292.

Thompson, J. R., Wiek, A., Swanson, F. J., Carpenter, S. R., Fresco, N., Hollingsworth, T., Spies, T. A. & Foster, D. R. (2012). Scenario studies as a synthetic and integrative research activity for long-term ecological research. BioScience 62 (4), 367–376.

UNEP/GPA (2006). Ecosystem-based management: Markers for assessing progress, Report commissioned by the Coordination Office of the Global Programme of Action for the Protection of the Marine Environment from Land-based Activities (GPA) of the United Nations Environment Programme (UNEP). The Hague, Netherlands.

Ward, V., House, A. & Hamer, S. (2009). Knowledge Brokering: The missing link in the evidence to action chain? Evidence and Policy 5 (3), 267–279.

Wilbanks, T. J., Leiby, P., Perlack, R., Ensminger, J. T. & Wright, S. B. (2007). Toward an integrated analysis of mitigation and adaptation: some preliminary findings. Mitigation and Adaptation Strategies for Global Change 12 (5), 713–725.

Wong, P. P., Losada, I. J., Gattuso, J.-P., Hinkel, J., Khattabi, A., McInnes, K.L., Saito, Y. & Sallenger, A. (2014). Coastal systems and low-lying areas. In Field, C. B., Barros, V.R., Dokken, D. J., Mach, K. J., Mastrandrea, M. D., Bilir, T. E., Chatterjee, M., Ebi, K. L., Estrada, Y. O., Genova, R. C., Girma, B., Kissel, E. S., Levy, A. N., MacCracken, S., Mastrandrea, P. R. & White, L. L. (Eds.), Climate Change 2014: Impacts, Adaptation, and Vulnerability. Part A: Global and Sectoral Aspects. Contribution of Working Group II to the Fifth Assessment Report of the Intergovernmental Panel on Climate Change (pp. 361–409). Cambridge: Cambridge University Press.

Rural resilience and renewable energy in North-East Groningen, the Netherlands: in search of synergies

Rozanne Spijkerboer, Elen-Maarja Trell and Christian Zuidema

Abstract

Using renewable energy (RE) for solving area-based socio-economic issues and by extension making peripheral, rural places more resilient is increasingly signalled in academic literature. Furthermore, governance capacity surrounding RE production is also being increased by a growing number of local initiatives around RE visible in various countries, including the Netherlands. It remains unclear, however, whether formal governments are recognizing and anticipating on the socio-economic potentials of RE. In this chapter we explore the perceived role of RE in government policy and planning in the Netherlands in connection to supporting the resilience of rural regions in the face of rural decline and peripheralisation. We focus on different governmental levels and the government as an institutional framework with official power and legitimacy to intervene in RE production. Our area of interest is the coastal and predominantly rural region of (North-East) Groningen. This area profiles itself as the 'energy portal' of the Netherlands while simultaneously dealing with a complex combination of opportunities and challenges related to both energy, as well as peripheralisation and rural decline.

Introduction

Social and economic benefits derived from natural resources can enhance the vitality of an area by contributing to "security of income, economic prosperity and the resilience of the region" (Karrasch et al. 2014, 528). RE production (e.g., wind energy, solar energy, energy crops) is a concrete example of natural resources offering social and economic returns. Previous research indicates that in an economic sense, RE has the potential to strengthen existing local and regional activities, such as farming (Muller 2009), or local industries (Avelino et al. 2012), while effects on generating employment also seem positive (Lehr et al. 2012). Finally, RE has additional benefits for local society, for example in contributing to citizen commitment and social capital (Seyfang and Haxeltine 2012). Consequently, RE production provides a window of opportunity for contributing to the resilience of the area in which they are developed.

De Boer and Zuidema (2015) argue that the 'local potential' of RE is conditioned by local circumstances. Physical circumstances condition opportunities for using, for example waste heat, biomass, and even solar, wind and hydro energy. The socio-economic context (e.g., local identities, social networks and groups, entrepreneurship and businesses) also condition opportunities for social support, initiatives and investments in RE production. Local circumstances, finally, are also relevant as RE production has a significant spatial impact. In contrast to most 'conventional' fossil fuel based energy extraction and production activities, RE production requires vast amounts of space and is highly visible. Its implications for the landscape make RE a highly contested matter prone to societal resistance (e.g., Strunz 2014). Hence, as Wüstenhagen et al. (2007) argue, without carefully embedding RE within both the local *spatial* and *social* context, RE initiatives are prone to produce societal aversion, impeding their development and implementation (see also Ohl and Eichhorn 2010).

If RE production is to strengthen the *resilience* of a region, it is crucial for plans and policies to recognize and capitalise on the socio-economic benefits that RE can potentially offer, while responding to the local conditioning factors just mentioned. Most notably, as De Boer and Zuidema (2015) argue, such plans and policies should therefore assist the integration of RE within the physical and especially socio-economic context of the area in which it is located. Such integration can be supported by an area-based approach (ibid.) which enables RE to be better co-located with other land-uses (Measham and Fleming 2014) and seeks local acceptance by applying planning approaches with local input (Ohl and Eichhorn 2010). However, we doubt whether this argument is also recognized and pursued in governmental plans and policies. Therefore, in this chapter we explore the extent to which RE[1] is 'formally' recognized as an opportunity to address socio-economic problems related to peripheralisation and rural decline and to what extent attempts are made to integrate RE initiatives in their spatial and socio-economic context.

Our exploration is set in the rural North-East of the Netherlands, where we analyse spatial planning and energy related policy documents of different governmental levels. We focus on formal government documents, as these documents have formal power and legitimacy to provide the institutional framework for the development of both RE and the region as a whole (Lloyd et al. 2013). We focus on the coastal and predominantly rural Province of Groningen and, more specifically, the so-called Eemsdelta region (home to about 65 000 inhabitants and comprising four municipalities: Delfzijl; Eemsmond; Appingedam; Loppersum). This area was chosen as, on the one hand, it has a large potential for the development of RE (e.g., favourable wind conditions; large waste heat

[1] When talking about renewable energy production we refer to the institutional framework related to the development of facilities for the production of renewable electricity on land. Other parts of renewable energy policy such as extension and redesign of networks and energy saving measures are not examined in the context of this paper.

potentials; socio-political interest in RE; enough space). On the other hand, this area deals with challenges related to peripheralisation, in the form of socio-economic challenges such as population decline, unemployment, and lack of services, where the benefits from RE production might increase the resilience of the region (Zuidema and Trell 2014).

Below we first discuss the connections we are looking for between peripheralisation, rural decline and renewable energy in the context of North-East Groningen. Second, we explain the theoretical position of this paper with regard to resilience and how we conceptualise this. Third we introduce our research approach, data and the methodology. Fourth, we interpret the results of the analysis and finish with a discussion and conclusion.

Renewable energy in the context of rural decline and peripheralisation: the case of North-East Groningen

"[U]ntil 2025, [...] in the eastern part of Groningen, Delfzijl and the surrounding area [...] the population will decline by 5 to 8%. In these regions, the population is on average older than in the rest of the country and this will lead to a higher mortality rate. Another factor which contributes to the population decline is that many young people leave these regions to settle in the large cities" (CBS 2013). Apart from population decline, also various other key indicators of the Dutch national statistics office illustrate North-East Groningen as an area in decline (CBS 2013). Unique to the area is also that gas extraction has created increasingly intense and frequent earthquakes that damage property and increase the risk of flooding. As a result, many home-owners and local businesses fear for their health and safety and doubt whether they are able to sell their property if they may wish to leave (Van den Voort and Vanclay 2015; de Kam and Spijkerboer 2015). Simultaneously, the problems related to natural gas extraction have created a sense of urgency to shift from fossil fuels to RE, particularly among younger inhabitants (de Kam and Spijkerboer 2015).

As illustrated by national indicators, North-East Groningen is a good example of an area subject to *rural decline* and peripheralisation (in the case of Groningen exacerbated by the earthquake problems). Literature provides three main indicators for rural decline that all also affect the North-East Groningen: (youth) out-migration; decline of human capital; and relatively low incomes (Maesham and Fleming 2014). The out-migration of mainly young people is often related to a lack of employment opportunities. This trend is connected to the problem of declining human capital since especially the 'best and brightest' tend to leave[2]. Furthermore, rural decline implies a growing gap between urban and rural incomes (Maesham and Fleming 2014).

According to Beetz et al. (2008) similar characteristics are typical to the process of peripheralisation. Moreover, an important aspect of peripheralisation processes is the lack of power of local institutions to advance their interests in responding to peripheralisation and hence, a dependence on political decisions made elsewhere that often lack sensitivity to the local context (Beetz et al. 2008). Although peripheral regions might sometimes be connected to supranational and even global networks, according to Beetz et al. (2008) they often derive only limited welfare effects from this connection. The Eemsdelta in North-East Groningen, but also more broadly the north of the Netherlands, is a perfect example of such a region (Figure 1), where despite the presence of the largest natural gas field in Western-Europe, the region is socio-economically lagging behind and largely depends on the central government for investment decisions. Similar to other rural areas, it is relatively scarcely populated, but simultaneously has an abundance of undeveloped space, with favourable wind conditions, and good agricultural land, while the ground prices are relatively low (Zuidema and Trell 2014). Moreover, the region itself sees RE production among the prime potentials for future development (State et al. 2014).

The emphasis on energy is illustrated in a vision document produced for the Eemsdelta region (Figure 1) with energy as one of the key opportunities. The vision positions the Eemsdelta region as a leader in biobased economy and in green energy; i.e., it is to become a so-called *'electric socket'* for the Netherlands (Stuurgroep Ontwikkelingsvisie Eemsdelta 2013). RE is also actively promoted by the province of Groningen as one of the main future economic basis for regional development. Partly built upon the existence of institutional networks and expertise from five decades of gas extraction, the Province of Groningen today invests in becoming the *'energy portal'* of the Netherlands (4Power 2012). Therefore, the province is part of a broader region consisting of the three northern provinces that is recognized as the so called 'energy valley' of the Netherlands (Figure 1). The Energy Valley Foundation, set up in 2003, unites over 200 public and private partners in the north of the Netherland. Governmental bodies, companies and knowledge institutions are working together on the growth of the northern energy sector (4Power 2012).

---

[2] Relevant with respect to these trends is to identify whether people stay away, or possibly return with new knowledge, in which case the temporary out-migration would not be a problem and could even be perceived as an opportunity.

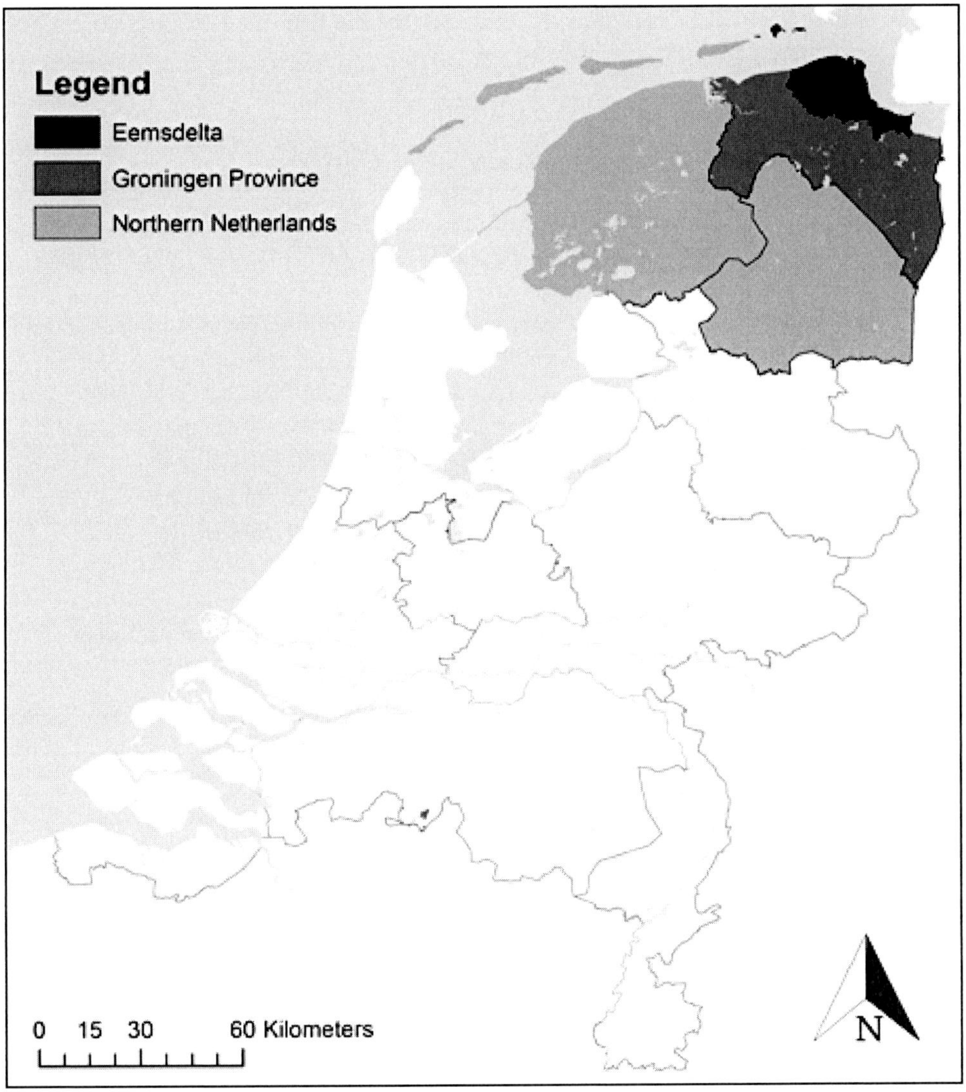

Figure 1: The location of the case study regions
Source: Authors' compilation

The importance of RE production is also clearly illustrated by a recent SWOT analysis in the multi-year infrastructure, spatial planning and transport agenda (MIRT agenda) for the north of the Netherlands (State et al. 2014). The summary of this SWOT in Table 1 illustrates that more than half of the recognized opportunities are related to energy. Simultaneously, many of the threats and weaknesses that are recognized in the SWOT are related to rural decline and peripheralisation. The discussion above, as well as the SWOT analysis, indicate that in the north of the Netherlands, particularly the province of Groningen and Eemsdelta, RE might indeed become a carrier for increasing the resilience of the region in light of rural decline and peripheralisation.

Table 1: SWOT from the MIRT Agenda northern Netherlands

Strengths	Weaknesses
Location on an European junction of gas and electricity infrastructure	Economic climate with limited resilience
Access to a high quality infrastructure to counter for imbalances in energy provision	Peripheral location from the point of view of Randstad
Soil conditions with storage capacity	Relatively high unemployment rates
Strong cities with economical clusters and knowledge institutions	Lower labor market participation and educational level then national average
Strong and well organized small and medium sized enterprises	Mismatch between educational level/-direction and labor market
Agricultural sector that contributes to the export strength of the Netherlands	Low percentage of R&D activities
Versatile economic cluster with niches	Big share of employment in declining sectors: industry, education, government
Fresh water reserve underneath Drents Plateau	
Opportunities	**Threats**
Earning capacity of the soil	Exhaustibility of gas
Increasing investments in energy	Social resistance against energy production and storage
Using residual heat	Demography: high vacancy rate; declining service base; declining transport value
Experimental space for innovations in energy and water technology	Decline of human capital
Space and facilities on land for offshore energy	Oversupply of modal working locations
Agriculture as producer of resources within a bio-based economy	Dependence on Ijsselmeer for the fresh water supply
Polycentric core structure with strong centrum-cities and location on the border	Soil subsidence due to peat oxidation with irreversible damage to peat lands
Technological developments in mobility	Negative influence of national policies, especially in the social domain and in the regions in decline
Space for experimentation	Increasing societal burdens and decreasing earnings capacity from area development
Flexible and efficient water system	

Source: State et al. 2014, 12

Towards resilient energy landscapes: a theoretical discussion

From a theoretical and conceptual perspective, the search for RE production as a strategy for coping with the effects of rural decline and peripheralisation can be supported by using the notion of 'resilience'. Resilience is a concept that is often used to analyse the way societies deal with, or respond to change (Chapin et al. 2009; Davoudi and Porter 2012; Steiner and Markantoni 2013). In this chapter we use an empirical definition of resilience[3] based on the so-called 'social-ecological' interpretation of resilience as provided by Folke (2006). He analyses the development of the concept of resilience as a shift from 'engineering-resilience' through 'ecological resilience' towards 'social-ecological resilience'. Most notably, this is a shift away from viewing resilience as the capacity of a system to 'bounce back' to its original state ('engineering resilience'), towards the capacity of a system to respond to contextual change by transforming itself to cope with changed conditions ('social-ecological resilience'). Folke (2006, 259–260) interprets social-ecological resilience as a combination of "(1) the amount of disturbance a system can absorb and still remain within the same state or domain of attraction; (2) the degree to which the system is capable of self-organization [...]; (3) the degree to which the system can build and increase the capacity for learning and adaptation". Resilience, as used in this chapter, therefore, directs our attention towards the capacity of regions and localities for responding to, and re-organizing themselves, in the face of larger socio-economic, demographic or physical challenges. In the case of North-East Groningen, the question then becomes how RE production might increase the region's capacity to use and transform its existing socio-economic and spatial characteristics so as to be better able to deal with rural decline and peripheralisation. In relation to the rural decline indicators discussed in the previous section, the question then is how RE production might be an opportunity to address these socio-economic trends in rural, peripheral areas (Table 2).

---

3 However, resilience is a contested notion with many different definitions. Some of these definitions are more normative focusing on how resilient systems ought to behave (e.g. Stokols et al. 2013; Chapin et al. 2009), whereas others are more empirical and concerned with explaining what can be observed in reality. These empirical definitions often allow for degrees of, or different forms of, resilience (e.g. Folke, 2006; Lloyd et al. 2013).

Table 2: Rural decline indicators as used in our analysis

Rural decline indicators	Questions
Migration	Is RE perceived as an opportunity to limit (youth) out-migration?
Skills and education	Is RE perceived as opportunity to retain and attract human capital?
Employment	Is RE perceived as an opportunity for employment options, both direct and indirect (spill-over)?
Community identity and social ties	Does RE production take into account and is it perceived as an opportunity to strengthen community identity and social ties?
Services	Is RE perceived as an opportunity to maintain or increase the available service facilities?
Income disparities	Is RE perceived as an opportunity to increase the local/regional income (and for whom?)

Source: Based on Koomen 2011 and Shucksmith et al. 2006

A resilient integrated energy landscape

In this chapter we argue that RE production might be a key to stimulate the resilience of regions and localities. De Boer and Zuidema (2015) suggest that the notion of the 'integrated energy landscape' illustrates how RE production could be integrated in the physical and socio-economic landscape. An integrated energy landscape is "a multifunctional physical and socio-economic landscape of which energy initiatives and systems are an integrated part" where "energy initiatives and systems become integrated through linkages, consisting of new relationships, interactions and movements with the local landscape" (De Boer and Zuidema 2015, 2). An area-based approach is important for achieving such integration for a number of reasons: (1) the development and implementation of RE is easier when based on available local potentials (e.g., the local physical or socio-economic fabric); (2) creating synergies between RE projects and other local land uses or socio-economic functions is crucial for enabling RE to have a clear impact on local socio-economic development (e.g., in supporting agriculture, nature management, entrepreneurship, liveability or mobility enhancement), and (3) the embedding of RE in the local context might reduce societal resistance and enhance societal support and trust while allowing for the inclusion of local knowledge in projects. When indeed integrated with local socio-economic functions and supported through societal engagement, local RE initiatives can also contribute to building social capital. Such social capital might not only help regions cope with their RE ambitions, but might well spread to other regional challenges where RE initiatives also address aspects such as flooding, food, mobility or local livability issues. In doing so, social capital also can create an improved local or regional governance capacity to act in the face of both RE and other regional challenges related to rural decline (De Boer and Zuidema 2015). Although not explicit in their work, these suggestions made by De Boer and Zuidema resonate with literature on increasing regional resilience. Their proposed integrated energy landscape is in its essence a framework for promoting RE production as a means to improve a regions' resilience by means of an area-based approach which closely knits RE projects to other local or regional activities. What remains important, though, is for the institutional dynamics at both the regional and higher scale to allow or even stimulate such knitting.

Addressing these institutional dynamics, Folke et al. (2010), suggest that resilience promotes "an adaptive approach, facilitating different transformative experiments at small scales and allowing cross-learning and new initiatives to emerge" (Folke et al. 2010, 5). Interactions between physical, social or economic innovations within and across spatial scales would subsequently contribute to the emergence of a new identity for the system; in our case a reconfigured physical and socio-economic regional structure. Strunz (2014) adds to this that such physical, social and economic innovations are indeed interdependent with the *institutional* system. He uses the example of the German *Energiewende* (energy transition) where the institutional context was very favourable to RE and large investment programs had paved the way for a system shift when a window of opportunity (the Fukushima disaster) presented itself (Strunz 2014). The above illustrates the importance of the existence of vertical interactions between localised and individual physical, social or economic innovations and the social, economic and institutional frameworks on higher scales. In other words, the resilience of regions is not only dependent on physical, social or economic innovations within and across spatial scales, but also on an institutional context allowing for and supporting local experimentation, learning and upscaling across institutional levels.

Drawing from the work of Alexander (2006, 2012; also Loorbach 2010) three, mutually influencing institutional levels can be identified: (1) the *macro*-level, where rules are set that apply to whole societies (i.e., constitution writing or major strategic programs) and that change very slow; (2) the *meso*-level, at which inter-organizational institutional change takes place, through governmental policies, regulations or the creation of (regional) organizations. The main purpose on this level usually is to create stability for current practices and actors, while protecting investments and existing values; (3) the *micro*-level focuses on intra-organizational change (i.e., organizational initiatives, committees and work groups) and operates on a lower scale where new initiatives are emerging and innovations are tested. Therefore, if RE production is to strengthen regional resilience in the face of rural decline and peripheralisation, it also requires an accommodating and stimulating institutional context on the micro-, meso- and macro-level.

Research approach, context and methodology

In this research we investigate whether and how the local, provincial, regional and national institutional context, here confined only to policy documents, recognizes and stimulates integration of RE production in the regional physical and socio-economic landscape. Rather than targeting specific institutional mechanisms such as legislation or subsidies, we aim to identify if regional and national policies recognize the need for developing an integrated energy landscape that potentially enhances regional resilience and develop *strategies* to stimulate this development.

This research draws on policy document analysis. We selected policy-documents ranging from the national to the local scale and targeted both energy policies and spatial planning policies. This choice is informed by the realisation that it is within the energy sector that many policies for RE production are made, while their localisation and regional embeddedness is, at least in a Dutch context, typically subject of spatial planning policies. Hence, it is the combined set of policies in the energy and spatial planning policies that allow for a comprehensive picture of the visions and strategies of the Dutch national, regional and local governments. The choice for governmental documents is informed by their power in setting the formal agenda for spatial planning, regional development and RE production. Furthermore, such documents are, specifically in the relatively strongly government led spatial planning context of the Netherlands, also crucial in setting the context for both governmental and often private investments. Especially in pursuing RE production, the Dutch depend relatively strong on government involvement (e.g., De Boer and Zuidema 2015; Rotmans 2011). However, in the Dutch context also societal partners are often involved in preparatory discussions for formal government documents and are sometimes also active in writing and endorsing the policies (Rotmans 2011). The choice for including national, regional and local policy documents is informed by their combined impact on local spatial and RE policies; in the Dutch context, regional and local spatial policies and RE policies are developed within a frame of constraints and stimuli provided by national and regional policies (EC 1997).

Choice of documents: RE and spatial planning in the Netherlands
Our choice of policy documents was informed by assessing how energy policies and spatial planning policies are both separated and linked in a Dutch context. The division of responsibilities in spatial planning is set down in the *Dutch Spatial Planning Act*. On the national level, the *Ministry of Infrastructure and the Environment* (MoIE) is responsible for the planning system and can exert power through national structure schemes and land-use plans. The national government is responsible for topics of national importance with consequences or gains that exceed the power or borders of municipalities and provinces or that are subject to international agreements. RE production is recognized as such a topic, illustrating that national spatial planning policies should address RE (MoIE 2012b). But it is important to note that the main responsibility for most policy regarding energy lies with the *Ministry of Economic Affairs* (MoEA) (Bosman 2013). Decisions regarding e.g., the location of onshore wind parks are consequently made in collaboration between the two ministries (MoIE and MoEA 2014).

On the national level the *National Policy Strategy for Infrastructure and Spatial Planning* (SVIR) was selected for analysis as it is a key document setting out the general spatial strategy for the Netherlands until 2040. The second national level document, the *Energy Agreement* is an agreement between the government and private parties regarding the direction of energy policy and actions that need to be taken to increase growth in this sector. These are the two main policy documents currently containing the national governments vision on RE energy production and spatial planning (Table 3).

Table 3: Analysed policy documents

Scale	Year	Name	Reference
National Netherlands	2012	Structuurvisie infrastructuur en Ruimte (SVIR) *[National Policy Strategy for Infrastructure and Spatial Planning (SVIR)]*	MoIE (2012a)
	2013	Energieakkoord (voor duurzame groei) *[Energy Agreement (for sustainable growth)]*	SER (2013)
Regional North of the Netherlands	2014	Energieke Noorderruimte — MIRT Gebeidsagenda Noord-Nederland *[MIRT Agenda Northern Netherlands]*	State et al. (2014)
	2013	Noorderzine — Noordervisie 2040 *[Northern vision 2040]*	Groningen Province et al. (2013)
	2014	SWITCH Noordelijke Energieagenda *[Northern Energy Agenda]*	BONN & Stichting Energy Valley (2014)
	2011	Green deal Noord-Nederland *[Green deal North-Netherlands]*	MoEA et al. (2011)
Provincial Groningen	2009	Provincie POP 2009–2013 *[Provincial area plan 2009–2013]*	Groningen Province (2009)
	2013	Lokale Energie — Programma Energie 2012–2015 *[Local Energy Program 2012–2015]*	Groningen Province (2013)
	2015	Ontwerp Omgevingsvisie *[draft Area Vision]*	Groningen Province (2015a)
Local Eemsdelta (North-East Groningen)	2013	Ontwikkelingsvisie Eemsdelta 2030 *[Development Vision Eemsdelta 2030]*	Stuurgroep Ontwikkelingsvisie Eemsdelta (2013)

Source: Authors' compilation

Provinces are mainly responsible for guiding regional development by creating integrated visions on spatial development while balancing different interests. Provinces can exert power through provincial structure schemes and land-use plans. The guiding strategic policy document for spatial development on that level in Groningen province is the *Provincial Area Plan* (MoIE 2012b), which was valid until June 2015; its successor, the *Area Vision,* is at the moment of writing this chapter under revision. Provinces play a key role in stimulating RE and collaborate closely with the national government on the allocation of space for wind energy. Furthermore, they formulate policy regarding the stimulation or constraining of more decentralised forms of RE such as solar energy and biomass (IPO 2015).

For the province of Groningen, the *Provincial Area Plan* provides the vision for both spatial planning and energy. The *Draft Area Vision* provides insights into the future of spatial planning and the allocation and stimulation of energy production, consumption and governance in the province. The *Energy Program* provides details on local energy initiatives. Other regional policies which have a less formal status are also influential in providing guidance to policy developments and choices for RE and spatial planning in Groningen. Although formally signed by the three northern Provinces, these documents are not part of the legally required policies Provinces need to develop. Hence, their contents and status can more easily be contested. A regional policy called the *Northern Vision 2040,* compiled by the three northern provinces, provides a shared vision on the development (including energy development) in the north of the Netherlands. The regional document *MIRT agenda* was selected for analysis as it provides regional information on the spatial challenges until 2040 and outlines the actions that will be taken to deal with them. The *Northern Energy Agenda* and the *Green Deal* are action programs in which the government and private parties have agreed on (investment) priorities and policies concerning sustainability and energy.

Municipalities are responsible for a safe and livable environment in which interests regarding mobility, the environment, water, economy and living are balanced by means of structure schemes and zoning plans (MoIE 2012b). However, on the municipal level in the North-East of Groningen there are no strategic policy documents that address the whole territory of a municipality. Therefore, the *Development Vision Eemsdelta 2030,* which collectively addresses four municipalities in North-East Groningen: Appingedam, Delfzijl, Eemsmond and Loppersum was analysed. The development vision has no formal legal status, but is relevant as it is supported by the Province of Groningen and the municipal councils. Therefore, this document was selected for analysis in order to gain insights into local vision and strategies for energy.[4]

---

4 In the remainder of the chapter the documents will be referred to using their English language names with the related governmental level provided in square brackets.

Methodology

The selected documents were analyzed focusing on whether they recognize and support the integration of opportunities for RE production with socio-economic issues and the spatial context of the region. The qualitative data analysis software Atlas.ti was applied to code the documents. With manifest coding concrete terms used in the documents were examined to gain an overview on the 'surface level' (Babbie 2010). Manifest coding established, for example, which types of RE production the documents discuss and which motives are provided for focusing on RE. Latent coding was used to analyze the underlying meanings and values in the documents and provided insights into strategies, narratives and concepts that are clearly advocated in the policy documents and those that are relatively marginalized (Aitken and Valentine 2006). These codes were divided in four code families: (1) rules and responsibilities; (2) area-based connections; (3) rural decline indicators (Table 2); (4) experimentation and innovation. These code families broadly reflect the structure of the result section, supported throughout by insights regarding the types of energy and motives. The result section starts with some observations regarding the responsibilities and priorities across governmental levels, providing insights into the institutional context. Second, the narratives and concepts relating to connections with the surrounding physical landscape (or area) are explained. Third, the results relating to the rural decline indicators provide insight into the extent to which connections are recognized between RE production and the socio-economic context. Fourth, the main strategies and institutional framework relating to local experimentation and innovation are discussed.

Findings and interpretations

Roles and priorities

A first important observation follows from the prioritization of RE production as a national interest. Following this prioritization, the national government has assigned itself the main responsibility for large energy projects, which in practice implies it takes a leading role in developing large wind parks of over 100 MW. Provinces are involved in selecting the areas where such projects are potentially allowed, but once a project is defined by investors and land owners (often farmers), the state takes over the planning (*SVIR* [national]; draft *Area vision* [province]). For such projects, the state applies a procedure called the *'National coordination scheme'*, implying that the Ministry of Economic Affairs, in consultation with lower levels of government, is responsible for the coordination of decisions regarding the project. This often includes the adoption of a *'National zoning scheme/spatial plan'* for the project area (RVO 2015). The choice for this strong top-down national coordination scheme is based on the need for swift and secure implementation of targets for wind energy. Local or regional interests are perceived as barriers that might frustrate such swift and secure implementation and are therefore often overlooked. This limits area-based innovations to link wind energy with regional or local development. It is therefore even debatable whether and how much space for innovation is left in such large wind energy projects (De Boer and Zuidema 2015). Currently, wind farms are, also in the province of Groningen, highly contested by local residents for producing only costs (visibility, noise nuisance, wind shadows) and having no visible benefits for the areas in which they are located. The national coordination scheme can thus be seen as impeding attempts at using RE production for supporting regional resilience.

In the meantime, a second observation on the provincial level confirms that RE production is not yet identified explicitly as a means for supporting regional resilience. As is shown in Figure 2, the *Draft Area Vision* [province] only sees RE production as connected to the regional economic structure and the production of clean energy. Alternative benefits, connected to the values and goals for 'thriving rural areas and strong cities' (and the attributes of attractive, livable and safe), or the core values of a distinctive, diverse and quiet landscape are not recognized.

On the surface it seems evident that national priorities to advance large scale RE production and provincial acknowledgement of its potential benefits for the region are hardly recognizing how RE production might support regional resilience. A more detailed analysis, however, is needed to assess whether in a more implicit sense, such recognition does exist. In the following paragraphs we will report on such detailed analysis.

Goals and values / Provincial interest	Sustainable economic structure			Strong cities and thriving rural areas			Core values			
	Competitive	Accessible	Future-proof	Attractive	Liveable	Safe	Distinctive	Diverse	Clean	Quiet
Provincial interest 1: Spatial quality			■	■			■			
Provincial interest 2: Attractive business climate	■	■	■	■						
Provincial interest 3: Space for energy	■		■						■	
Provincial interest 4: Strong agricultural sector	■		■		■				■	
Provincial interest 5: Protect the landscape			■	■			■	■	■	
Provincial interest 6: Increase biodiversity			■	■	■		■			
Provincial interest 7: Water safety			■			■	■			
Provincial interest 8: Clean and sufficient water	■		■						■	
Provincial interest 9: Accessibility	■	■	■		■					
Provincial interest 10: Prevent environmental disturbance			■		■	■			■	■
Provincial interest 11: Use of the sub-soil			■		■	■				

Figure 2: The perceived connection between provincial interests and overall goals and values

Source: Draft Area Vision [province], 12

Fitting RE into the local spatial context

The policy documents recognize that RE requires more space compared to tradetional energy sources. They therefore emphasize that RE is to be carefully 'fitted into' the physical landscape (*inpassing*), implying that RE has to be adapted to the physical landscape to minimize changes to the landscape. However, how such *inpassing* is to take place remains ambiguous and is not operationalised.

Wind energy receives most attention throughout the government documents and is perceived at the local and provincial levels as something that should be developed based on national targets, for which the local and provincial level refer to 'deals' made with the national government. Illustrative for this attitude are statements by the Province of Groningen, when they declare that that "space will only be reserved for the amount of wind energy that is nationally required" (*Draft Area Vision* [province], 37). Wind parks are perceived to have an industrial function and should therefore be located near industrial areas. In general the province of Groningen has pursued a policy in which three 'concentration areas' for wind energy have been determined (D*raft Area Vision* [province]). This concentration policy "has *saved* the province from many solitary turbines" (ibid., 24). However, under previous policy schemes some solitary turbines were located in the province of Groningen, which are now perceived, as stated in the *Provincial Area Plan*, as a 'prominent disturbance to the landscape' (87) that needs to be relocated to the concentration areas and on the long term preferably to the North Sea. The question about how large scale, concentrated wind parks can 'fit into' the landscape structure is not posed. Neither do these provincial documents recognize the possible contribution of solitary turbines for supporting villages or communities to generate financial gains to be used for the support of local services. The use of solitary turbines to do exactly that is, however, a development visible in both neighbouring countries and even neighbouring provinces (e.g., Maesham 2012). Wind energy, it seems clear, is in Groningen only pursued 'because the national government tells us'.

In their *Draft Area Vision* the Province of Groningen thus aims for alternative means of RE production (as compared to wind energy) as well. For example, the province also claims to stimulate biomass production. Surprisingly, though, biomass installations are only allowed in areas that are zoned for agricultural buildings to 'protect the landscape'. Again, rather than searching for how these installation might link to the socio-economic landscape, it is the protection of the existing physical landscape that is prioritized. A similar approach is visible when it comes to solar parks. They are only allowed in urban areas while in rural areas they are only allowed as an exception (D*raft Area Vision* [Province]). These regulations emphasize that RE and landscape are not seen as an integrated whole but rather RE has to be adapted to the landscape in a way that minimizes the impact to it.

Overall, RE seems to be perceived as a threat to the physical landscape, including cultural historical assets. Opportunities for the physical landscape in the form of enhancing nature and recreation vis-à-vis biomass production, the use of urban design with solar panels to create new cityscapes or wind energy to strengthen landscape structures are not recognized or contested. As a result, RE production is seen as conflicting with the landscape, narrowly defined in physical terms (e.g., in the *Development Vision Eemsdelta* [local]). Despite this perception, in the *Development Vision Eemsdelta* [local] by 2030 the landscape around the Eemshaven is expected to reflect the status of the region as an international player with regard to RE. It is then noteworthy that at present, integration of RE (especially wind energy) with the landscape is not a priority. Rather than searching for opportunities to achieve integration, the focus in the policy documents seems to lie on the restrictions.

Furthermore, what is perceived as the landscape here, is largely (if not solely) confined to the physical landscape. The need for integration with the socio-economic landscape are all but overlooked. When it comes to the socio-economic landscape, all governmental levels emphasize the need for social acceptance. Again, however, this is very broad and it remains unclear how social acceptance is to be achieved. Since literature suggests that connecting RE production to the socio-economic fabric of an area could potentially enhance social acceptance (Maesham 2012; Seyfang and Haxeltine 2012), in the next paragraph we examine whether, more implicitly, such connections between RE production and socio-economic indicators (in our case we specifically looked at rural decline indicators) are recognized at all.

Rural decline indicators: connections with the socio-economic context

The document analysis reveals that links between RE production and rural decline indicators exist, but they are not abundant and not all indicators are equally addressed. The rural decline indicators that are most often addressed in all analysed documents in connection to RE production are employment, and skills and education. RE production is expected to create opportunities for employment in the energy sector and the related sectors such as the installation- and construction sector (e.g., *Energy Agreement* [national]).

With regard to skills and education, the focus within the documents lies on the development of knowledge centres, innovation, and R&D activities, which are presented as factors that can help increase the competitiveness of the region. When considering skills and education as a factor to increase rural resilience, perhaps the most relevant aspect is whether it is perceived as a factor that might help create, attract and retain human capital. We indeed found that RE production was considered as a means to do exactly that in only six instances, but across all governmental levels (in the *Development Vision Eemsdelta* [local], the *Draft Area Vision* [province], the *Northern Vision 2040* [regional], and the *Energy Agreement* [national]) (Box 1).

Box 1: Example of RNE addressing rural vitality indicators

"Another important point of attention within our vision is to profile the North as a pioneer in the field of sustainable energy production and energy consumption and focus on the innovation sectors. These sectors and the sustainability ambition should give a boost to employment in the North. That will attract and keep companies and highly educated inhabitants in the area."

(Northern Vision 2040, 121 [regional])

Other rural decline indicators are hardly mentioned in connection to RE production in any of the analysed documents. On the regional scale a few remarks are made which demonstrate that close knit communities are seen as an important prerequisite for the successful development of local energy initiatives, but the reverse is typically not identified. Only the *Northern Energy Agenda* [regional] explicitly states that the new economic activities related to RE might bring new perspectives to communities, thereby contributing to community-building.

Migration in relation to RE is addressed once in the *Draft Area Vision* [province] and twice in the *Northern Vision 2040* [regional], with a reference to *attracing* new inhabitants. Box 1 contains one of the few examples where RE is connected to attracting 'highly educated inhabitants'. How RE production might help to keep and attract youths is not addressed in any of the analysed documents.

Only once, a reference is made to the generation of income. The *Development Vision Eemsdelta* [local] states that the development of biomass industries is "economically beneficial, because connections are made between sectors in a manner that creates new employment opportunities and income" (63). However, nothing is explicitly stated about reducing income disparities, which is also related to the debate on rural decline.

Box 2: RNE helps to retain services in the area

"When the yields from energy production are bundled in a cooperative or fund, it provides perspective to be used in the interests of the respective neighbourhood or village. For example facilities such as libraries or sports clubs could be supported from revenues from own production."

(Northern Vision 2040, 133 [regional])

Only the *Northern Vision 2040* [regional] makes a connection between the loss of services and RE (Box 2). In this case local production of energy by cooperatives is perceived as an opportunity to retain services. Also the draft *Area Vision* [province] contains a reference to retainment of services, albeit more implicitly. This document states that that in 2014 the province of Groningen approved a framework which sets out a reorganized policy for wind energy, which includes area based funds (Groningen Province 2015b). When looking into this specific policy in more detail, as illustrated in Box 3, it turns out these area based funds oblige developers to deposit money in this fund for every Megawatt they realise, which subsequently can be used to compensate people living in the direct vicinity of the wind parks and for community projects that increase the livability of the area. However, this procedure is only valid for the parks that are realised by the province, not for those realised under the '*National coordination scheme'* (Groningen Province 2015b). Currently, Groningen does not yet have such parks under development.

Box 3: Area-based funds

"When the yields from energy production are bundled in a cooperative or fund, it provides perspective to be used in the interests of the respective neighbourhood or village. For example facilities such as libraries or sports clubs could be supported from revenues from own production."

(Northern Vision 2040, 133 [regional])

Literature suggests that an explicit connection between RE production and local needs can enhance the resilience of both the initiatives and the region itself. The examples above demonstrate that governments do identify the possible contribution of RE production to the economic resilience of the area. However, linkages between RE production and social aspects of rural decline are (still) rarely made. These results support the initial observations that there is only limited recognition of the opportunities offered by RE production for supporting the resilience of the region beyond regional economic development.

North of the Netherlands as a 'living lab'

According to the concept of an integrated energy landscape, synergies on the local level are crucial for making energy initiatives less vulnerable, enhancing societal support and building social capital, and by extension resilience (de Boer and Zuidema 2015; Shucksmith et al. 2006). The relative lack of integration of RE production into the physical and especially socio-economic landscape, as well as the rather negative attitude towards the spatial impacts of RE revealed in our analysis, raise the question how the different governmental levels aim to realise the vision of the north of Netherlands as Energy Port. The following paragraph will therefore focus on the main strategy presented in the policy documents to achieve this ambition: the ambition to make the North of the Netherlands a 'living lab' in which experimentation and local energy initiatives play a central role.

The regional and provincial level documents present the region as a so-called 'living lab' for energy-related experiments. According to the *Northern Energy Agenda* [regional], the *Green Deal* [regional], and the *MIRT Agenda* [regional], the strength of the living labs lies in closer connections between knowledge and industries within the area and the testing of innovations in practice which should give the area an image of 'energy experimentation region'. This, in turn, is expected to attract more capital and talent to the region thus connecting the ambitions to the rural decline indicator of skills and education.

Box 4: Experimenting with RNE in declining regions

"In connection to the approach to population decline the North [of the Netherlands] is divided into so- called 'anticipatory regions' (North-East Friesland; North-West Friesland and East Drenthe) and 'decline regions' (Eemsdelta; East Groningen and De Marne). In these areas the specific complex task is to combine restructuring, renovation and demolishing with energy saving measures and locally produced renewable energy. In these areas there is space to experiment and, among other things, develop new earning-back models"

(MIRT Agenda, 14 [regional])

Box 5: 'Energy Port' image

"Drenthe, Groningen, Friesland and North-Holland have in the past years proven with Energy Valley that they are able to develop RNE on the basis of close public private cooperation and successfully connect 'green' and 'growth'. This success has been acknowledged in national policy and illustrated also by the recent nomination of North-Netherlands as the Energy port (SVIR). This strong national energy position will be given form in collaboration with, among others, North-West Germany with the goal to grow and become the 'European Region of Energy Excellence'."

(Green Deal North NL, 4 [regional])

Box 4 illustrates that especially the declining rural areas are encouraged to develop energy initiatives and initiate experiments. This shows that, although not explicitly connected to rural decline indicators, opportunities of RE production in declining areas are recognized at the regional level. However, in light of the regulations restricting RE production based on its fit into the physical landscape, it remains questionable to what extent such local initiatives are actually supported or even possible at the different governmental levels.

The provincial and regional levels have taken an active role in creating the image of the north of the Netherlands, and particularly Eemshaven, as an energy port. Box 5 illustrates that such efforts are recognized at the national level and have contributed to the nomination of the north of the Netherlands as Energy port in the national SVIR. This nomination demonstrates that initiatives at the meso-level are capable of influencing the higher level institutional framework. Moreover, in 2015 the national government adopted the 'experimentation regulation' (*Experimenteer AmVB*) to support 'living labs'. This regulation allows for deviations from the Gas and Electricity Act by local energy initiatives in order to test policy adaptations which, if successful, may lead to permanent changes in policy and legislation. A successful experiment is defined as the one leading either to significant increase in the use of REs, better balanced networks, or large citizen involvement (Energeia 2014). While this regulation does shows that at the national level there is support for local experimentation, it also demonstrates that these experiments need to be initiated by local parties.

In general, local initiatives are considered an important aspect of experimentation in living labs at all governmental layers. Such initiatives are to be supported "when necessary and possible by municipalities, provinces and the State" (*Energy Agreement*, 19 [national]) and it is recognized that they can provide benefits for citizens through investments in the local economy (*Local Energy Program* [Province] and local services (*Northern Vision 2040* [regional]) (see, e.g., Box 2). However, it is noteworthy that compared to the other governmental levels, the *Development Vision Eemsdelta* [local] seems to place the least emphasis on experimentation. Moreover, with regard to local initiatives, the *Development Vision Eemsdelta* [local] states that such initiatives need to fit in the spatial structure of an area and be embedded into the physical landscape. In line with the argumentation earlier in this chapter, this illustrates a reluctant attitude towards changes in the physical landscape on the local level, even when initiated by citizens and although encouraged on paper by other levels of government. Therefore, in our analysis, the government level that is considered to be the closest to the citizens seems to be the least considerate of local initiatives in its policy documents compared to other governmental levels.

The emphasis on experimentation in 'living labs' demonstrates that the importance of micro-level innovations is recognized across governmental levels. Moreover, some measures are proposed on the meso-level to encourage and enable such experiments. Simultaneously, the discourse surrounding the 'fitting in' of RE production demonstrates that the visible impact of these energy-innovations is problematic, especially at the local level. Moreover, except for the *national experimentation regulation*, no explicit strategies are mentioned to support these experiments. The *Green Deal* [regional] states that in order to realise a living lab in the north of the Netherlands, "energy needs to be structurally embedded in spatial policies and permit procedures, with transparent regional and national tasks and responsibilities" (5). This statement demonstrates that the private and public parties that together created the *Green Deal* [regional] acknowledge the fact that energy is not yet embedded in spatial policies and are aware of the necessity to do so. How this connection between energy and spatial planning is going to be improved is not clarified. In other words, while the intention of becoming a living lab exist and is even connected to alleviating the consequences of rural decline, current spatial policies are not yet accommodating or even allowing for such a living lab to thrive.

Discussion and conclusion

In this chapter we analysed different governmental levels' strategies and visions regarding RE production in relation to spatial planning. Specifically, we focused on whether these strategies and visions recognize and support opportunities for RE production to alleviate challenges caused by rural decline and peripheralisation and enhance regional resilience. We did so by drawing from the notion of an integrated energy landscape, conceptualised as a multifunctional physical and socio-economic landscape with energy initiatives and systems forming an integrated part of them (de Boer and Zuidema 2015). RE is in the context of an integrated energy landscape seen as both being conditioned by the physical and socio-economic landscape and as a means to stimulate resilience through possible positive spin-offs from RE production to strengthen the regions' socio-economic fabric and prospects to cope with future changes. Specifically, we noted that for RE production to be a contributor to regional resilience, it was also essential that the existing institutional framework on a national, regional, provincial and local level would identify this possible contribution.

Our analysis of policy documents in the fields of energy and spatial planning does show that RE production is recognized at all analysed governmental levels as an important opportunity for regional economic development. Nevertheless, it has not yet been translated into a more comprehensive strategy that enables RE production as an integrated part of the physical and socio-economic landscape in the area under investigation. Despite the prospects for using RE production to alleviate pressures caused by rural decline and peripheralisation in the north of the Netherlands, our analysis demonstrates that at all government levels, RE is hardly considered as a key opportunity for rural development or an enabler of a more resilient region. Rather, RE production is often conceptualised as a threat to the physical landscape.

Despite the rhetorical claim made that RE production can benefit the socio-economic development of the region, the documents show little vision upon how such benefits might be materialised. The idea of a regional living lab that might become the basis for the development of a new regional socio-economic landscape might be the only example of a strategy to move towards this vision. Nevertheless, it is an idea that is advanced in a context where RE production is largely seen as a threat to the physical landscape. RE production is not only actively isolated from its regional and local context by mechanisms such as the national coordination scheme, it is also in national and provincial visions largely seen as a disturbance to this landscape. The resulting prudence and resistance to embed RE production in the region is a logical consequence. If the region is to become the kind of Energy Port it aims to become, such prudence and resistance are counterproductive. We therefore argue for looking beyond the 'negative' implications of RE production for the landscape. Instead, we opt for a reconceptualisation of RE production as a possible means for supporting a strategy to respond to rural decline and its consequences. It is such a reconceptualisation we hope might provide a breeding ground for a true vision on the future of North East Groningen, where RE production become an integrated element in defining a resilient region.

In this research we did not explicitly focus on the parties involved in setting the RE agenda. Therefore the research includes only a very specific section of the total governance framework involved in the dynamics concerning RE in the (north of the) Netherlands. Further research could examine the role and power of other actors involved in setting the agenda. Especially on the local level, research could provide additional insights into the institutional dynamics since strategic policy documents — especially in cases with smaller, more rural municipalities — appear to be rare. Furthermore, it would be interesting to examine in more detail the findings regarding the role of informal cooperation between governments in spreading more innovative approaches (in our case particularly on the regional scale). Moreover, it would be interesting to compare the findings in the Netherlands with neighbouring countries who seem to be further advanced in their attempts towards energy transition. The North of Germany would form an interesting case study, in particular in the context of ongoing cross-border cooperation, with similar issues and opportunities in the rural regions across the border and the shared vision of becoming the *'European Region of Energy Excellence'*.

References

Aitken, S. & Valentine, G. (Eds.) (2006). Approaches to Human Geography. London: Sage.

Alexander, E. R. (2006). Institutional Design for Sustainable Development. The Town Planning Review 77 (1), 1–27.

Alexander, E. R. (2012). Institutional Design for Value Capture and a Case: The Tel-Aviv Metropolitan Park. International Planning Studies 17 (2), 163–177.

Avelino, F., Bressers, N. & Kemp, R. (2012). Transition management as new policy making for sustainable mobility. In Geerlings, H., Shiftan, Y. & Stead, D. (Eds.), Transition towards Sustainable Mobility: The Role of Instruments, Individuals and Institutions (pp. 33–52). Hampshire: Ashgate.

Babbie, E. (2010). The Practice of Social Research. 12th edition. Wadsworth: Belmont.

Beetz, S., Huning, S. & Plieninger, T. (2008). Landscapes of Peripherization in North-Eastern Germany's Countryside: New Challenges for Planning Theory and Practice. International Planning Studies 13 (4), 295–301.

BONN & Stichting Energy Valley (2014). SWITCH Noordelijke Energie Agenda. www.energyvalley.nl/uploads/bestanden/37bc3c14-50e4-46a1-bf4a-d1229eef4143. Accessed 17 June 2015.

Bosman, R. (2013). Big ambitions, little action — Energy Transition in the Netherlands. Renewables International. www.renewablesinternational.net/energy-transition-in-the-netherlands/150/537/62014/. Accessed 20 May 2015.

CBS (2013). Grote gemeenten goed voor driekwart van bevolkingsgroei tot 2025. www.cbs.nl/nl-NL/menu/themas/bevolking/publicaties/artikelen/archief/2013/2013-062-pb.htm. Accessed 17 June 2015.

Chapin, F. S. III, Folke, C. & Kofinas, G. P. (2009). A framework for understanding change. In Chapin, F.S. III, Kofinas, G. P. & Folke, C. (Eds.), Principles of Ecosystem Stewardship: Resilience-Based Natural Resource Management in a Changing World (pp. 3–28). New York: Springer.

Davoudi, S. & Porter, L. (Eds.) (2012). Applying the Resilience Perspective to Planning: Critical thoughts from Theory and Practice. Planning Theory & Practice 13 (2), 299–333.

De Boer, J. & Zuidema, C. (2015). Towards an integrated energy landscape. Urban Design and Planning 168 (5), 231–240.

de Kam, G & Spijkerboer, R (2015). Wat willen de Groningers zelf?: Een overzicht van de uitkomsten van een onderzoek naar de opvattingen van leden van de Vereniging Eigen Huis over de oplossingen voor de aardbevingsproblematiek in Groningen (onderzoeksrapportage). Faculteit Ruimtelijke Wetenschappen, Rijksuniversiteit Groningen, Groningen.

EC (1997). The EU compendium of spatial planning systems and policy. Luxembourg: Office for Official Publications of the European Communities.

Energeia (2014). Experimenten lokale energie kunnen veel sneller tot wetswijziging leiden. www.hieropgewekt.nl/actueel/experimenten-lokale-energie-kunnen-veel-sneller-tot-wetswijziging-leiden. Accessed 20 May 2015.

Folke, C. (2006). Resilience: The emergence of a perspective for social-ecological systems analysis. Global Environmental Change 16 (3), 253–267.

Folke, C., Carpenter, S.R., Walker, B., Scheffer, M., Chapin, T. & Rockström, J. (2010). Resilience thinking: Integrating Resilience, Adaptability and Transformability. Ecology and Society 15 (4), 20. www.ecologyandsociety.org/vol15/iss4/art20/ Accessed 20 May 2015.

Groningen Province (2009). Provinciaal Omgevingsplan 2009–2013. Bedum: Scholma Druk.

Groningen Province (2013). Programma Energie 2012–2015: Projectplan Lokale Energie. www.provinciegroningen.nl/fileadmin/user_upload/Documenten/Downloads/Verbeterd_Definitief_Samenvatting_Projectplan_Lokale_Energie.pdf. Accessed 17 June 2015.

Groningen Province (2015a). Ontwerp Omgevingsvisie Groningen maart 2015. www.provinciegroningen.nl/omgevingsvisie. Accessed 6 May 2015.

Groningen Province (2015b). Parkfonds. www.provinciegroningen.nl/uitvoering/klimaat-en-energie/windenergie/parkfonds/. Accessed 3 June 2015.

Groningen Province, Friesland Province & Drenthe Province (2013). Noorderzine: een co-creatieve verkenning van de toekomst van Noord-Nederland. www.ikdrenthe.nl/pers/ontwerp-noordervisie-2040-gepresenteerd. Accessed 6 May 2015.

IPO (2015). Energie en IPS2E. www.ipo.nl/milieu-energie-klimaat/energie. Accessed 17 June 2015.

Karrasch, L., Klenke, T. & Woltjer, J. (2014). Linking the ecosystem services approach to social preferences and needs in integrated coastal land use management — A planning approach. Land Use Policy 38, 522–532.

Koomen, E. (2011). Indicators of rural vitality. A GIS-based analysis of socio-economic development of the rural Netherlands. Research Memorandum 2011–50. degree.ubvu.vu.nl/RePEc/vua/wpaper/pdf/20110050.pdf. Accessed 17 June 2015.

Lehr, U., Breitschopf, B., Diekmann, J., Horst, J., Klobasa, M. et al. (2012). Renewable energy deployment — do the benefits outweigh the costs? GWS Discussion Paper 2012/5. www.gws-os.com/discussionpapers/gws-paper12-5.pdf. Accessed 17 June 2015.

Lloyd, M. G., Peel, D. & Duck, R. W. (2013). Towards a social-ecological resilience framework for coastal planning. Land Use Policy 30 (1), 925–933.

Loorbach, D. (2010). Transition management for sustainable development: a prescriptive, complexity-based governance framework. Governance 23 (1), 161–183.

Measham, T. (2012) . Renewable Energy: Community Benefit and Ownership. SPICe Briefing 12/71. www.scottish.parliament.uk/ ResearchBriefingsAndFactsheets/S4/ SB_12-71.pdf. Accessed 18 June 2015.

Maesham, T. G. & Fleming, D. A. (2014). Impacts of unconventional gas development on rural community decline. Journal of Rural Studies 36, 376–385.

MoEA, MoIA, MoIE, Drenthe Province, Friesland Province et al. (2011). Green Deal Noord-Nederland. www.energyvalley.nl/uploads/bestanden/ e32eb386-aee4-476c-be76-60414b3fd5e8. Accessed 17 June 2015.

MoIE (2012a). Structuurvisie Infrastructuur en Ruimte — Nederland concurrerend, bereiekbaar, leefbaar en veilig. www.rijksoverheid.nl/ documenten-en-publicaties/rapporten/2012/03/13/structuurvisie-infrastructuur-en-ruimte.html. Accessed 19 May 2015.

MoIE (2012b). Kamerbrief betreffende definitieve structuurvisie infrastructuur en ruimte (SVIR). www.rijksoverheid.nl/documenten-en-publicaties/ kamerstukken/2012/03/13/definitieve-structuurvisie-infrastructuur-en-ruimte.html. Accessed 19 May 2015.

MoIE (2012c). Summary National Policy Strategy for Infrastructure and Planning. www.government.nl/documents-and-publications/publications/ 2013/07/24/summary-national-policy-strategy-for-infrastructure-and-spatial-planning.html. Accessed 19 May 2015.

MoIE & MoEA (2014). Structuurvisie Windenergie op land. www.rijksoverheid.nl/onderwerpen/duurzame-energie/documenten-en-publicaties/rapporten/2014/03/31/bijlage-1-structuurvisie-windenergie-op-land.html. Accessed 19 May 2015.

Muller, A. (2009). Sustainable agriculture and the production of biomass for energy use. Climatic Change, 94 (3–4), 319–331.

Ohl, C. & Eichhorn, M. (2010). The mismatch between regional spatial planning for wind power development in Germany and national eligibility criteria for feed-in-tariffs — A case study in West Saxony. Land Use Policy, 27 (2), 243–254.

4Power (2012). Province of Groningen. www.4-power.eu/about-4power/ partners/province-of-groningen.aspx. Accessed 17 June 2015.

Rotmans, J. (2011). Staat van de energietransitie in nederland. Message posted Augustus 2011 to www.janrotmans.blogspot.nl/2011/08/staat-van-de-energietransitie-in.html. Accessed 29 January 2015.

RVO (2015). De Rijkscoordinatieregeling. www.rvo.nl/subsidies-regelingen/de-rijksco%C3%B6rdinatieregeling. Accessed 10 June 2015.

SER (2013). Energieakkoord voor duurzame groei. Den Haag: Sociaal-Economische Raad.

Seyfang, G. & Haxeltine, A. (2012). Growing grassroots innovations: exploring the role of community-based initiatives in governing sustainable energy transitions. Environment and Planning C: Government and Policy 30 (3), 381–400.

Shucksmith, M., Cameron, S., Merridew, T. & Pichler, F. (2006). First European Quality of Life Survey: Urban-rural difference. Luxembourg: Office for official Publications of the European Communities.

State, Drenthe Province, Friesland Province, Groningen Province, Municipality of Assen, et al. (2014). Energieke Noorder ruimte: MIRT-Gebiedsagenda Noord-Nederland 2040. www.tweedekamer.nl/kamerstukken/detail?id=2014D42200. Accessed 6 May 2015.

Steiner, A. & Markantoni, M. (2013). Unpacking community resilience through Capacity for Change. Community Development Journal, 49 (3), 407–425.

Stokols, D., Lejano, R. P. & Hipp, J. (2013). Enhancing the resilience of Human-Environment Systems: a Social Ecological Perspective. Ecology and Society 18 (1), 7–18.

Strunz, S. (2014). The German energy transition as a regime shift. Ecological Economics 100, 150–158.

Stuurgroep Ontwikkelingsvisie Eemsdelta (2013). Ontwikkelingsvisie Eemsdelta 2030. Groningen: WM Veenstra Druk BV.

van den Voort, N., & Vanclay, F. (2015). Social impacts of earthquakes caused by gas extraction in the Province of Groningen, The Netherlands. Environmental Impact Assessment Review 50, 1–15.

Wüstenhagen, R., Wolsink, M. & Bürer, M. J. (2007). Social acceptance of renewable energy innovation: An introduction to the concept. Energy Policy 35 (5), 2683–2691.

Zuidema, C. & Trell, E. (2014). Proeftuin voor Europa. Noorderbreedte. www.noorderbreedte.nl/themas/landschap_archeologie/proeftuin-voor-europa. Accessed 17 June 2015.

Climate change mitigation and energy concepts: also an issue for rural areas? Experiences of Communities in Northwest Germany

Peter Schaal

Abstract

Rural areas play an important role for the German "Energiewende". Renewable energy production in the form of wind energy plants, biomass facilities and photovoltaic farms are located largely in rural areas. In the last 15 years the number of plants has grown enormously, and with it there are many opportunities and even threats within this massive structural change. Renewable energies are one main pillar of the German Energy Concept 2010, the national climate change mitigation strategy. One important instrument to bring the strategy to its roots is the "Municipal Directive" (Kommunalrichtlinie), a national funding initiative for municipalities that started in 2008. Meanwhile, more than 8000 projects have been established in around 3.000 communities. Central funding objects of the initiative are "integrated climate change mitigation concepts" (ICCMCs). This expert article shows, from the perspective of a consultant, that even in the "Energy Region Northwest rural communities" haven't utilised the opportunities hidden in this funding initiative to prepare their future development.

Introduction

Climate change mitigation is currently a politically and ecologically high-level issue in Germany. The Intergovernmental Panel on Climate Change (IPCC) has calculated the expected magnitude of climate change in its climate models and scenarios and has described the expected impacts. The First Assessment Report of the IPCC dates from 1990 and served as the basis for the United Nations Framework Convention on Climate Change (UNFCCC). Since the publication of the expected extent of global warming in the 2001 IPCC Report, politicians can no longer ignore the facts about climate change (IPCC 1990; IPCC 2001; UN 1992).

The German federal government under Chancellor Angela Merkel has gladly assumed the role of the forerunner in the climate change mitigation negotiations and has launched programs in Germany to realise the agreed upon goals. The national climate change mitigation program includes a schedule for the reduction of greenhouse gas emissions in Germany, which addresses all sectors at all levels of planning (Deutscher Bundestag 2000; BMU 2015). Particular importance is given to the municipal level with respect to the adoption of suitable measures for the reduction of greenhouse gases in Germany. The largest cities and municipalities with more than 50 000 inhabitants have long since reacted and developed their own concepts and action strategies, which should be integrated into the federal government's climate change mitigation program.

With their well-staffed and differentiated authorities, these large municipalities are able to voluntarily take on further tasks in addition to their mandatory tasks, and moreover they are able to acquire funding from federal programs for climate change mitigation. But what role will rural areas take within national climate change mitigation? 94% of all communities in Germany have fewer than 20 000 inhabitants, and a large portion of these communities is located in rural areas. A vast number of these small cities and villages with populations under 20 000 are equipped with municipal authorities that can just barely manage their obligatory tasks. And by no means is climate change mitigation among the obligatory tasks.

On the other hand, the German energy transition (*Energiewende*) is taking place for the most part in rural areas. Wind energy, solar energy, and biogas production require space and are therefore predominantly produced in the countryside. And the resulting conflicts with the populace and nature conservation about the construction of the renewable energy production systems need yet to be resolved. Solely on the basis of these problematic conditions, it is already worthwhile to examine the topic more closely and to pose the question, as to whether or not issues of climate change mitigation and energy policies have reached rural areas and how significant this topic has become within politics at the local level. Another aspect of this question is to what extent climate change mitigation has been conceptually promisingly laid out in Germany, or if the concept's definition of objectives possibly misses its target audience.

National climate change mitigation and instruments

But what strategy is the German federal government pursuing to motivate the general population at lower levels of planning to more efficiently address the issue of energy? The climate change mitigation on the federal level is currently based on the Energy Concept 2010 (*Energiekonzept* 2010; Bundesregierung 2010). The concept intends to lower greenhouse gas emissions by at least 40% — relative to 1990 — by 2020. In the long term the Energy Concept follows the recommendations of the IPCC: in order to stay under the 2-degree limit, greenhouse gas emissions must drop at least 80 to 95% by 2050. Interim goals were also set for the years 2030 and 2040.

These goals shall be achieved by means of developing energy production using renewable resources and through considerable increases in energy efficiency. The share of renewable energy in respect to the gross energy consumption should increase to 18% by 2020. By 2050 the share of renewables should reach

60% (Bundesregierung 2010). The share of electricity produced from renewable resources should contribute 80% of the gross electricity consumption by 2050. Energy efficiency shall be improved through the minimization of primary energy consumption and electricity production. The annual rate of energy efficiency building renovations is expected to double from the current rate of 1% to 2%, with the aim to make all buildings virtually climate-neutral by 2050. In the transport sector the goal is to reduce final energy consumption, compared to the consumption in 2005, by 10% by 2020 and up to 40% less by 2050. At least one million electric cars should be on the road by 2020 — which is currently a nearly utopian ambition, because as of 2015, not more than 19 000 electric cars were licensed in Germany (Kraftfahrt-Bundesamt 2016). By 2030 the number of electric cars should reach six million. In the logistics and freight transport sector the ambition is to decouple traffic volume and energy consumption. As it soon became clear that the set goals cannot be achieved with the current instruments, in 2014 the Action Program for Climate change mitigation 2020 (*Aktionsprogramm Klimaschutz* 2020; BMUB 2014) was launched, which designates additional activities in order to save nearly 80 million tons of CO_2 equivalents. The most important political measures include, for example, the National Energy Efficiency Action Plan (NAPE) and especially the strategy "Climate-friendly Building and Housing" (*Klimafreundliches Bauen und Wohnen*).

Nevertheless, important national instruments were also created in Germany between 2000 and 2010 in order to reinforce climate change mitigation, expand renewable energy capacity, and improve energy efficiency. Among these measures was the Renewable Energy Act (EEG), which constitutes the central instrument to manage the development of renewable energy sources. With set remuneration rates, guaranteed purchasing, and priority feed-in for electricity produced from renewable energy sources, the act established sustainable technologies, such as wind and solar power on the market. The success of the act is evident based upon the establishment of wind, solar, and biogas technologies in Germany: over the last couple of years the energy supply from renewable energy sources has rapidly grown (Figure 1). In 2014 renewable energy sources became Germany's most important source of electricity — even ahead of lignite coal. Since its introduction the Renewable Energy Act has been continuously developed.

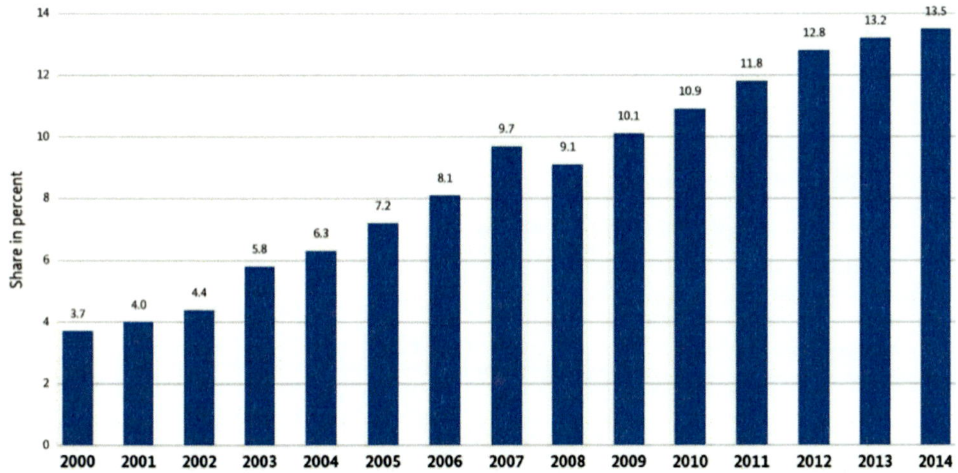

Figure 1: Development of renewable energy share of gross final energy consumption in Germany (Calculated without applying specific calculation rules of EU-Directive 2009/28/EC; BMWi based on Working Group on Renewable Energy Statistics (AGEE-Stat); as at August 2015; all figures provisional)

Source: BMWi 2015

With the funding for renewable energy based upon the EEG, electricity prices for consumers have greatly increased, which has sparked intense discussion about the amount of remuneration rates and the orientation of the funding policy. As a result the last amendment to the EEG in 2014 readjusted the funding rates in order to increase the overall cost effectiveness and economic efficiency. In the future the further development of the renewable energy sources shall be achieved within a designated corridor of expansion for the types of energy generation biogas, both on and offshore wind energy, as well as solar energy. Correspondingly the rate of guaranteed funding is oriented towards the development goals.

At the sub-national level according to the Action Program for Climate Action, the newly realigned European Structural and Investment Funds and, in particular the European Regional Development Fund (ERDF) should make an important contribution toward climate change mitigation with the federal states of Germany. In the funding period from 2014 to 2020, the states intend to annually utilise 2.9 billion euros from the European Structural and Investment Funds for climate change mitigation throughout all sectors of the economy. At least 20% of the funds from the ERDF, approximately 2.4 billion euros, will be employed Germany-wide towards the greenhouse gas reduction goal.

Primary objectives include:

- the sustainable production of energy from renewable energy sources,
- the improvement of energy efficiency and the use of renewable energy in businesses as well as in public infrastructures, public buildings and residential buildings,
- sustainable urban mobility and energy research (including pilot and demonstration projects).

With the funding from the ERDF programs, annual greenhouse gas emissions are expected to decrease by 4.25 million tCO_2eq by 2023 (BMUB 2014, 72). However, with respect to the 2013 calculation of 953 million tons of total annual emissions in Germany (UBA 2015), this declared goal is only like a drop in the ocean.

Climate change mitigation and communities

The majority of the measures mentioned by the federal government in the Action Program for Climate change mitigation 2020 and the National Energy Efficiency Action Plan (NAPE) are aimed at specific sectors and refer to electricity generation, building and housing, transportation, industry, trade, services, waste management, as well as agriculture and forestry. However, the energy transition ultimately takes place at the municipal level, where federal incentive tools influence the production of renewable energy (Figure 2). Local conflicts of use arise due to the amount of land and space required for the production of energy from renewable resources. Therefore it is important to increase energy efficiency through the use of intelligent solutions, such as heat and power co-generation, and to turn to energy saving measures at the municipal level also of planning and coordination. But in the end, residents must be convinced that energy transition is worthwhile and have to bear the retrofitting of the system and the resulting burden. Hence, local climate change mitigation must undertake an important task for the realisation of national and international targets for climate change mitigation.

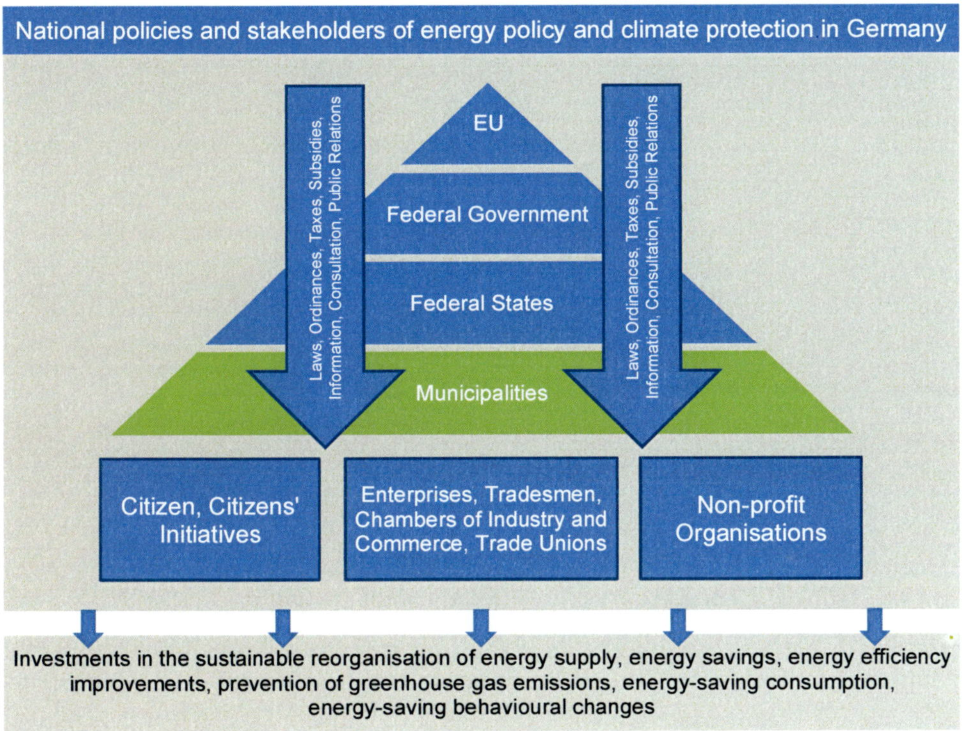

Figure 2: Energy policy and climate change mitigation in Germany
Source: KfW (2010, 15), modified

Following this line of thought, in 2008 the German Federal Ministry for the Environment enacted the so-called Municipal Directive (*Kommunalrichtlinie*) with a funding program for municipal climate change mitigation projects. Since 2008 over 8 000 projects in around 3 000 communities have contributed to the reduction of greenhouse gas emissions and provided an extensive approach for addressing the topic of climate change mitigation. The directive is amended on an annual basis. The current version aims to cut greenhouse gas emissions by 80 to 90% by the year 2050 (BMUB 2015). The core components of the funding are municipal climate change mitigation concepts and sub-concepts, in which 65% of the costs are subsidized by the federal government. Financially weak communities can be compensated for up to 91% of the costs.

Integrated climate change mitigation concepts: new opportunities for rural areas?

The Municipal Directive uses the term "integrated climate change mitigation concept" (*Integriertes Klimaschutzkonzept*) in its funding regulations, which implies a concept that recognizes all relevant areas of action within municipal climate change mitigation politics. On the contrary, the climate change mitigation sub-concepts, which are also subsidized, only refer to a single climate relevant sector like transportation or local heating plants. The objective of the integrated climate change mitigation concept (ICCMC) is for the municipality to define concrete measures in order to meet realistic climate change mitigation goals and to integrate climate change mitigation as an inherent element of municipal action. Consequently, an ICCMC presents a strategic decision-making tool and planning guide for communities.

But does this strategic approach also lead communities to accept this objective, and are these ICCMCs elements of municipal development strategies? Do the required contents in the ICCMCs even reflect the interests of the municipalities? After all, municipalities are generally financing 30% of the costs of ICCMCs and are subsequently expected to realise the concept's intended measures partly with their own budgetary resources. As previously explained, the vast majority of German communities have fewer than 20 000 inhabitants and merely have poorly-staffed municipal administrations compared to mid-sized and bigger cities. Precisely these are the rural municipalities, in which the energy transition in power generation primarily takes place. Are these communities even participating in the funding program? This question can only be answered in a scientifically correct manner by a statistical evaluation of the ICCMCs that have been funded thus far. My following statements are based upon the experience that we gained as community consultants during the creation of the integrated climate change mitigation concepts for smaller communities in the Weser-Ems region in Lower Saxony, in the northwest of Germany.

The northwest of Germany is deemed the region of competence for the expansion of renewable energy. In 2013, a study was published that even referred to the region as an "energy hub" (OLEC 2013). There are 55 000 small power stations in the domain of the regional energy provider EWE. 4 000 facilities were newly constructed in 2013. Wind energy has already reached a high level of expansion in the districts with 2 700 wind turbines connected to the energy grid in 2013. The number of biogas facilities is extremely high especially in districts with a high density of livestock, such as Cloppenburg and Emsland. In 2013 a total of 1 100 energy facilities fed electricity into the grid, and also the level of expansion in the photovoltaic system is enormous with 51 700 single systems. The grid company of the regional energy provider EWE recorded an average share of 70% green energy of the total feed-in to the energy grid already in the year 2013 and this almost meets the German 2050 target to have an 80% share of green energy (NOZ 2014).

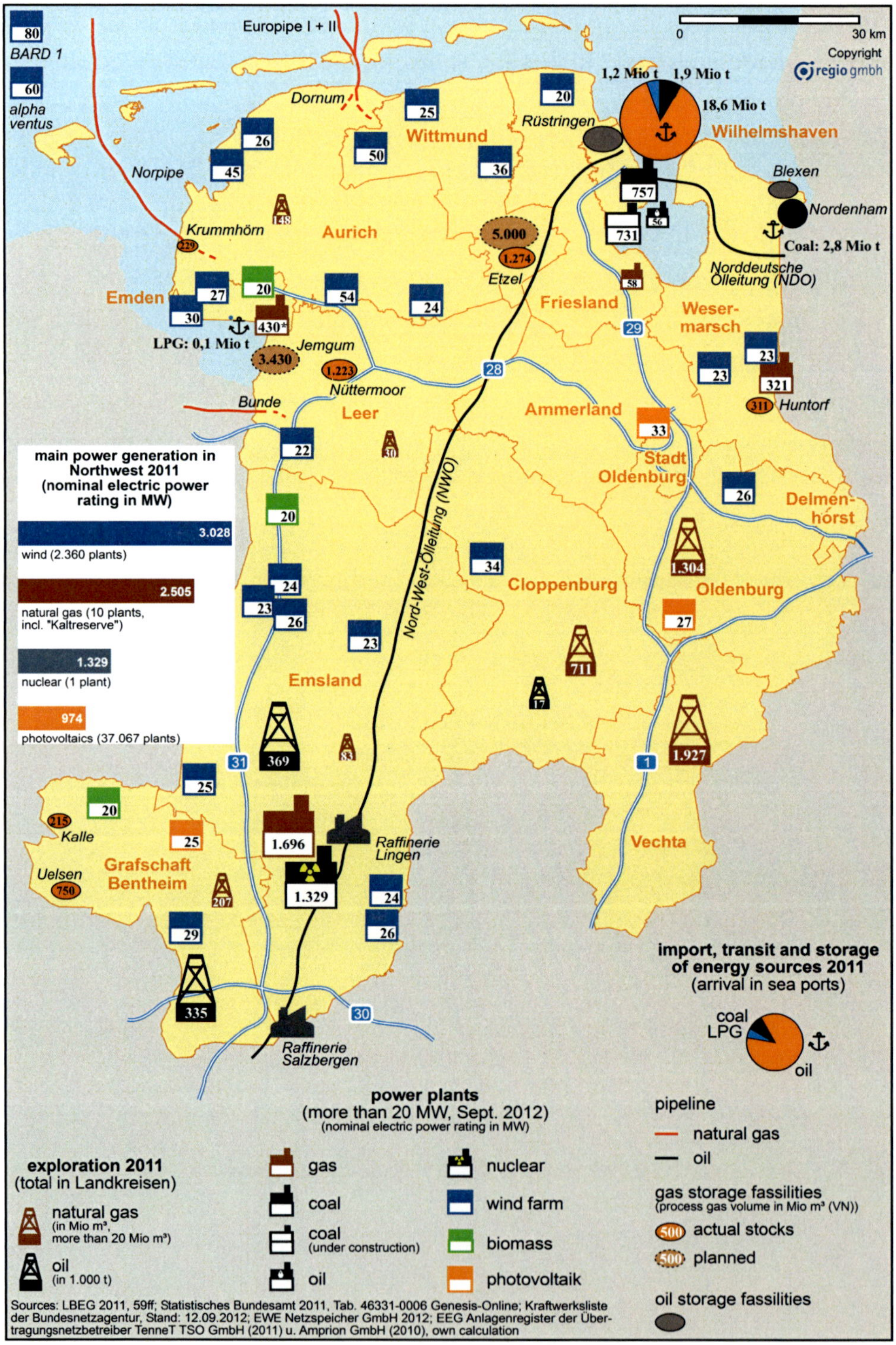

Figure 3: Energy Region Northwest — Electricity Generation and Energy Hub

Considering these numbers it is clear that the existing competition for land will increase due the area required for renewable energy production. At the same time, the energy production enables added value for the region and tax revenue for the communities, which is gladly utilised in structurally weak rural areas and would be used even more intensively if could be. Yet, frequently the conflicting interests of participating parties lead to substantial conflicts concerning the construction of wind parks and biogas plants in the communities. Municipal politics must search accordingly for conceptual solutions to deal with this issue in order provide appropriate responses for the individual questions and proposals of facility operators. The development of integrated climate change mitigation concepts, which are comparably cost-effective due to the federal funding program, offers these municipalities the opportunity to purchase consulting services from correspondingly specialised firms.

The contents of integrated climate change mitigation concepts

It is worthwhile to take a look at the contents of the climate change mitigation concepts. The Federal Ministry for the Environment has highly standardized the planned contents of the ICCMCs. With the comprehensive Practical Guide for Climate change mitigation in Municipalities (*Praxisleitfaden Klimaschutz in Kommunen*, DIFU 2011) the ministry has given communities consulting assistance. The contents of the concepts are featured in leaflets published by Project Management Jülich (PtJ), which are updated on a regular basis. On behalf of the Federal Ministry for the Environment, PtJ is responsible for the realisation of the Municipal Directive and manages the funding proposals and support for the individual initiatives. In the current version from September 2015, the leaflet regarding the development of climate change mitigation concepts defines the following working steps:

1) Energy and Greenhouse Gas Audit
2) Potential Analysis
3) Stakeholder Participation
4) Catalogue of Measures
5) Continuation Strategy
6) Controlling Concept
7) Communication Strategy

The *energy and greenhouse gas audit* is a mandatory component for all ICCMCs. In smaller municipalities this balancing is performed in a highly simplified form. For this process the surveying offices often use the internet-based tool "EcoRegion" from Ecospeed Climate Software Solutions. In most cases, the balance from this audit is used as the so-called opening balance. The balancing process uses socio-economic values such as population and employment data taken as time series and regional consumption data to be able to approximately calculate the energy consumption within a municipality. This method is pragmatic yet does not provide any sensitive information that could help pinpoint and assess local climate change mitigation action in the balance. Another weakness turns out to be the fact that the local energy production from renewable resources is not accounted for in the balance, in that it only considers the end consumption. The auditing tool "Ecoregion" does allow updates of the local consumption data, which can therefore improve the balance's sensitivity regarding local climate action. However, the collection of the local consumption data tends to be quite time-consuming.

The *analysis of potentials* assesses the energy-saving potential and the possibilities for producing energy from renewable resources within a municipality. This evaluation is broken down by sector and focal points specific to the community. This method analyses municipal properties in terms of their energy consumption as well as the sectors private households, economy, and transport in regards to possibilities for energy savings. On the side of energy production, there is an analysis of the potential possibilities for the expansion of wind and solar energy, biomass and local heating, as well as geothermal energy. Theoretical possibilities for improvements in the municipal greenhouse gas balance have resulted from the investigation. These theoretical considerations tend to be subject to criticism from community representatives, since in particular the analysis of expansion options for biogas production and considerations for the expansion of wind energy deals with very sensitive issues regarding their location, which local residents and the potential facility operators view very differently.

Stakeholder participation refers to the part of the climate change mitigation concept, in which the populace is involved in the process. Consulting firms methodically handle participation in quite different ways. Normally, they conduct acceptance surveys, expert interviews, round table meetings, workshops, and extensive rounds of discussion with citizens. Interactive online public forums may also accompany the process. Though these web forums represent a useful addition for large municipalities and in the case of highly interested citizens, they are able to incorporate stakeholders into the process who may not be reach through other means. The results of these surveys and participation processes include evaluations of acceptance for the measures and local climate change mitigation politics on one hand, and on the other hand citizen suggestions for local measures for climate change mitigation. An additional significant side effect of participation is the increased awareness within the population for climate change mitigation and private energy use.

Responsible local politicians have a great interest in the incorporation of the population in energy and climate change mitigation politics. However, the population in rural areas does not reciprocate this thematic interest. In the more than 20 communities that we studied, participation is concentrated among the operators of renewable energy production facilities (wind and biogas) and among the local council representatives who are active in committees. The involvement of additional groups such as homeowners and local commercial and industrial businesses has proved to be very difficult, even though it is a topic from which the private sector can make financial gains. Only occasionally does one find citizens who have discovered climate change mitigation as an issue worth committing to.

While drafting ICCMCs communities and experts focus their work on creating a coordinated *catalogue of measures*, which provides a basis for future climate change mitigation policy and energy policy. This catalogue presents all of the measures that will help to improve the local climate balance and achieve the self-imposed climate change mitigation objectives. The report must describe the measures and their content, include an estimation of the costs, name the parties necessary for the implementation, and include the contributed amount of emissions reductions. Additionally, a prioritization of the measures should follow and it should designate an implementation timeframe. The catalogue of measures also provides a basis for the application for funds for the implementation of climate change mitigation measures and for the application for climate change mitigation managers, who work on the realisation of the community climate change mitigation strategies.

Additional binding elements to develop for an ICCMC are the *continuation strategy*, *controlling concept*, and *communication strategy*. These working steps should be used to ensure that the ICCMC is not a discrete report but instead as a progressive process that is integrated into the municipal activities.

Integrated climate change mitigation strategies: innovation strategies for communities in rural areas?

Energy transition and climate change mitigation politics are providing a strong stimulus for development in rural areas. In particular the decentralisation of the energy supply through renewable energy presents opportunities and risks for the populations of rural communities. With the support for the ICCMCs the federal government has provided an instrument for a conceptual approach towards the issue. But are climate change mitigation energy strategies recognized and used as innovation strategies in rural areas?

If we take a look at the contents of climate change mitigation concepts that were presented above, the community's content-related interest is largely focused on the potential analysis, the stakeholder participation, and the catalogue of measures. The approaches of the municipal politics and administration concentrate primarily on the exploration of local chances for the development of renewable energy sources (wind, solar, and biogas). In many communities it is a matter of coordination between municipal development and the facility operators' desire to expand, i.e., primarily regarding issues of location and land use. Additionally, municipalities themselves are developing into the operators of renewable energy facilities. Cooperative models for the operation of wind parks, photovoltaic facilities, biogas facilities, and heating networks enable the involvement of residents in the creation of local added value and generate tax revenues for the municipality. At the same time, there is an increase in the level of acceptance for local projects due to the participation of a portion of the population.

In contrast, an important aspect of the ICCMCs takes a backseat in municipalities, namely the reduction of greenhouse gas emissions in the private household, economic, municipal property, and transport sectors. At the same time the energy-savings potential of the municipal properties are pursued with high levels of interest. This is an important incentive that prompts the possibility of applying for funding for climate change mitigation managers and climate change mitigation measures as long as they are listed in the municipal climate change mitigation concept as one of the measures. A glimpse at the potential funding determines the consent for the financing of the concepts for many representatives in the municipal councils. Otherwise, the thoughts of community council members are sceptical towards the conceptual plans, and this scepticism then manifests itself in the result that only a few advanced communities are willing to invest their own finances in climate change mitigation concepts and their implementation. These pioneer communities recognize among other advantages the image boost provided by an offensive climate change mitigation policy. The model of the "bioenergy villages" stands as a representative for the positive perception of such municipalities and shows the profitability of a community's public image.

A look at the approach towards the possibilities that are associated with an active climate change mitigation policy for rural communities shows the following contradictory tendencies. Communities with growth trends utilise the opportunity and tend to actively work with climate change mitigation and energy concepts and form the concepts into strategies for community development. In contrast, peripheral rural communities with structural problems rarely recognize the utility that these concepts might offer them. It is evident that climate change mitigation offers opportunities for development and new income for rural areas with an entrepreneurial culture. Here is a substantial difference between financially well-situated communities and communities in a critical financial situation. The former are more willing to take new directions, includeing new operator models for regenerative energy production, which generate addition financial resources in the intermediate term and expand the financial leeway for communities. Communities with financial problems, on the other hand, must focus on their obligatory tasks and are less prepared to take risks. Moreover, the politicians and administrations must be capable of attracting funding for their communities. For the application for funding from the climate change mitigation directive, municipal administrations require a certain professionalisation of their work.

Conclusion

From the findings of the considerations of the aforementioned questions, it is noted that a precise alignment and a reinforcement of the climate change mitigation action is necessary in communities in rural areas in order to be able to reach the ambitious climate goals set by the federal government within the intermediate term. Rural areas are main setting for the *Energiewende* with its transition to renewable energy sources. This fact presents great opportunities but also substantial risks for rural areas (Kaphengst et al. 2009, 11–14). In addition to the growth in added value in rural areas, there is an increase in the number of considerable land use conflicts that arise between agriculture, forestry, nature conservation, and the residents that live there, and these conflicts will need to be solved in order to accomplish the energy transition and support climate change mitigation action. These solutions must be found conceptually, and integrated climate change mitigation and energy concepts according to the climate change mitigation directive provide an appropriate starting point.

The form of the climate change mitigation directive must take the conditions and interests in rural communities more seriously into consideration than it has thus far. Especially the financial situation of the communities have to be taken into account. According to the new directive, financially weak communities may receive up to 91% funding. Furthermore, these communities have a higher quota for the funding of investment measures and for the creation of climate change mitigation sub-concepts. Nevertheless these prospects of funding are not enough motivation for many communities. Climate change mitigation awareness is poorly represented in the municipal councils and reservations about the development concepts are significant. In this case the largely standardized form of the ICCMCs presents a challenge. Not all of the required contents of the ICCMCs reflect the key concerns and problems in the communities. Furthermore, the conception of the climate change mitigation directive implies a kind of "self-fulfilling prophecy" in local climate change mitigation policy. Funding goes primarily to measures and the employment of climate change mitigation managers that are based upon the previously funded climate change mitigation concepts. Municipalities without conceptual climate change mitigation are left out of the system.

In order to reach the ambitious German climate objectives, climate change mitigation must be defined as an obligatory task for municipalities. It is the only possible way to include all communities in an active climate change mitigation strategy. The rural areas that have already recognized climate change mitigation and the energy transition as an innovation strategy are already profiting and creating numerous incentives for the regional economy as well as actively supporting environmental awareness in the general population.

References

BMU — Bundesministerium für Umwelt, Naturschutz und Reaktorsicherheit (Ed.) (2005). Nationales Klimaschutzprogramm 2005. Beschluss der Bundesregierung vom 13.Juli 2005. Sechster Bericht der Interministeriellen Arbeitsgruppe „CO2-Reduktion". Berlin.

BMUB — Bundesministerium für Umwelt, Naturschutz, Bau und Reaktorsicherheit (2014): Aktionsprogramm Klimaschutz 2020. Kabinettsbeschluss vom 3. Dezember 2014. Berlin.

BMUB — Bundesministerium für Umwelt, Naturschutz, Bau und Reaktorsicherheit (2015). Richtlinie zur Förderung von Klimaschutzprojekten in sozialen, kulturellen und öffentlichen Einrichtungen im Rahmen der Nationalen Klimaschutzinitiative (Kommunalrichtlinie) vom 22. September 2015. BAnz AT 06.10.2015 B4.

BMWi — German Federal Ministry for Economic Affairs and Energy (2010). Energiekonzept für eine umweltschonende, zuverlässige und bezahlbare Energieversorgung.

BMWi — German Federal Ministry for Economic Affairs and Energy (2015). Development of Renewable Energy Sources in Germany 2014. Berlin.

Deutscher Bundestag (2000): Nationales Klimaschutzprogramm. Fünfter Bericht der Interministeriellen Arbeitsgruppe "CO_2-Reduktion". Drucksache 14/4729. Berlin.

DIFU — Deutsches Institut für Urbanistik (Ed.) (2011). Klimaschutz in Kommunen. Praxisleitfaden. Berlin.

International Panel on Climate Change — IPCC (1990). Climate Change. The IPCC Scientific Assessment. Report prepared for Intergovernmental Panel on Climate Change by Working Group I. Edited by J.T. Houghton, G.J. Jenkins and J.J. Ephraums .Cambridge, New York, Melbourne: Cambridge University Press.

International Panel on Climate Change — IPCC (2001). Climate Change 2001: Impacts, Adaptation & Vulnerability: Contribution of Working Group II to the Third Assessment Report of the International Panel on Climate Change.PCC. Edited by J. J. McCarthy, O. F. Canziani, N. A. Leary, D. J. Dokken and K. S. White. Cambridge: Cambridge University Press.

Kaphengst, T., Friedrich, S. & Herbert, S. (2009). Zukünftige Herausforderungen für den ländlichen Raum in Deutschland und Europa — die Megatrends. Hintergrundpapier zur Veranstaltungsreihe von BMU und BfN: "Nachhaltig und innovativ: Zukunft im ländlichen Raum — Perspektiven für naturverträgliches Wirtschaften nach 2013". Bonn.

KfW Bankengruppe (2010). Förderung von Klimaschutz und nachhaltiger Energiepolitik für die Kommune der Zukunft. Abschlussbericht des Expertenkreises. Frankfurt/M.

Kraftfahrt-Bundesamt (2016). Bestand an Pkw in den Jahren 2007 bis 2016 nach ausgewählten Kraftstoffarte. www.kba.de/DE/Statistik/Fahrzeuge/Bestand/Umwelt/b_umwelt_z.html?nn=663524. Accessed 22 June 2016.

NOZ — Neue Osnabrücker Zeitung (2014). 70 Prozent Ökostrom. EWE: Stromnetz durch Energiewende stark belastet. Report from 02/24/2014.

OLEC — Oldenburger Energiecluster e.V. (Ed.) (2013). Potenzialstudie Energieregion Nordwest. Studie im Rahmen des Projektes Hansa Energy Corridor HEC, Oldenburg.

UBA — Umweltbundesamt (2015). Emissionsentwicklung 1990 bis 2013. www.umweltbundesamt.de/daten/klimawandel/treibhausgas-emissionen-in-deutschland. Accessed 29 October 2015.

UN (1992): United Nations Framework Convention on Climate Change, New York 9 May 1992, in force 21 Mar 1994 [1771 UNTS 107; S. Treaty Doc No. 102-38; U.N. Doc. A/AC.237/18 (Part II)/Add.1; 31 ILM 849 (1992)].

Local mobility for local supply: Recommendations for municipalities and local suppliers in rural peripheries

Werner Dietl, Andreas Friedwagner, Benedikt Hahn, Martin Heintel, Thomas Langthaler, Alice Wanner and Norbert Weixlbaumer

Abstract

The overarching goal of this paper is to analyse and discuss the reciprocal effects of local supply and local mobility on the example of a structurally weak peripheral rural region in the most eastern province of Austria. In our context local supply can be defined as the supply of the population with goods for daily or short-term needs in the residential surroundings, including retail and artisanal grocery and culinary establishments with local supply characteristics.

Crucial questions of the empirical investigation are firstly the requirements of the population for sustainable mobility during errands, secondly the ability of a suitable transportation service in contributing to securing local supply and thirdly synergy effects between these two fields in the context of effects that are able to assure viability and financial feasibility of the services. The findings show, that the incorporation of interests and target groups are able to create optimistic scenarios for the future. In general the paper shows that solutions to problems, such as the current lack of small scale local supply, are in sight. At the base of a broad solution to such issues is the underlying point that politics and society must come together with coordinated approaches and methods. Some best practice examples addressed this by demonstrating effective solutions.

Introduction: issue and posing questions

The province of Burgenland is considered to be a peripheral rural region in Austria. In addition the Burgenland marked the border between East and West until 1989. Burgenland and especially the Southern regions illustrated here are classified as "structurally weak peripheral rural regions" (Mose 2005, 577–78) including all of the attributed typical characteristics. The distance to centres and lacking connection efficiency to these, deficient infrastructure on various levels, loss of population and work migration (particularly during the week), small-scale agriculture and niche offers of sustainable tourism (cycling, hiking and wine tourism) characterize this region.

As Austria proceeded to join the European Union in 1995, Burgenland was the sole recipient region of Objective-1 in Austria within the structural funds policy. To this day Burgenland can draw funding from Phasing-Out support and is considered a model for success in regard to many aspects of European development. Nevertheless, it is difficult to compensate for the peripheral location of southern Burgenland. The global framework and situational issues that have sprung up in these peripheral rural regions have resulted in debates on equal living conditions. This article aims to address some aspects of this issue. It proposes possible forms of sustainable mobility and local supply.

Securing the local supply locations will not depend solely on local mobility services. Just as the adaptation of the public transport services to the needs of consumers will not be the only way to secure business operations. Successful examples have shown that effective local development can be achieved through interplay of both of these fields and appropriate commitment. The following investigation and evaluation are based on this underlying concept.

301 people were questioned in four survey clusters in Burgenland (Figures 1 and 3). 275 valid interviews were obtained. The survey, which was conducted in October 2014 in households of the municipalities, was a set of 31 mainly closed questions. The average length was 20 minutes. Simultaneously, the research project "Pilot ERRAM HU-AT NaVer — Nahversorgung Oberwart" (Verracon GmbH 2015b), flowed into this contribution.

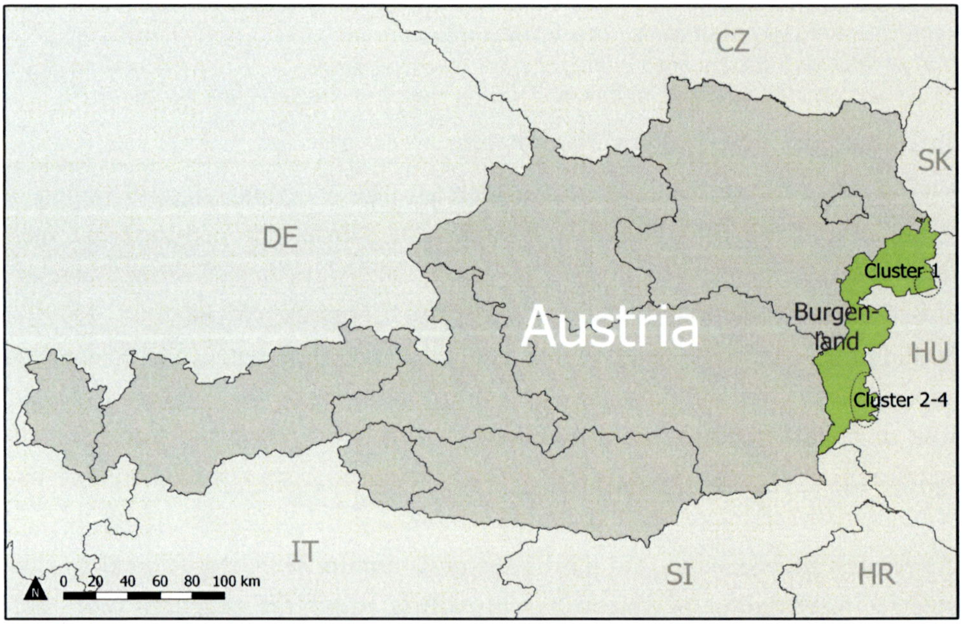

Figure 1: Austria and the province of Burgenland (with survey clusters)
Source: Authors' compilation

The rising percentage of elderly people is fundamental to changing consumer habits, budget restrictions in public local mobility and demographic processes. This is especially true for rural peripheral municipalities and has been the basis for the development of several strategies and concepts[1]. As examples, the study on local supply in Burgenland (Reinswald et al. 2013) or the guideline "nah& versorgt in Tyrol (ATL 2011) need to be mentioned. These examples mainly address issues of securing local supply away from major centres. In addition, there are many studies on mobility in rural areas (e.g., VRÖ 2012).

The aim of this paper is to investigate the reciprocal effects of local supply and local mobility. The following crucial questions were established:

- What are the requirements of the population for sustainable mobility during errands?
- Can suitable transportation services contribute to securing local supply?
- Can synergy effects be identified between these two fields? And would these effects assure viability and financial feasibility of the services?

At the end of the paper areas of action for bicycle traffic and flexible public transport service are discussed.

Local supply and local mobility: definitions

According to the local supply study of Burgenland (Reinswald et al. 2013), *local supply* can be defined as follows: In its most basic form it is the supply of the population with goods for daily or short-term needs in the residential surroundings (including retail and artisanal grocery and culinary establishments with local supply characteristics). More broadly it includes facilities of basic supply and services that support social, cultural or economic life. In this context local suppliers are those offering the population goods of daily needs and are in either the close residential surroundings or in the centres of municipalities. Aside from grocery retailers this would include bakeries and confectioneries, butchers, businesses with additional postal services and culinary establishments with local supply characteristics. Other examples may also include mobile local supply or innovative concepts.

---

1 The presented results were devised by SmaShMob (see www.b-mobil.info/projekte/internationale-projekte) They were aided by ETC through the AT-HU program.

Local mobility is mobility for short distances in the residential, work or shopping environment. In our case the focus should be on eco-friendly forms of mobility such as public transport, cycling or walking. In addition to the classic forms of scheduled public transport special attention should be paid to new demand oriented services, also known as flexible public transport services (e.g., the "Gmoabusse"[2]). Due to its customized approach to individual needs and results, flexible public transport services have varying forms and configurations. Figure 2 illustrates an overview of different categories of flexible public transport services and their relations to other transport services. These services combined with non-motorized transportation (cycling and walking) create the framework for eco-friendly mobility in local supply.

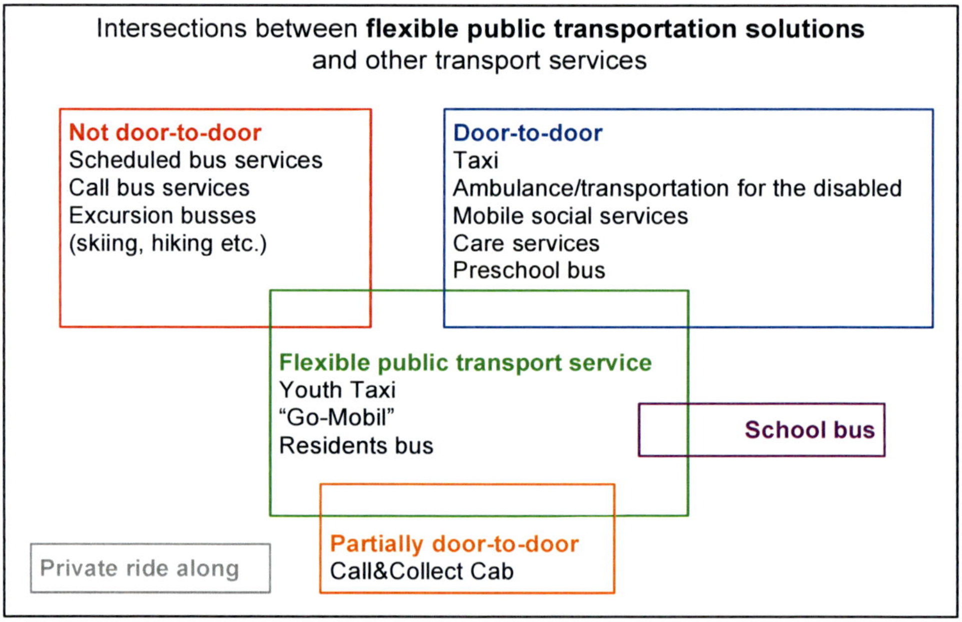

Figure 2: Overview of flexible public transport services in relation to other transport services

Source: Wolf-Eberl et al. 2011

[2] This term refers to demand-driven bus services in the municipalities Breitenbrunn, Mörbisch, Pöttsching and Purbach in Burgenland.

Ways of securing local supply: the research area

The trends of the past decades have led to the development of more and larger supermarkets at the edges of towns and along the axes of private (road) transport. This trend is criticized in spatial planning. It is an ever growing competition for local suppliers in the town centres. This has also led to a deficit in services in many smaller municipalities.

A growing market transparency and aspiration levels are further contributors to the challenges of local suppliers. While quality and exclusiveness are in demand, discounter products are also in high demand. As a result, the loyalty of consumers to their suppliers has dropped dramatically. At the same time there has been a growing consciousness towards regional economic cycles and appreciation of proximity to producers. This awareness is both a positive development and an opportunity for local suppliers.

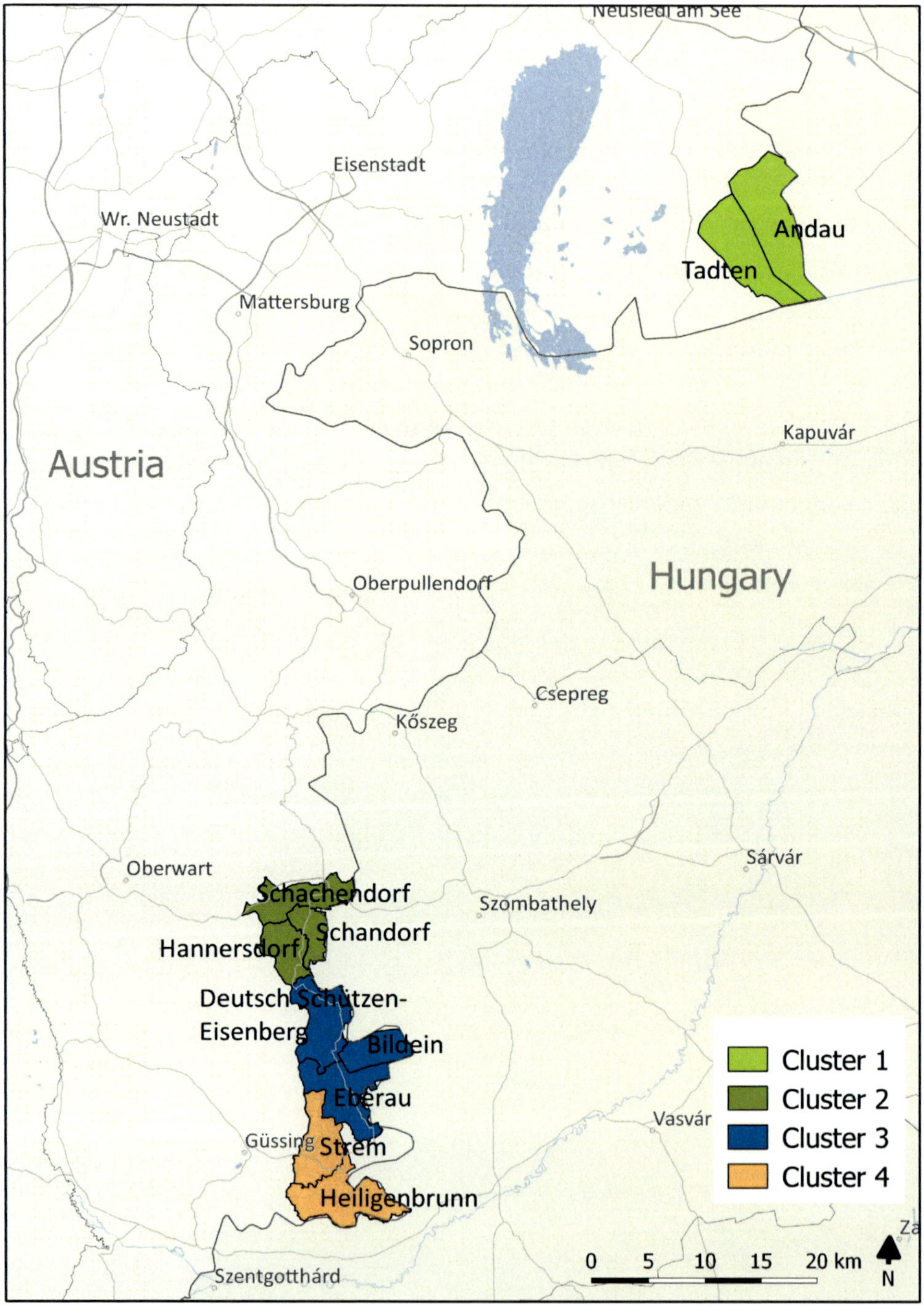

Figure 3: SmaShMob project area with municipality clusters

Source: Authors' compilation

Mobility and shopping under empirical investigation: results from SmaShMob[3]

Specific needs are associated by users for shopping traffic. In this situation the two following points should be under special consideration:

- Flexibility and frequency: A mode of transportation that requires the user to wait two hours after a 15 minute errand is unacceptable, especially for daily shopping. Therefore, basic characteristics of a public transportation system should be short intervals and flexibility.
- Mobility options: When speaking about shopping mobility, special attention has to be drawn to the options to transport the purchased goods. This point is usually given as one of the main reasons individuals do not want to opt for a more eco-friendly alternative and use their own private vehicle.

Regardless of whether or not a local supplier is present, the trend was the same: The existing public transport systems play a minor role for the population and their daily shopping routines (Figure 4).

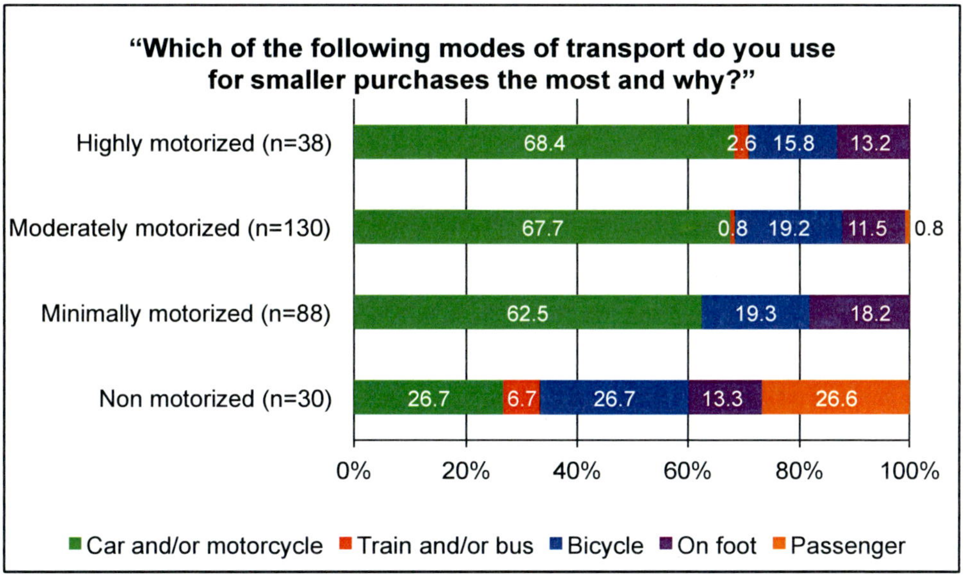

Figure 4: Mode of transport used for small purchases

Source: SmaShMob survey 2015

[3] SmaShMob = Smart Shopping Mobility. The cited results were taken from the following study: www.b-mobil.info/projekte/internationale-projekte.

Therefore, it must be clarified if this lack of use is based on the absence of service offers, the insufficiency regarding the needs and desires of shoppers, or if there is just a general lack of willingness amongst shoppers to use public transport. Simultaneously there is also a part of the population without access to a car at all. This is most prevalent amongst the elderly. Consequently it is important to investigate how public transport can offer an autonomous and independent lifestyle for these people.

Concerning the first aspect, the SmaShMob survey has a clear indication of the population's attitude. The willingness to use public transport for shopping trips is relatively low (Table 1). For the age group 36–49 years, it is out of the question for roughly three quarters of those questioned.

Table 1: Willingness to use public transport for shopping trips by age

"Would you use public transport services for shopping trips if the offer was improved?"	≤ 35 years (n=31)	36–49 years (n=61)	50–64 years (n=112)	65+ years (n=87)
True	35.5%	27.9%	42.9%	33.3%
Not True	64.5%	72.1%	57.1%	66.7%

Source: SmaShMob survey 2015

Why is there a lack of willingness to switch to public transport even if the offer has been improved? One of the main issues identified in the survey was that it was inconvenient to transport purchases home by bus or train. This is an issue where action can be taken.

On the other hand, it is also of interest how the public transport service would have to be modified to fulfil the demands of those willing to switch. One important aspect here is surely the frequency of services. This aspect was also addressed in a survey for the General Transport Strategy Burgenland (ABL 2014).

Table 2: Desired frequencies of public transport

"At what intervals would public transport have to run to increase your interest in its use for shopping?"	≤ 35 years (n=33)	36–49 years (n=62)	50–64 years (n=112)	65+ years (n=90)
Multiple times per hour	9.1%	6.4%	6.3%	2.2%
Hourly	39.4%	9.7%	8.0%	2.2%
Every 2 hours	9.1%	12.9%	17.0%	6.7%
Semi-daily (morning, afternoon)	6.1%	8.1%	13.4%	13.3%
Weekly	3.0%	0.0%	7.1%	8.9%
Not worth considering at all	33.3%	62.9%	48.2%	66.7%

Source: SmaShMob survey 2015

It has been shown that it is predominantly younger people who are willing to switch mode of transport, however, only under the condition that the services are frequent and flexible (Table 2). This is difficult to achieve, since potential demand in rural areas and financial feasibility of classic public transport services are very difficult to combine.

Among the elderly there is a notably smaller percentage willing to switch, but for that their demands are distinctively lower. About 9% would be content with a weekly option. This alone demonstrates that this specific age group has a need for basic supply.

This leads us to the second aspect addressed above: Dependency on family members or others rises with increasing age (and declining car ownership) and no local suppliers within walking distance, if public transport services are unavailable (Figure 5). This is highly relevant for an autonomous lifestyle among this age group, as shopping is one of the most common purposes of trips. In addition it offers the chance for social interaction in daily life.

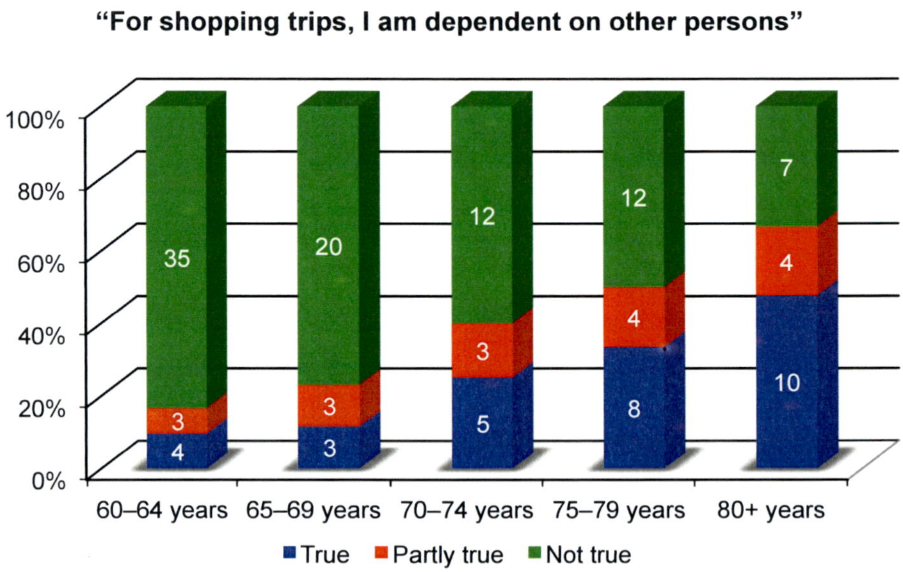

Figure 5: Dependency of the elderly on others for shopping trips
Source: SmaShMob survey 2015

In summary it can be said, that the expected demand in shopping traffic will be too low to create and maintain the desired service offer in a financially feasible manner. At the same time it is evident that there is a need for access to basic supplies for age groups that do not possess a car. It is reasonable to assume that there is a certain potential to pull in new clientele (e.g., younger population), if the service offer was flexible enough.

With regard to the demand of population of rural areas it is safe to assume, that the solution will not be found in classic public transport systems. Moreover, the focus must lie on demand-driven, flexible offers. These will be addressed in the areas of action.

The bicycle: "a friend of local supply"

Cyclists are often loyal customers of local shops. Using bicycles for shopping trips is usually preferred in residential or working environments. This presumption has been confirmed by studies in the past (CIMA 2007). The purchasing power remains in the municipality, and does not move to the shopping centres in the outskirts. Therefore, promoting cycling should be considered an instrument to secure local supply. In addition, it was determined that cycling customers not only visited stores more often but also lingered longer. Shop staff can then build a better relationship with their customers and target specials for regulars more precisely. Aside from these positive aspects, the question must be raised whether or not the population is even willing or able to cycle to shops.

To begin with, it must be said that generally in SmaShMob (and other studies as well) there is a very high availability of bicycles. Statistically speaking nearly every citizen of Burgenland owns a bicycle (ABL 2014). And these are actually frequently used in municipalities with local suppliers. In the municipalities studied in SmaShMob over 30% of the trips for smaller purchases were conducted by bicycle. Strong differences in proportions among the municipalities suggest that these numbers are in relation to available infrastructure and topography.

The potential increase for the bicycle is shown in the survey (Table 3). Answers to questions about willingness to walk or cycle for smaller purchases, and the general interest in cycling for everyday trips (ABL 2014) suggest a lot of room for growth. This has also been shown to be true for towns without local suppliers. There is a willingness to cycle or walk even in such cases, which is a positive sign for revival of local supply.

Table 3: Willingness to use a bicycle for shopping trips

"I could imagine cycling to do my shopping, if the distance was not too great and bicycle paths were safe."	≤ 35 years (n=33)	36–49 years (n=58)	50–64 years (n=110)	65+ years (n=91)
True	57.6%	48.3%	55.5%	42.9%
Not True	42.4%	51.7%	44.5%	57.1%

Source: SmaShMob survey 2015

According to the household survey of General Transport Strategy Burgenland, safer bicycle infrastructure and a growing favour for cycling was given as driving reasons for their choice of use. However, it is also important not to forget the influence of weather and season of the year on willingness to cycle. Even if in some cases more than 50% of people asked in the survey (CIMA 2007) admitted their choices were not in favour of cycling, there are possible measures to counteract these opinions. Winter road clearance would be a notable example in this case. Other examples are addressed in the "Areas of Action" portion of this paper.

"Horses for courses": possibilities of intervention and target groups

The previous chapters have addressed cycling as an everyday, flexible and demand-driven mode of transport. Cycling should be considered sustainable local mobility for shopping trips. But it is not a fitting solution for all municipalities or population groups. To assure effectiveness, possible measures must be integrated and adapted to the specific municipality and target group. The target groups are defined and then readdressed in "Areas of action".

Concerning the townships and municipalities there is one major distinguishing criterion: are there local suppliers in the town centre (or in its proximity) or are there none? Therefore it is necessary to differentiate between towns/ municipalities with local suppliers on the one hand and towns/municipalities without local suppliers, on the other.

Concerning the population the target groups for local mobility for shopping can be categorized into three groups (Figure 6):

- People who incorporate shopping with other trip purposes. Especially commuters are a difficult group to incorporate into local mobility (not-only shoppers)
- People who do not incorporate shopping trips into others and have a car available to them (car-receptive shoppers)
- People who do not incorporate shopping trips into other trips and do not have a car available to them (independent shoppers)

The group of *not-only shoppers* is very large. It is difficult to reach the entire group with the following measures. However, their mobility behaviour can be changed by offering appropriate services. For example flexible transport services may allow them to omit certain drop-offs and pick-ups, which would form a new prerequisite for other trips such as shopping.

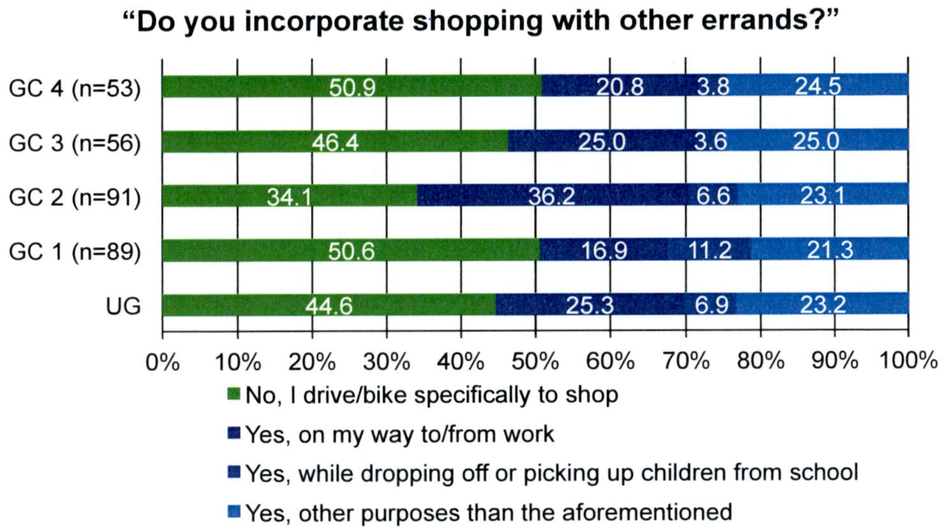

Figure 6: Linking shopping trips with other trip purposes.
GC 1–4 = Municipality cluster 1–4; UG = all interviewees

Source: SmaShMob survey 2015

The group of *car-receptive shoppers* is found predominantly in rural areas. Members of this group always have a car at their disposal. Their car is perceived to give higher comfort and better transport for purchases. Generally speaking, it is very difficult to win over members of this group for eco-friendly mobility options. Nonetheless, it is possible to create potential for switching mode of transport through convincing arguments, good service offers and corresponding marketing and awareness strategies.

The group of *independent shoppers* consists of elderly people, who no longer use a car, young people without driving licenses, and people who have made a deliberate decision to forgo the use of a car. They are the easiest to win over for eco-friendly mobility services but still require relevant basic supply.

Areas of action for bicycle traffic and flexible public transport service

Bicycle traffic: recommendations for municipalities

In many ways bicycles are the best suited mode of transport for local supply. Local structures are preserved, while simultaneously cycling saves money and is healthy for the user. In the following passage measures for municipalities, consumers and local suppliers are discussed. These measures should help create positive shopping experiences for cyclists.

For municipalities there are several reasons to promote cycling for shopping:

- Location security: Cyclists enjoy shopping close to where they live or work. They prefer the local supplier in the town centre to the supermarkets on the edge. Therefore, the purchasing power remains in the municipality itself. Hence cycling itself secures local retail business. A great advantage of cycling is found in areas with a low-density development: The distances to the shops may be too far to walk, but the sparsely driven-on roads offer plenty of routes for cycling.
- Saving money: The infrastructure required for cycling is comparatively low in many cases. Oftentimes it is enough to apply certain traffic laws to make cycling more attractive and securer.
- Environmental protection: The bicycle is the most eco-friendly mode of transport. Supporting it does not just contribute to climate protection but also creates a positive image for the municipality.
- Positive image: A growing popularity and acceptance for cycling in everyday trips has been observed over the course of the past years. The General Transport Strategy Burgenland survey was able to determine that it is a common desire of the population to increase the offers and services for daily cycling. In return municipalities have responded with initiatives.

The following initiatives can be implemented by municipalities:

- Giving the cyclists space and safety: For daily paths, the connections should be safe and as direct as possible. In areas with barely any motorized traffic certain laws could be put to use. For example implementing a 30km/h zone or speed limit could be sufficient to supply ample feeling of security. On through roads a separate bike path is more sensible.
- Bike Stands: Bike stands in central areas and close to shops offer a safe location to store bicycles while also making it a more visible part of the overall transport system. The main requirement is to have them as close as possible to the destination, giving users a safe and sturdy resting point for the bicycle. Best case scenario would to offer a design that allows frame and front wheel to both be locked securely. Metal frames as shown in the picture are often the best option. Further options that are appreciated would be roofed, dry and well lit.

- Public services for bicycles: Offerings services for daily bicycle use shows the appreciation of the municipality and encouragement for the use of eco-friendly mobility options. It also eases use for cyclists. Some examples for this include pump stations or (free) rental of trailers to transport purchases more easily.
- Taking cycling into account during planning: A bicycle friendly development of the local structures and local transport network can contribute a lot towards increasing the share of cyclist. The needs of cyclists should be considered while redesigning streets or squares or new designations for business or residential areas. In doing so, later investments can be avoided and the required structures can be implemented.
- Reduction of weather and seasonal obstacles: Winter weather conditions are a reason many people opt against cycling in daily life. This hurdle can be overcome through winter road cleaning services on relevant routes. This must also be communicated to the cycling population to encourage use, even in bad weather conditions.
- Communication, public relations and raising awareness: This type of work is never a one way street. The population should always be incorporated as much as possible in these considerations. Information, marketing and awareness should be used to create a better image of cycling and to plant the ideas of alternative modes of transport into the populations head. Examples for such measures include:
  - Incorporating cyclists in the planning by inviting them to discussions or creating a committee
  - Including local press and local papers to communicate the on goings
  - Having politicians set an example by cycling themselves
  - Dedicate certain campaign days towards cycling: test runs on bicycles with trailers, special sales or closing the town for cars for an afternoon, day or weekend.

Bicycle traffic: recommendations for local suppliers

Local businesses also have many good reasons to encourage cycling for local mobility, for example:

- Connections to customers: Statistically, cyclists do spend less on purchases each trip, but they do come more often and are loyal customers. Investing in bicycle infrastructure will pay off quicker than starting pricing competitions.
- Positive image: Bicycles continue to symbolise a healthy lifestyle. This can be used to the advantage of local suppliers who value regional products additionally. Combining these aspects could be a good marketing strategy.
- Spatial and financial aspects: Bike stands take up far less space than parking spaces and are cheaper to implement

The following initiatives can be implemented by local suppliers:

- Good, sheltered bicycle stands: Cyclist appreciate good and safe stands that are directly next to store entrances. Having roofed bicycle stands is the cherry on top.
- (Luggage) lockers: Cycling customers have no possibility to leave previous purchases or bags locked up safely in a vehicle. Therefore, a storage space is highly appreciated.
- Service: Additional services such as air pumps, repair tools, or lending trailers not only demonstrate the support of local suppliers but also give the cyclist a welcoming feeling.
- Delivery service for larger purchases: For most shopping trips baskets, saddlebags or trailers are sufficient for transporting purchases. Some items, however, are large or bulky and cannot be transported by bicycle. In such cases delivering should be available to encourage people to stay away from car use and travelling to the larger shopping centres.
- Specials for cyclists: Local suppliers themselves must also promote cycling amongst their customers. Introducing sales or campaign days for cyclists are known to be an effective form of encouragement.
- Communication and cooperation: Cooperation between local suppliers has a stronger external perception and makes marketing bicycle services easier.

Bicycle traffic: recommendations for consumers

Most trips are under 5 kilometres, particularly for local suppliers (Figure 7). On these stretches bicycles show their strengths as a quick and cheap mode of transport. Also important to mention are the positive health effects everyday cycling has on a person.

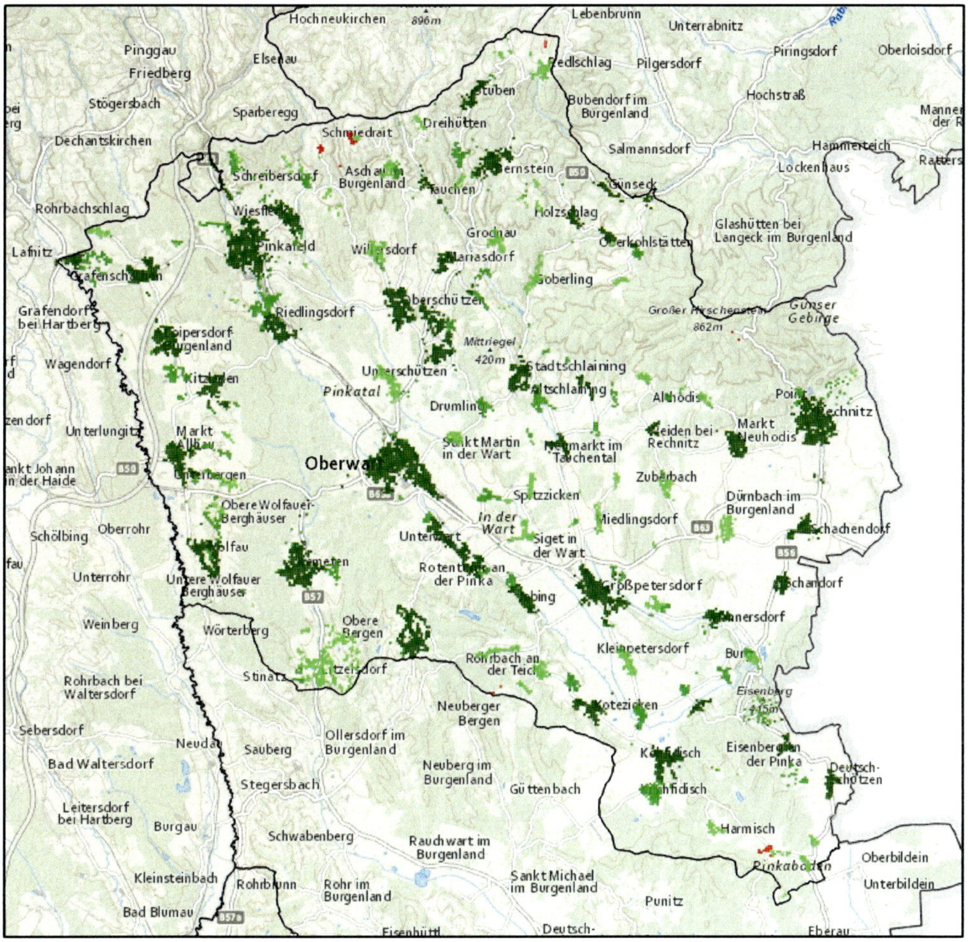

Figure 7: Accessibility of local suppliers by bicycle
Source: Verracon GmbH 2015b

Just as in public transport the main problem with cycling is perceived to be the transportation of purchases. Actually though most purchases weight and volume would be quite easily transportable by bicycle. And there is usually a fitting possibility to transport nearly anything[4]:

- Bike baskets: For transporting smaller, daily purchases. They can be mounted at the handlebars or on the rear baggage carrier. It is a cheap and simple solution.
- Saddlebags: Modern models are easy to use and perfect for transporting daily goods. They can be easily attached to the rear baggage carrier. Some are waterproof, allowing dry transportation even on rainy days.
- Bike Trailers come in two different variations. One type can be used to transport children, while the other type can be used to transport larger purchases.

Flexible public transport services: recommendations for municipalities

As shown, public transport is essential to people without access to a car. These people too must be able to attain basic necessities. It is hard to imagine other groups of people switching to public transport. If they did it would only be through an offer that was highly flexible and frequent.

For rural and peripheral areas a classic public transport system could not fulfil these desires or is not financially feasible in the long run for state authorities. Therefore, the following area of action will concentrate on flexible public transport service solutions.

Flexible public transport systems can take many different forms. They can be run by licensed operators, as a communal service or in a volunteer network. The municipality will always play a crucial part regardless which form the system is developed as. The following recommendations are aimed at the creation of a flexible transport system that is geared towards shopping trips. This is not a development solely to profit local suppliers, but this should also create synergy effects with a greater potential for transport services as a whole.

The flexible public transport service handbook (Wolf-Eberl et al. 2011) recognized four external factors and nine components that create flexible systems and determine their individual success (Figure 8).

---

4 A good compilation of various transport option can be found at Sicheres Vorarlberg: www.sicheresvorarlberg.at/sport/sommer/fahrrad/transport/

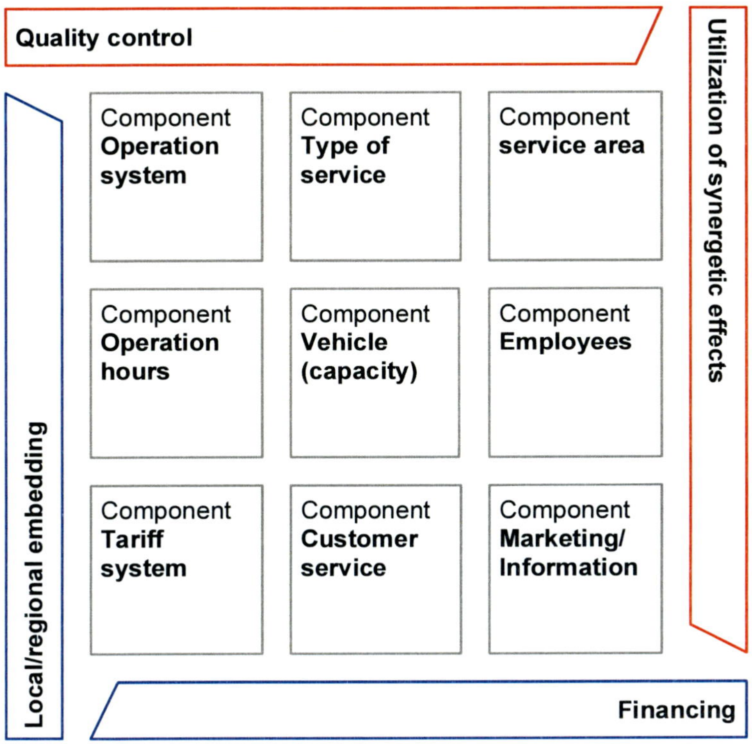

Figure 8: External factors and components of flexible public transport services
Source: Wolf-Eberl et al. 2011

The following recommendations are based on the components and factors that are relevant to shopping and the shoppers' needs.

- Local and regional identity and embeddedness: Cooperating with local suppliers can contribute enormously to embedding the flexible public transport system into the community. Local businesses are amplifiers in their communication with others. The population can identify with the system better, if their needs where addressed and solutions were given. "Passive supplies" such as solo trips for picking up prescriptions and having the medications delivered are one area to consider. Another possible service worth considering is transportation of bulky goods or options for transporting bicycles on the way home from shopping trips.
- Type of service: If the system is not a door-to-door service, but rather a service from stops or collection points, then the local suppliers should be considered in the planning of their locations.
- Service area: Taking both the population and local shops into consideration, an appropriate service area should be determined.

- Operating hours and shopping habits of the population should be taken into account while planning the service hours for a flexible transport system.
- Vehicle: The choice of vehicle should also consider that shoppers may also have bulky objects to transport. This should be made possible. Some cyclists would also appreciate the option of being able to transport their bicycles after larger shopping trips (bicycle trailers /racks).
- Tariff system: Creative solutions for tariffs of flexible public transport systems are useful in fulfilling the wishes of the population and to support the local economy. For example, one could offer free rides as of a certain amount spent at the local shops. Inexpensive multiple ride or time dependent tickets encourage more trips to be made with the flexible transport system.
- Marketing and information: Cooperation between local suppliers and transport operators can emerge as strong partnership in this area. Targeted marketing strategies or combined information days can be very effective. Campaign days could include offering customers using public transport specials at the shops, or offering shop customers free rides with public transport.
- Financing: Local businesses can financially support the flexible transport systems by sponsoring them. An additional option would be to advertise in or on the vehicles.

Flexible public transport service: recommendations for local suppliers

Many successful examples have demonstrated that flexible public transport services strengthen local and regional economy. People who do not own a car or rely on another person with a car truly understand the benefits of shopping locally and remain loyal customers. Securing local supply is often linked to offering additional services. In many cases this has taken the form of creating seating or coffee corners, thus making the shops a social gathering point of the townspeople. However, it is essential for success to assure that the target groups (usually the elderly) can reach this destination regularly and independently. For this reason, flexible public transport services are a sensible idea for local suppliers.

Local suppliers have the following options:

- Delivery services: Combining delivery services with the flexible transport operators would allow home deliveries without additional employees or vehicles. Not only elderly or the mobility-impaired would appreciate such an offer, but also those who have been shopping on foot or by bicycle.
- Cooperation in marketing and awareness: Joint marketing can give the entire regional local supply a more easily perceptible image. Information or advertising in vehicles and at the stops or stations can distribute the information to the local suppliers target groups directly and easily. Discounts, cheap tickets for shoppers or other campaign days could be organized together. Joint campaigns can help create success for both shops and operators.

Mobile local supply

For a long time and in many areas there have been successful approaches to delivering goods of daily requirement to people, for example, bakers that deliver fresh bread to the doorstep every morning. With the slow decline of local suppliers in smaller rural municipalities and the simultaneous wish to maintain basic requirements, new and alternative services have been pushed for. At this point a few examples of such services and their effects on local mobility and synergy potential are investigated.

- Delivery services: Traditional delivery services, such as the aforementioned bakery example, could make shopping trips obsolete, depending on the range of products. Deliveries could step into the local supply function. They cannot, however, replace the social functions of shopping trips which are quite important for the elderly. Local suppliers could offer delivery of purchases from the shop to the customers' doorstep as an extra service, an especially good offer for those on foot. And a cooperation that can benefit both businesses and the transport system, if transport volume can be accommodated.
- Local suppliers on wheels — mobile shops: This too could make shopping trips obsolete and offer basic supplies in municipalities without stationary local supplier. Social functions are limited. Synergy potential is nearly non-existent.
- Neighbourhood assistance: In existing town structures neighbourhood assistance can fill the gaps in local supply and local mobility and secure

basic supply. At the end of the day, people are still dependent on others. This option should only be introduced if no other sustainable options can be found that would secure local supply and local mobility offers. Dependency should not be institutionalised.

Good practice examples in Austria

FLEXIBLE PUBLIC TRANSPORT SERVICE
Ernsti-Mobile

This example from Ernstbrunn, Lower Austria, aimed to strengthen local economy and assure car free target groups an independent shopping experience in the town. Volunteer drivers take people (who have registered with the association) to shops, doctors or restaurants. Nowadays this is done by eco-friendly electric vehicles.

For more information:
www.ernstbrunn.gv.at/system/web/sonderseite.aspx?menuonr=222529618&detailonr=222529618

FLEXIBLE PUBLIC TRANSPORT SERVICE
Shopping bus Traisen

While most public transport services try to address different trip reasons, there are also examples of affordable (for operator and customer) services to support shopping trips and basic supply. One example of this is the shopping bus in Traisen. Twice a week (once in the morning and once in the evening) it connects the entire municipality with the town centre. The service is offered by the officials to secure basic supply and strengthen local economy. It is free of charge for the population.

For more information:
www.traisen.com/buergerservice_verkehr.asp?style=

FLEXIBLE PUBLIC TRANSPORT SERVICE
Shopping bus Bad Erlach

A further example of a shopping bus can be found in Bad Erlach, Lower Austria. It aims to connect the elderly with the local suppliers twice a week. Retirees volunteer as drivers, which adds an additional social component.

For more information:
www.baderlach.gv.at/system/web/sonderseite.aspx?menuonr=221679749&detailonr=221679749

Good practice examples in Austria

BICYCLE
Bicycle Spring — A Campaign by the eco-region Kaindorf

An attractive sweepstakes showed the advantages of cycling for short distances and also demonstrated what the region has to offer. The participants received a sticker for their sweepstakes passport if they arrived on bicycle to one of the bicycle partner shops. A turned in and full passport offered the chance to win prizes related to cycling.

For more information:
www.oekoregion-kaindorf.at/index.php/archiv/2012/radfruehling-2012/gewinnspiel

BICYCLE
Shopping – loading – cycling

In cooperation with ARGUS Steiermark and the state of Styria campaign days at various supermarkets were organized as part of a roadshow. It was geared at showing the benefits of cycling for shopping trips. Information materials were handed out, free technical inspections for personal bicycles and free bicycle trailer trial runs were offered.

For more information:
www.verkehr.steiermark.at/cms/dokumente/10553958_11163140/c55c948e/argus_folder_roadshow.pdf

LOCAL SUPPLY
„Unser G´schäft“ in Strem

„Unser G'schäft“ („Our shop“) in Strem in Burgenland is an example of an association that offers a different approach towards local mobility. Not only do they have a delivery service, they also do pick-ups for mobility-impaired residents in an electric car.

For more information:
www.unser-gschaeft.at/cms/index.php/bestellservice

Conclusion

This contribution discusses concepts of local supply and local mobility by showing application in peripheral rural areas. A classic peripheral area of Austria was used for this specific example — the province of Burgenland, which used to lay next to the Iron Curtain. Empirical studies from 2014 and 2015 were used to illustrate the concepts discussed. Building on the examples given, some areas of action were established for rural peripheral municipalities. These concentrated on components that would support sustainable flexible public local mobility and cycling.

The findings and recommendations show that the incorporation of interests and target groups creates optimistic scenarios for the future. Solutions to problems, such as the current lack of small scale local supply, are in sight. At the base of a broad solution to such issues is the underlying point that politics and society must come together with coordinated approaches and methods. The best practice examples addressed this by demonstrating effective solutions. Greater diversity in delivery services on one hand, and flexible mobility services on the other, are the base for synergetic coordination between local mobility and local supply.

References

ABL (Amt der Burgenländischen Landesregierung, Gesamtverkehrskoordination) (Ed.) (2014). Gesamtverkehrsstrategie Burgenland — Gemeinsam mehr erreichen. Eisenstadt.

ARGUS Steiermark (Ed.) (2015). Einkaufen, Aufpacken, Losradeln. www.verkehr.steiermark.at/cms/dokumente/10553958_11163140/c55c948e/argus_folder_roadshow.pdf. Accessed 22 February 2016.

ATL (Amt der Tiroler Landesregierung, Abteilung Raumordnung und Statistik) (Ed.) (2011). nah&versorgt — Die Zukunft der Nahversorgung in Tirol. Innsbruck.

Bauer-Wolf, S., Payer, H. & Scheer, G. (Eds.) (2008). Erfolgreich durch Netzwerkkompetenz. Handbuch für Regionalentwicklung. Wien, New York: Springer.

CIMA (2007). Nichtmotorisierter Einkauf in Niederösterreich: Untersuchung des Verhaltens, des Einkaufsvolumens sowie des Potenzials von nicht motorisierten Einzelhandelskunden. Ried im Innkreis. wko.at/wknoe/handel/homepagehandel/Einkaufen_und_Radfahren.pdf. Accessed 22 February 2016.

Marktgemeinde Bad Erlach (Ed.) (2015). Einkaufsbus. www.baderlach.gv.at/system/web/sonderseite.aspx?menuonr=221679749&detailonr=221679749/. Accessed 22 February 2016.

Marktgemeinde Ernstbrunn (Ed.) (2015). Bürgerbus "Ernsti-Mobil". www.ernstbrunn.gv.at/Buergerbus_Ernsti-Mobil_/Fahrbetrieb. Accessed 22 February 2016.

Marktgemeinde Traisen (Ed.) (2015). Einkaufsbus. www.traisen.com/buergerservice_verkehr.asp. Accessed 22 February 2016.

Mose, I. (2005). Ländliche Räume. In Akademie für Raumforschung und Landesplanung — ARL (Ed.), Handwörterbuch der Raumordnung (pp. 573–579). Hannover: ARL.

Ökoregion Kaindorf (Ed.) (2015). Radfrühling 2012 — Das Gewinnspiel. www.oekoregion-kaindorf.at/index.php/archiv/2012/radfruehling-2012/gewinnspiel/. Accessed 22 February 2016.

Reinswald, F., Damyanovic, D. & Schönemann, M. (2013). Sicherung der Nahversorgung in den burgenländischen Gemeinden — Grundlagenstudie zu Strategien, Instrumenten und Gute-Praxis Beispielen zur Förderung der Nahversorgung im Burgenland. Wien, Eisenstadt.

Unser Strem — Verein zur Sicherung des Grundversorgung (Ed.) (2015). Unser Gschäft — Zustellung. www.unser-gschaeft.at/cms/index.php/bestellservice. Accessed 22 February 2016.

Verein Regionalmanagement Österreich (Ed.) (2012). Mobilität im ländlichen Raum. Nachhaltig und den Mobilitätsbedürfnissen angepasst. Tagungsbroschüre. Öhling. www.rm-austria.at/itemacms/content/files/Downloads/Veranstaltungen/Broschuere-Mobilitaet_14-12-2012.pdf. Accessed 22 February 2016.

Verracon GmbH im Auftrag des Amts der Burgenländischen Landesregierung (Ed.) (2015a). SmaShMob — Smart Shopping Mobility Nahmobilität und Nahversorgung — Empfehlungen für eine nachhaltige Mobilität im lokalen Einkaufsverkehr. Wien.

Verracon GmbH im Auftrag des Amts der Burgenländischen Landesregierung (Ed.) (2015b). Pilot ERRAM HU-AT NaVer — Nahversorgung Oberwart. Wien.

Wolf-Eberl, S., Koch, H., Estermann, G. & Fürdös, A. (2011). Ohne eigenes Auto mobil. Ein Handbuch für Planung, Errichtung und Betrieb von Mikro-ÖV Systemen im ländlichen Raum. Wien: Klima- und Energiefonds.

How to counteract the increasing discrepancy between cost causers and payers in the total costs of individual mobility in rural peripheries? A case study in Germany

Henry Lewerentz

Abstract

Using two case studies in the rural periphery of Mecklenburg-West Pomerania, Germany, it is shown how the share of private costs of spatial mobility compared to the total household consumption will increase in the future. Additionally, these areas are regions with significant losses in basic public services and infrastructure. One consequence is that the daily privately covered distances of travel are rising. This results in an increasing discrepancy between the attempts of saving administrative expenses and the efforts and costs of private households.

The chapter shows that alternative engine technologies (e.g., electromobility) can create economic, ecological and social advantages for the households and the region. Therefore, we need a consistent implementation of these technologies in the future. A combination of several engine technologies can help to reduce or stabilise the rate of cost increases. Especially rural peripheries are very good test locations for the integration of these "new" technologies, but actually, they do not play a major role in research on rural areas.

Introduction

Nowadays, mobility and the overcoming of space are of great importance for social and economic participation. Travelling and mobility promise a life-style, which is equivalent to freedom and self-determination. However, for many years the private costs for individual mobility increased. This is proven by the rising part of fuel costs in total household spending (Statistisches Bundesamt 2014) which in the following is called mobility budget. It is expected that the daily privately covered distances will rise because of the centralisation of local administration as well as the reduced offers of public services in rural areas. As a result, an increasing discrepancy occurs between the attempts of saving administrative expenses and the efforts and costs of private households. This is also in contradiction to the premise, "care for payable, qualitatively high-quality fuels" which has been published by the Federal Ministry of Transport, Building and Urban Development (BMVBS 2013, 62) in its mobility and fuel strategy (*Mobilitäts- und Kraftstoffstrategie der Bundesregierung*). By the rapidly growing amount of private car traffic, especially in the threshold countries, the demand on fossil energy sources will rise further. Worldwide, the same will apply to oil in the mid-term (Birol and Fatih 2010). The natural resources and therefore the offer of these non-renewable materials are very restricted. In the classical economic sense the prices for a demanded good will rise more and more.

In the following, it will be shown how a combination of existing and new engine technologies can stabilise the rising mobility budget. By this, the individual mobility can remain payable. In addition, other advantages are discussed briefly. In this work the Liquefied Petroleum Gas (LPG), Compressed Natural Gas (CNG), electric and the hybrid based engine technologies are summarized as alternative engine technologies. All others like petrol and diesel-based technologies are called conventional technologies. In Germany, more than 99% of all passenger cars have the latter types of engines.

Case study and research areas

The two research areas (RAs) are located in the north-east part of Germany and are shown in Figure 1. The RA 1 belongs to the district of Vorpommern-Greifswald and RA 2 to the district of Mecklenburgische Seenplatte. Both RAs lie between the metropolitan areas of Hamburg, Berlin and Szczecin on the Polish side. Each RA consists of several municipalities. The RA 1 consists of 24 and RA 2 of 22 municipalities around an arrangement of three central towns in each case. On the one hand, the three towns in the RA 1 are Ueckermünde, Torgelow, Eggesin and in RA 2 Waren, Malchow, Röbel, respectively. The population of both areas were about 42 000 (RA 1) and 44 000 (RA 2) inhabitants in 2010. The population density is about 40 inhabitants per square kilometre. The lowest value is reached for a municipality with about five inhabitants per square kilometre. This indicates the rural characteristics of the research area. Due to their low population densities, these areas retained their natural character. The surrounding landscape is formed and characterised by the ice age. Therefore, the regions have relatively low altitude differences of up to 40 metres between lowest and highest point.

In the local context the medium regional centres of Ueckermünde and Waren are particular important for their individual RA. Also, the other four towns are smaller regional centres or basic centres. Due to their local facilities, they offer the supply function for items of everyday necessities in the rural areas.

It is important to note that the RAs have a big potential for the production of renewable energies. In 2012, in the RA 1 were 368 power plants installed according to the German renewable energy law (EEG). These had a nominal output of 58 359 kWpel and delivered about 78 GWh electricity. At the same time, the RA 2 had 268 power plants with a nominal output of 31 744 kWpel and generated 53 GWh electricity (DGS n.d.). Together, the power plants could cover the electricity consumption of more than 65 000 people. That is three quarters of the inhabitants of both RAs.

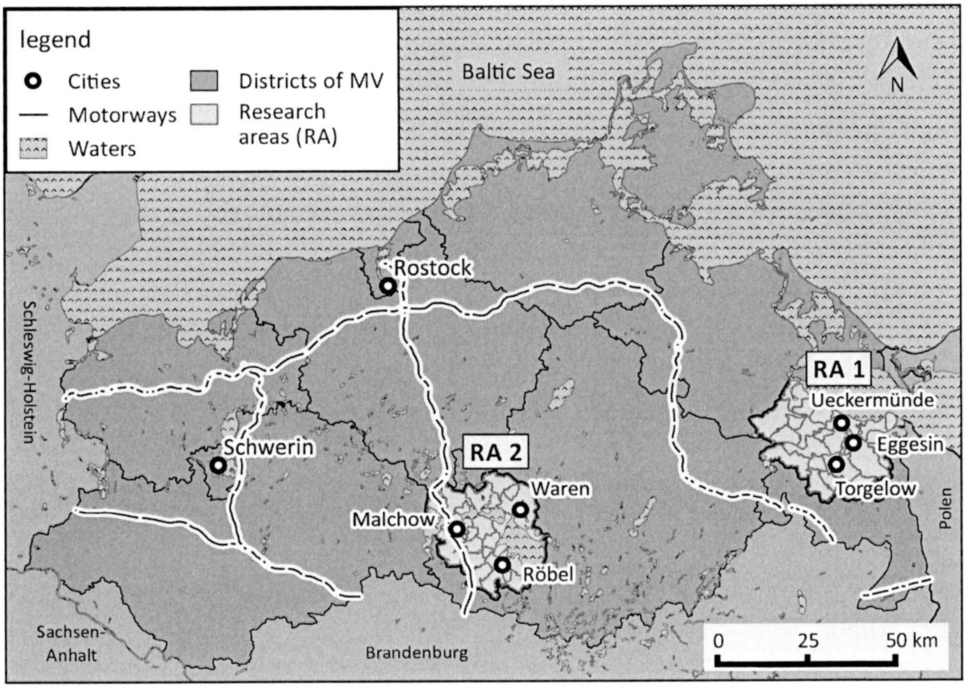

Figure 1: The location of the two research areas (RAs) within MV is shown
Source: Author's compilation

State of research

Many authors consider a move away from carbon-based energy sources in the 21st century unavoidable (e.g., Rothfuß 2012). With this in mind Wagner and Rothfuß (2012) describe a high correlation between the spatial structure and the choice of means of transportation. Therefore, different kinds of mobility and energy forms in rural peripheries should be introduced as the means of transportation (Wagner and Rothfuß 2012), e.g., electromobility.

Researching rural peripheries in light of alternative engine technologies is rewarding from several points of view. On the one hand — depending on the definition — in nearly 70% of the area of Germany more than 30% of inhabitants live in the countryside (BBSR n.d.). On the other hand, in studies about alternative engine technologies, e.g., electric engines, rural peripheries do not play a major part (Proff et al. 2012; Slupetzky and Stroj 2012). If rural regions in the context of alternative engine technologies are considered as locations for production, then almost exclusively areas in or close to metropolitan areas or big urban centres are thought of (e.g., *Modellregionen Elektromobilität* (model regions of electromobility) and the *Schaufenster Elektromobilität* (showcase electromobility), BMVI n.d.). Especially, rural municipalities in the periphery as areas of energy production play a major role for the energy turn (Hirschl et al. 2010). Also the RAs produce to a big part more electric energy than they use for themselves (DGS n.d.). Thus, it can be noted that alternative engine technologies are particular sustainable in the areas between the metropolitan areas. Furthermore, numerous studies show economic advantages for rural municipalities, which build renewable energy plants or promote the settlement of enterprises in the field of renewable energy plants. In addition, car ownership per 1 000 inhabitants and the share of the motorised private traffic in rural areas are higher than in urbanised regions (infas and DLR 2008). The logical consequence is that the cost of using a private car is by the far biggest share in the traffic expenses of private households. Further, the car parking situation in the rural peripheries is special. Many cars can be parked on private property. That can be an important criterion for the development of electromobility, because this offers the possibility for easy recharging of the cars at home (infas and DLR 2008).

Methodology

The results of our research are obtained from a combination of interviews with key stakeholder and from the development of a computer model. All steps of research and the project logic are shown schematically in Figure 2.

The first step of research is an analysis of the actual situation and also future prospects. This is done through interviews with experts and local actors from the region as key stakeholders (Atteslander 2010). It comprises a quantitative and qualitative approach. Mobility, public services and general characteristics are the main topics addressed by the questionnaire survey. The survey with 26 experts (mayors, entrepreneurs, representatives of organizations) took place in the period of autumn and winter 2013. The interviews are analysed and evaluated for later use.

Further, statistical data has been analysed to describe and estimate the mobility budget in order to define the research object as well as the timescale involved. Initially, the dependencies of mobility and traffic in both RAs are described and documented (research step two). For all forecasts, the base year of the data is 2010. For all municipalities the necessary data (population, amount of cars, renewable energy plants, etc.) is collected, if possible including also historical data. When no data exists at the local level in certain areas, data of the next higher administrative level is used.

Based on the first two research steps we can build a model, which illustrates the complex system of mobility and traffic. This step can be summarized as system description (research step four, illustrated in Figure 2). In particular, the model includes variables of regional population, local car market structure, and data of human capital. The complex model is divided in six modules. The main components are shown in the flow chart in Figure 3. The model is numerically implemented in Matlab (The MathWorks 2009).

After model validation, the specific simulation runs are generated. This step is closely connected with the evaluation, discussion and representation under the view of the research questions.

All simulations are carried out at the local level, the level of the municipalities (spatial component).

Another important aspect is the temporal component. Here, the predicted errors increase with time. That is why the simulation is limited to short-term or to medium-term periods. The data until 2010 exists. The model then advances the modules successively on a year per year basis until 2030.

Each module of Figure 3 can be considered as a closed system with its own results. The combinations of the modules to the resulting model increase the complexity. The advantage of this approach is that this kind of procedure is fast and easy extendable. The parameters (shown in Figure 3, middle part) are determined by the results of the other modules, e.g., variable "energy consumption of a household" is calculated as a result from module "demography". The aim of this procedure is a consistent description and linkage of the implemented modules.

Changes within the modules (e.g., different price trends or demographic changes) are estimated by available statistical data or plausibly fixed according to the process "what would be if". Not all details of the forecast method will be discussed in detail here. To give just one example: the forecasting of the fuel prices is based on the development of this value during the last years, which is shown in Figure 4 (on the left-hand side of the figure). With few exceptions the fuel prices grew yearly since 1986. The prices increased by an average of 3.6% for petrol, 4.5% for diesel and 4.9% for electricity annually. The difference between petrol and diesel remains steady at the end of the forecast period, because a minimum distance of seven cents is implemented. This is because the government supports the diesel and therefore the taxes are lower than for petrol.

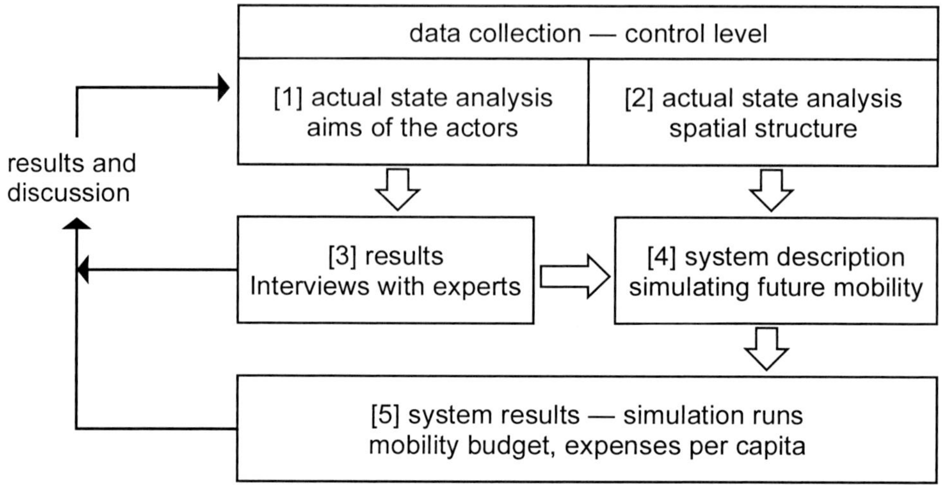

Figure 2: The research steps and main research modules
Source: Author's compilation

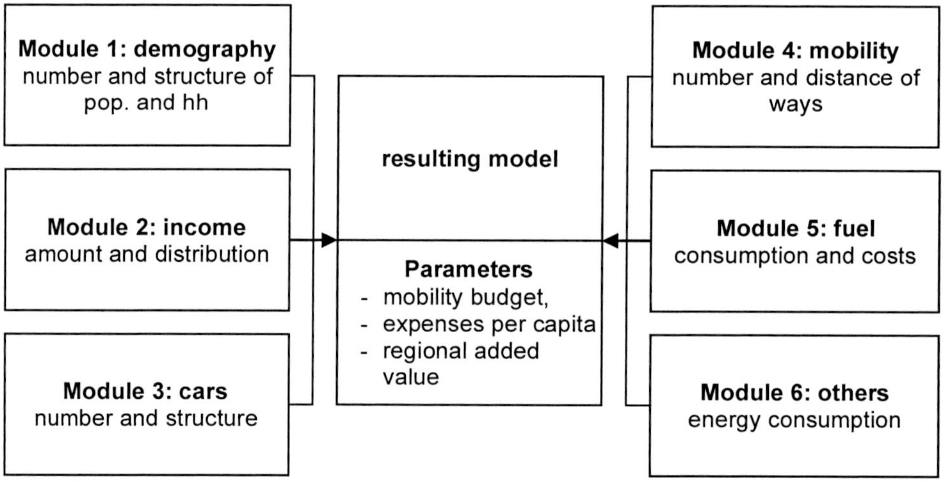

Figure 3: The simulation modules
Source: Author's compilation

To analyse alternative possible developments of each module the scenario technology is used. The different scenarios are based on assumptions using literature and some time series analysis. Time series analysis is a frequently used method to explore short-term associations.

The most important part is module 3, which models a change in the engine technologies from fossil to renewable ones. But how is the implementation of the mobility change done in the model? By the study of real historical data of several successful technology market penetrations (e.g., radio, TV, CD, fridges, microwave), a s-shaped curve is characteristic for various examples (shown in Figure 5, Vanston and Hodges 2004, 32–34). Also, this model is often used in forecasts in the car sector. The market penetration increases quickly and then it is slowing down as it progresses. To describe such a process mathematically it is common to use (Vanston and Hodges 2004) the Gompertz equation (Equation 1), which describes an asymmetric s-shaped curve. In the present work, the same function is used to model the adoption of alternative engine technologies.

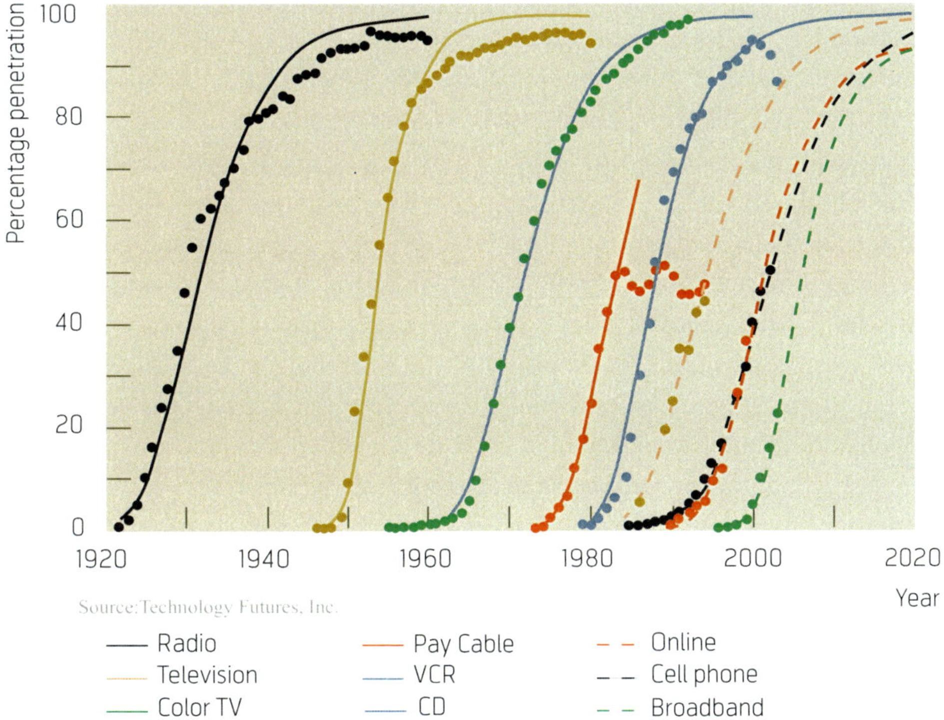

Figure 4: Gompertz examples

Source: Vanston und Hodges 2004, 34

Equation 1: Gompertz-Equation

$$f(t) = M * e^{-ae^{-bt}}$$

Where,

f(t) = depending variable

t = independent variable, time step.

M = lower asymptote, market share

a, b = fitting coefficients

The current situation: results from the field research

In the stakeholder interviews, general development trends and main problems were identified as obstacles for further regional development.

In September 2011, a comprehensive administrative reform and reorganisation of the districts took place in Mecklenburg-West Pomerania. Twelve administrative districts and six independent cities were re-organized to six administrative districts and two independent cities. Since then, each of the considered two groups of municipalities is in the same two administrative districts. The results of the reform and reorganisation have far-reaching effects on the inhabitants. The linear distance from Penkun at the southern end of the administrative district of Vorpommern-Greifswald (which is the district of RA 1), up to the county seat Greifswald is larger than 100 km. Such long travel can last a whole day due to the badly developed public transport in the countryside. In addition, a number of smaller organisational reforms regarding the municipalities have been carried out during the last years. Relating to the RAs there were six local reorganisations of municipalities since February 1, 2012.

Further, the district structure of the courts was also changed and reorganised. This had a particular impact. One district court in the RA 1 (Ueckermünde) was dissolved.

In the communication with experts, it has been recognized that one of the most important obstacles in the RAs is the financial situation. Because of their bad finance situation, most municipalities already work according to a budget consolidation concept. In the future, the cost pressure will further rise and this could lead to other attempts from the administrative level to save money.

However, also beyond the administrative level namely in the private sector one can observe a big withdrawal of suppliers of public services. Especially visible is the vacancy rate of former supermarket buildings and the empty Schlecker buildings, a major chemist chain all over Germany that went bankrupt in 2012. Once more, the results are rising distances for the daily care. The longer distances go hand in hand with negative consequences for environment and climate.

The expected rise in distances with the corresponding rise in the mobility budget will affect a region, which is currently among the most underdeveloped regions of Germany. One reason is the very low income level. The average income is very low and about 1 250 euro per month (year 2010; Mecklenburg-West Pomerania about 1 360 euro, Germany about 1 860 euro per month (AK VGRdL 2013). The unemployment rate is up to 15%. Higher rates are no rarity in the municipalities considered.

Another critical point is that the research areas are regions with a very high average age of more than 46 years. Official forecasts assume that in 2030 about 30% of the local population will be older than 65 years (Scholz et al. 2008).

Another statement from the interviews is that a strong solidarity between the inhabitants compensates many negative aspects. At the moment, some laws restrict the acting of the local actors. Most of them would like to get more rights of self-determination and more financial power allocated from higher administrative levels.

Results of the model

As expressed before, the mobility budget is in the centre of interest of this chapter. That is why the share of the fuel costs compared to the total household consumption expenses is calculated. From the model, especially from the demographic module, a shrinking number of inhabitants and a rising average age of the population become evident. In the income module, a rising net income and household consumption expenses are predicted. In addition, we see, in spite of considerable population losses, the sum of the net income of the region to be expected to increase. This is based on assumptions of using literature (Kunert et al. 2008). The trend is the same for the regional buying power.

A rising number in passenger cars in relation to a sinking population number leads to an increase of the motorisation grade (private car/population). Besides, petrol and diesel vehicles have a share of almost 99% of all vehicles.

In the following three scenarios are generated. The first one is the "status quo scenario" (SQS). In this scenario, the data from 2010 is consistently extrapolated. That means that the proportional distribution (column "MV 2010 SQS", in Table 1) also looks the same in 2030. In the "Good case scenario" (GCS), a realistic market penetration of alternative engines is assumed. Realistic means from the perspective of experts, studies and literature. Here, the shares are given for the target year (2020 and 2030). In the third and last scenario, the "Best case scenario" (BCS), an excessive market penetration of alternative engines is implemented (Table 1). Some penetration assumptions were twice as high as the GCS.

Table 1: Status and distribution of engine technologies in M-V

engine technology	*share in % 2010*	*MV 2010; SQS per piece*	*target share GCS 2020 / 2030 in %*	*amount in 2020 / 2030 per piece*	*target share BCS 2020 / 2030 in %*	*amount in 2020 / 2030 per piece*
Petrol	76.3	622 675	not specified*	not specified*	not specified*	not specified*
Diesel	22.5	183 423	not specified	not specified*	not specified*	not specified*
LPG	1.0	7 972	3.35 / 5.0	27 400 / 40 800	5.7 / 9.0	46 600 / 73 400
CNG	0.1	1 088	3.7 / 6.0	30 200 / 49 000	7.0 / 12.0	57 100 / 97 900
Electric	0.0	13	2.4 / 11.8	19 300 / 96 400	7.0 / 23.6	57 100 / 193 000
Hybrid	0.1	718	3.0 / 4.7	11 600 / 38 600	10.0 / 15.0	81 600 / 122 400
Other	0.0	17	not specified*	not specified*	not specified*	not specified*
Total	100.0	815 906	100.0	815 906	100.0	815 906
	Source SqS: (Kraftfahrt-Bundesamt (KBA) 2011)		Source GcS: (Stiller et al. 2010; Dudenhöffer und Pietron 2010; Die Bundesregierung 2009)		*not specified (is calculated from the model)	

Further, a loss of facilities in the public services can be expected from the interview results presented above and therefore a rising mobility effort. This fact will not be taken into consideration and the average of kilometre per path is assumed to be constant in the prediction. In the model, all rising efforts and distances from mobility within a municipality arise only from demographic effects. These assumptions are the same for every scenario.

In 2010 people in the RA 1 moved about 24.1 km/d in the motorised private traffic (RA 2 about 24,2 km/d). In 2030 this value will rise, mainly because of the rise in mobility with age, about +7.1% (25,8 km) or +4.8% (25.4 km), respectively.

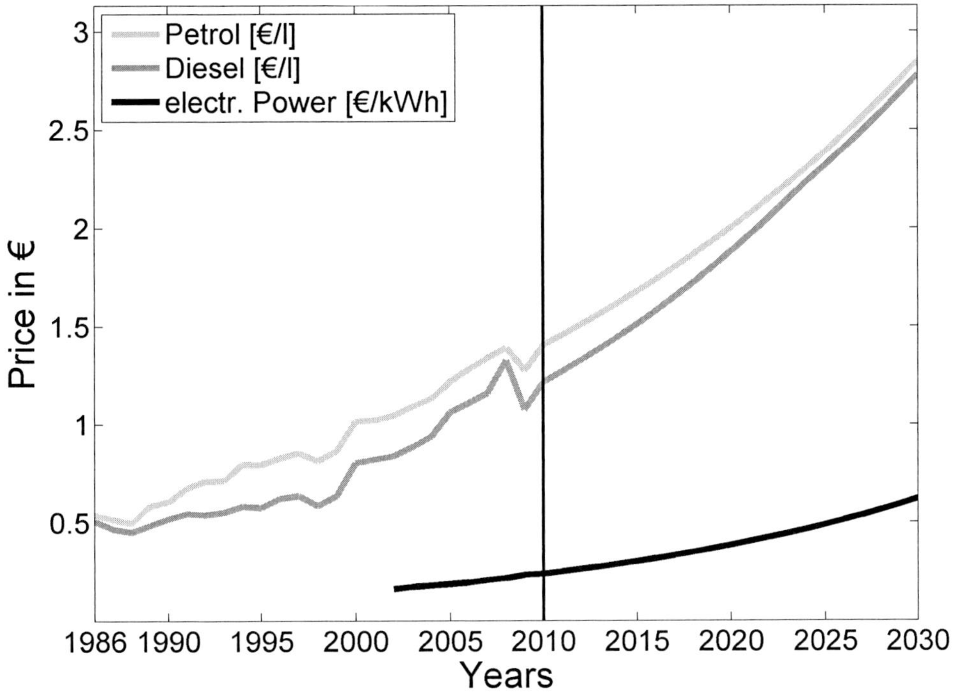

Figure 5: Fuel prices

Remark: From 2010 onwards, the model is based on data extrapolation.
Source: Author's compilation

Prices for petrol and diesel of about three euros may look absurd at first, nevertheless, similar values are found in comparison with other model calculations (Adolf et al. 2014; DEKRA and IFA 2008). Comparing historical oil price forecasts Altenburg and Peter (2014, 30–31) state that these often underestimate the actual price trends. They predict for 2025 a cost of petrol of 3.50 euro/l and for electricity of 0.40 euro/kWh. The resulting fuel price in this work is in a similar range (Figure 5). All model results are compared to available data from the base year as well as to other model calculations. The forecast results are plausible and understandable. From simulations, the mobility budget can be estimated. That is the most important number for households, stakeholders and local actors.

Between the years 1995 and 2012 the share of the fuel costs to the household income was between 3% and 3.5% (Statistisches Bundesamt 2008, 2014). At the end of this period, we can find the highest values. In the present model the parts for 2010 are already higher, on average about 7.1% (Figure 6). This seems realistic, because in the rural area many short distances are done by private car and the prize-intensive motorised private traffic is often the only possibility for mobility. In addition, the RAs are one of the weakest regions in Germany in terms of structure and income. For 2030 an increase of the mobility budget is calculated in scenario SQS to 10.7% (RA 1) or 10.4% (RA 2). In the second scenario (GCS) the share of the mobility budget rises to 9.6% (RA 1) and 9.3% (RA 2), respectively, until 2030. In the third scenario (BCS) the mobility budget in 2030, with 7.2% (RA 1) and 7.0% (RA 2) is on a similar level as in 2010. Due to the extreme market penetrations of the electric vehicle and hybrid electric vehicle the mobility budget even decreases in the middle of the 2020s. The lower percentages in the scenarios two and three is due to the more favourable price of alternative fuels. Overall, in the BcS (second scenario) the costs per capita rise further, although the mobility budget remains steady. The reason for this is the strong decrease of the population and simultaneously the rising household income expectation at a regional level. It is evident that the mobility budget in RA 2 is lower than in RA 1 (Figure 6). The main reason is the different composition of the population.

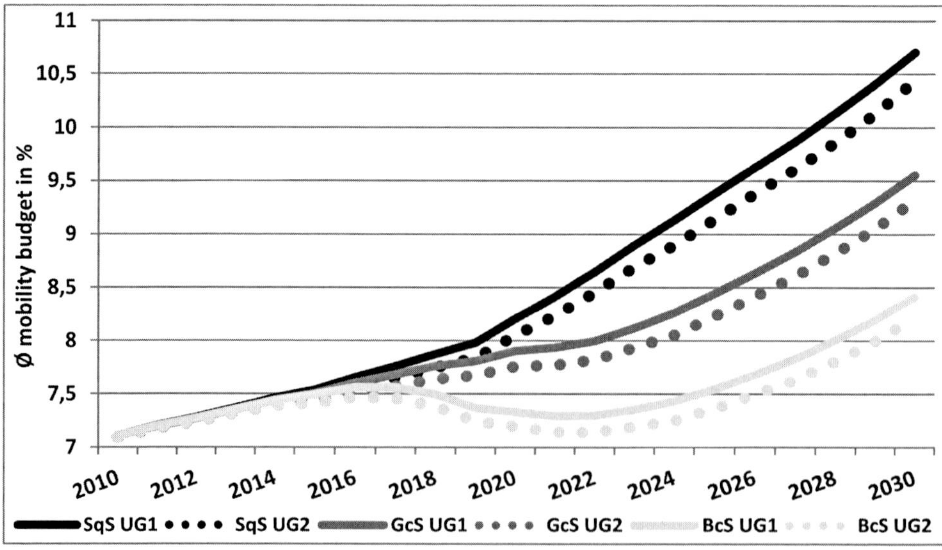

Figure 6: Estimated development of the mobility budget in different scenarios from 2010 to 2030

Source: Author's compilation

In the RAs the number of passenger cars decreases, also the population and the overall daily covered kilometres. Because the population itself has the strongest (82.6%) decrease, whereas the covered kilometres show a much less strong decrease (88.1%), the infrastructures (e.g., streets, ways) need less costs to be retained. Therefore, the cost per capita will further rise and the cost pressure as well the saving attempts will increase.

When the energy for the electromobility comes from renewable energy power plants, it is not only a decrease of the mobility budget, rather there is also a decrease of CO2 emissions.

Conclusion

The analysis of the increasing discrepancy between cost causers and cost payers in the total costs of individual mobility has revealed that demographic, technological and economic developments will change the motorised private traffic in the rural peripheries analysed in this paper. New alternative engine technologies are suitable for these areas, but they play only a minor part both in everyday life and present research.

On the one hand, alternative engine technologies offer chances and many advantages for sustainable economic, ecological and social improvements of the private households and the municipalities. On the other hand, one must not forget the high initial costs for the household to buy a private car with an alternative engine and the financial risks for companies or municipalities to build up the filling and charging stations.

The share of private mobility costs within total household expenses will further rise. This effect is due to the increasing prizes of fuels, especially conventional fuels. On the other hand, it is to be expected that by centralisation of the administrative as well as the public services the daily covered distances will further rise. There is a discrepancy between the attempt of the political-administrative level to save money and the raised cost burden of the private households. The search for solutions leads to turn our focus on alternative engine technologies. The rate of price increases can be reduced by the described combination of different alternative engine technologies. Private mobility remains payable. Exploding costs are avoided and the disparity between the cost causers and the cost payers of mobility does not diverge further. In summary, only if the consequent turn in engine technologies succeeds, the mobility may be payable for people in rural peripheries in the future. This transformation requires the development of alternative engine technologies. With the growing use of alternative engine technologies on the local level the efforts of actors will rise.

According to the theory of endogenous development (Schätzl 2003, 155–158), it can be expected that alternative engine technologies have big economic advantages for the local economy and thus local municipalities. Further studies must clarify in which way or to what amount the existing potentials can be enhanced and expanded in order to increase the benefit in the region.

From the point of view of the local actors and municipal representatives, the absence of self-determination rights and financial resources are big problems. Therefore, teamwork between investors, customers and local actors is necessary for the realisation of the mobility change. It highlights the geo-economic and geo-political efforts. This means the establishment of new business relations, which can be (financially) positive for all partners.

Local politicians should be a trailblazer regarding the support of electromobility and the development of renewable energy plants (Anonymous 2014).

References

Adolf, J., Krämer, L. & Rommerskirchen, S. (2014). PKW-Mobilität am Wendepunkt. Bedeutung des demographischen und des Verhaltenswandel für den PKW-Verkehr in Deutschland bis 2040. Internationales Verkehrswesen 66 (4), 64–67.

AK VGRdL — Arbeitskreis „Volkswirtschaftliche Gesamtrechnungen der Länder“ im Auftrag der Statistischen Ämter der 16 Bundesländer, des Statistischen Bundesamtes und des Bürgeramtes, Statistik und Wahlen, Frankfurt a. M. (Eds.) (2013). Einkommen der privaten Haushalte in den kreisfreien Städten und Landkreisen der Bundesrepublik Deutschland 2000 bis 2013. Reihe 2, Band 3.

Altenburg, S. & Peter, M. (2014). Steigende Energiepreise und ihre Wirkung auf planerische Strategien. Erfahrungen aus einem transdisziplinären Forschungsprojekt. Internationales Verkehrswesen 66 (3), 30–32.

Anonymous (2014). Stadtbedienstete sind jetzt mit Elektroflitzer unterwegs. Elektroauto gehört ab sofort zum Fuhrpark der Stadt, 12.06.2014. www.malchin.de/index.php?option=com_content&view=article&id=725: elektroauto&catid=1:latest-news&Itemid=57. Accessed 30 May 2016.

Atteslander, P. (2010). Methoden der empirischen Sozialforschung. 13th ed. ESV basics. Berlin: Schmidt.

BBSR (Bundesinstitut für Bau-, Stadt- und Raumforschung) (n.d.). Raumabgrenzungen. www.bbsr.bund.de/BBSR/DE/Raumbeobachtung/Raumabgrenzungen/raumabgrenzungen_node.html. Accessed 30 May 2016.

Birol, F. (2010). World Energy Outlook 2010. Paris: International Energy Agency (IEA). www.worldenergyoutlook.org/media/weo2010.pdf. Accessed 30 May 2016.

BMVI (Bundesministerium für Verkehr und digitale Infrastruktur) (n.d.). Modellregionen Elektromobilität. www.bmvi.de/SharedDocs/DE/Artikel/G/modellregionen-elektromobilitaet.html?nn=36210. Accessed 4 March 2015.

BMVBS (Bundesministerium für Verkehr, Bau und Stadtentwicklung) (2013). Die Mobilitäts- und Kraftstoffstrategie der Bundesregierung (MKS). Energie auf neuen Wegen. Berlin. www.bmvbs.de/SharedDocs/DE/Anlage/UI-MKS/mks-strategie-final.pdf. Accessed 30 May 2016.

DEKRA & IFA (DEKRA Automobil GmbH and Institut für Automobilwirtschaft e.V. (2008). Gemeinsam im Dienste der Automobilwirtschaft. Die Ergebnisse der DEKRA/IFA-Studie “Das Management der Cost-of Ownership“. Stuttgart, Geislingen an der Steige.

DGS (Deutsche Gesellschaft für Sonnenenergie e.V.) (n.d.). EnergyMap – Mecklenburg-Vorpommern. www.energymap.info/energieregionen/DE/105/115.html (as of 21 February 2014). Accessed 30 May 2016.

Die Bundesregierung (2009). Nationaler Entwicklungsplan Elektromobilität der Bundesregierung. www.bmub.bund.de/service/publikationen/downloads/details/artikel/nationaler-entwicklungsplan-elektromobilitaet-der-bundesregierung/. Accessed 30 May 2016.

Dudenhöffer, F. & Pietron, K. (2010). CNG as automotive fuel for Europe/CEE is it possible to achieve 5% + x market share for CNG? necessary steps and actions to achieve? Duisburg: CAR-Center Automotive Research. www.uni-due.de/~hk0378/studien/2010/CAR-Studie_CNG.pdf. Accessed 30 March 2016.

Hirschl, B., Aretz, A., Prahl, A., Böther, T., Heinbach, K., Pick, D. & Funcke, S. (2010). Kommunale Wertschöpfung durch erneuerbare Energien. Studie im Auftrag der Agentur für Erneuerbare Energien (AEE). Schriftenreihe des IÖW. Berlin: Institut für Ökologische Wirtschaftsforschung.

infas & DLR (Deutsches Zentrum für Luft- und Raumfahrt e.V.) (2008). Mobilität in Tabellen. Webanwendung zur Auswertung der Erhebung Mobilität in Deutschland 2008 (MiD 2008). www.mobilitaet-in-deutschland.de/02_MiD2008/mit2008.htm (as of 01.07.2010). Accessed 30 May 2016.

Kunert, U., Horn, M., Kalinowska, D., Kloas, J., Ochmann, R., Schulz, E. Zinsmeister, F. (2008). Mobilität 2025. Der Einfluss von Einkommen, Mobilitätskosten und Demographie. Anhang 2: Die Einkommen der privaten Haushalte im Jahr 2025. Berlin: Institut für Mobilitätsforschung (ifmo). www.ifmo.de. Accessed 8 January 2015.

Proff, H., Schönharting, J., Schramm, D. & Ziegler, J. (Eds.) (2012). Zukünftige Entwicklungen in der Mobilität. Betriebswirtschaftliche und technische Aspekte. Wiesbaden: Gabler.

Rothfuß, R. (Ed.) (2012). Elektromobilität aus raumwissenschaftlicher Perspektive. Vom technologischen zum integrierten Wandel. Global Studies Working Papers of the Tübingen Institute of Geography 1. Universität Tübingen, Geographisches Institut.

Schätzl, L. (2003). Wirtschaftsgeographie 1. Theorie. Stuttgart: UTB.

Scholz, R., Rößger, F., Kreft, D., Steinberg, J. & Doblhammer-Reiter, G. (2008). Bevölkerungsprognose für Mecklenburg-Vorpommern auf Kreisebene bis zum Jahr 2030. Diskussionspapier 22. Rostock: Rostocker Zentrum zur Erforschung des Demografischen Wandels.

Slupetzky, W. & N. Stroj (2012). Chancen und Potentiale von Elektromobilität im ländlichen Raum am Beispiel des Forschungsprojekts eMORAIL. e & i Elektrotechnik und Informationstechnik 129 (3), 123–127.

Statistisches Bundesamt (2008). Energieverbrauch der privaten Haushalte. Wohnen, Mobilität, Konsum und Umwelt. Wiesbaden. www.destatis.de/DE/PresseService/Presse/Pressekonferenzen/2008/UGR/pressebroschuere_ugr.pdf. Accessed 30 May 2016.

Statistisches Bundesamt (2014). Wirtschaftsrechnungen. Laufende Wirtschaftsrechnungen. Einnahmen und Ausgaben privater Haushalte. Haushalte - Fachserie 15 Reihe 1 – 2014. www.destatis.de/DE/Publikationen/Thematisch/EinkommenKonsumLebensbedingungen/EinkommenVerbrauch/EinnahmenAusgabenprivaterHaushalte2150100147004.pdf. Accessed 30 May 2016.

Stiller, C., Schmidt, P.,Weindorf, W. & Matra, Z. (2010). CNG and LPG for Transport in Germany. Environmental Performance and Potentials for GHG Emission Reductions until 2020. München: Ludwig-Bölkow-Systemtechnik GmbH. www.bdew.de/internet.nsf/id/2EAD46D327C5D769C125783E005C0E7D/$file/LBST_CNG%20and%20LPG%20for%20Transport_Studie.pdf. Accessed 30 May 2016.

The MathWorks (2009). Matlab. Version 7.9.0.529. Natick, Mass.): The MathWorks. www.mathworks.com. Accessed 30 May 2016.

Vanston, L. K. & Hodges, R. L. (2004). Technology forecasting for telecommunications. Telecommunications Forecasting. Telektronikk 100 (4), 32–42.

Wagner, U. M. & Rothfuß, R. (Eds.) (2012): Einsatzpotenziale der Elektromobilität im Hinblick auf Raumstrukturen. Global Studies Working Papers of the Tübingen Institute of Geography 1. Universität Tübingen, Geographisches Institut.